Achieving sustainable cultivation of cassava

Volume 2: Genetics, breeding, pests and diseases

It is widely recognised that agriculture is a significant contributor to global warming and climate change. Agriculture needs to reduce its environmental impact and adapt to current climate change whilst still feeding a growing population, i.e. become more 'climate-smart'. Burleigh Dodds Science Publishing is playing its part in achieving this by bringing together key research on making the production of the world's most important crops and livestock products more sustainable.

Based on extensive research, our publications specifically target the challenge of climate-smart agriculture. In this way we are using 'smart publishing' to help achieve climate-smart agriculture.

Burleigh Dodds Science Publishing is an independent and innovative publisher delivering high quality customer-focused agricultural science content in both print and online formats for the academic and research communities. Our aim is to build a foundation of knowledge on which researchers can build to meet the challenge of climate-smart agriculture.

For more information about Burleigh Dodds Science Publishing simply call us on +44 (0) 1223 839365, email info@bdspublishing.com or alternatively please visit our website at www.bdspublishing.com.

Related titles:
Achieving sustainable cultivation of cassava Volume 1: Cultivation techniques
Print (ISBN 978-1-78676-000-5); Online (ISBN 978-1-78676-002-9, 978-1-78676-003-6)

Achieving sustainable cultivation of grain legumes Volume 1: Advances in breeding and cultivation techniques
Print (ISBN 978-1-78676-136-1); Online (ISBN 978-1-78676-138-5, 978-1-78676-139-2)

Achieving sustainable cultivation of grain legumes Volume 2: Improving cultivation of particular grain legumes
Print (ISBN 978-1-78676-140-8); Online (ISBN 978-1-78676-142-2, 978-1-78676-143-9)

Chapters are available individually from our online bookshop: https://shop.bdspublishing.com

climate•SMART•publishing
IN AGRICULTURAL SCIENCE

BURLEIGH DODDS SERIES IN AGRICULTURAL SCIENCE

NUMBER 21

Achieving sustainable cultivation of cassava

Volume 2: Genetics, breeding, pests and diseases

Edited by Dr Clair Hershey, formerly International Center for Tropical Agriculture (CIAT), Colombia

burleigh dodds
SCIENCE PUBLISHING

Published by Burleigh Dodds Science Publishing Limited
82 High Street, Sawston, Cambridge CB22 3HJ, UK
www.bdspublishing.com

Burleigh Dodds Science Publishing, 1518 Walnut Street, Suite 900, Philadelphia, PA 19102-3406, USA

First published 2017 by Burleigh Dodds Science Publishing Limited

Library of Congress Control Number: 2017936826

British Library Cataloguing in Publication Data
A catalogue record for this book is available from the British Library

ISBN 978-1-78676-004-3 (print)
ISBN 978-1-78676-006-7 (online)
ISBN 978-1-78676-007-4 (online)
ISSN 2059-6936 (print)
ISSN 2059-6944 (online)

Typeset by Deanta Global Publishing Services, Chennai, India
Printed by Lightning Source

Contents

Series list

Title	Series number
Achieving sustainable cultivation of maize - Vol 1 From improved varieties to local applications Edited by: Dr Dave Watson, CGIAR Maize Research Program Manager, CIMMYT, Mexico	001
Achieving sustainable cultivation of maize - Vol 2 Cultivation techniques, pest and disease control Edited by: Dr Dave Watson, CGIAR Maize Research Program Manager, CIMMYT, Mexico	002
Achieving sustainable cultivation of rice - Vol 1 Breeding for higher yield and quality Edited by: Prof. Takuji Sasaki, Tokyo University of Agriculture, Japan	003
Achieving sustainable cultivation of rice - Vol 2 Cultivation, pest and disease management Edited by: Prof. Takuji Sasaki, Tokyo University of Agriculture, Japan	004
Achieving sustainable cultivation of wheat - Vol 1 Breeding, quality traits, pests and diseases Edited by: Prof. Peter Langridge, The University of Adelaide, Australia	005
Achieving sustainable cultivation of wheat - Vol 2 Cultivation techniques Edited by: Prof. Peter Langridge, The University of Adelaide, Australia	006
Achieving sustainable cultivation of tomatoes Edited by: Dr Autar Mattoo, USDA-ARS, USA & Prof. Avtar Handa, Purdue University, USA	007
Achieving sustainable production of milk - Vol 1 Milk composition, genetics and breeding Edited by: Dr Nico van Belzen, International Dairy Federation (IDF), Belgium	008
Achieving sustainable production of milk - Vol 2 Safety, quality and sustainability Edited by: Dr Nico van Belzen, International Dairy Federation (IDF), Belgium	009
Achieving sustainable production of milk - Vol 3 Dairy herd management and welfare Edited by: Prof. John Webster, University of Bristol, UK	010
Ensuring safety and quality in the production of beef - Vol 1 Safety Edited by: Prof. Gary Acuff, Texas A&M University, USA & Prof.James Dickson, Iowa State University, USA	011
Ensuring safety and quality in the production of beef - Vol 2 Quality Edited by: Prof. Michael Dikeman, Kansas State University, USA	012
Achieving sustainable production of poultry meat - Vol 1 Safety, quality and sustainability Edited by: Prof. Steven C. Ricke, University of Arkansas, USA	013
Achieving sustainable production of poultry meat - Vol 2 Breeding and nutrition Edited by: Prof. Todd Applegate, University of Georgia, USA	014
Achieving sustainable production of poultry meat - Vol 3 Health and welfare Edited by: Prof. Todd Applegate, University of Georgia, USA	015

Acknowledgements

We wish to acknowledge the following for their help in reviewing particular chapters:

- Chapter 1: *Professor Mabrouk El-Sharkawy, formerly CIAT, Colombia; Dr James Cock, formerly CIAT, Colombia; and Emeritus Professor David Connor, University of Melbourne, Australia*

Introduction

1 The roles, challenges and opportunities for cassava in development

For the last century agricultural scientists have embraced and met the challenges of providing technologies that gave farmers the capacity to stay ahead of the growing food demands from increasing populations. Hunger and malnutrition still exist -- and cause far too much misery and livelihood risk -- but UN statistics show clear progress in reducing food deficits at the global level. Agricultural science has ever more tools at its disposal, but at the same time an expanding panorama of expectations from its users. While the Green Revolution of the 1960s and beyond made landmark progress by focusing mainly on productivity of cereal grains under high input conditions, the world rightly demands more attention to environmental impact of agriculture (water quality, soil erosion, genetic resources), quality and nutrition of food, dwindling inputs for expanding productivity, and social justice issues such as equity and opportunity for women. Agricultural sciences are slowly adapting to these new realities, but not quickly enough, and with too few resources committed to success, especially in developing countries. This is the broader context within which we consider the cassava story as a key target for sustainable improvement.

Although cassava is little-known outside the tropics, globally it is the fourth most important calorie source after wheat, rice and maize. It is especially renowned as a crop that supports farmers in more marginal conditions and with poor access to inputs, due to its ability to yield well under low-fertility soils and with periodic drought where more sensitive crops would be devastated. Although it was domesticated from its wild ancestors in the Americas, over half of global production now comes from Africa where it has a leading role in food security. Cassava is largely an energy crop. Early in the growth cycle, beginning at about 2-3 months, some of the roots convert to starch storage organs. From a physiological standpoint, this is a mechanism to store energy through periods of stress, and for recovery when the stress is alleviated. In agriculture, farmers typically aim to minimize stress so that as much of the root starch as possible can be recovered for human use. Although cassava is often referred to as a *subsistence crop*, this is in fact only a small part of present-day production. There is a growing dichotomy between cassava as a low-input, low-yield crop of the poor, and cassava as a vibrant and highly flexible crop driven by industrial demand. Even in Africa where it is nearly entirely a crop of small farmers, mainly for human food, a large majority of production is marketed for off-farm use. In Asia, most of production is already for industrial uses, and with the exception of a few countries, especially Indonesia, India and the Philippines, human consumption is relatively low. Even in industrial markets, most of production is on small farms. Nonetheless, large-scale industrial systems are emerging around the globe, driven by robust demand for cassava products and by technologies that facilitate production and processing at scale. Some facets of cassava production are not easy to industrialize, i.e. to mechanize or implement at large scale, for several reasons. These constraints are gradually being overcome. Planting and harvesting can be mechanized. But there are still many aspects of production that favor small farmer management, and cassava production is likely to be dominated by small farmers for decades to come. Just as the crop is grown primarily by poor farmers, consumers also tend to be from the lower income strata. For governments and development agencies who want to differentially benefit the poor, cassava is an ideal target for sustainable improvement.

2 Science progress and the needs for supporting cassava's future roles

Cassava's features, especially its minimal production in developed countries, have often meant that research investment has long been far less than for most important food crops. This situation was reversed to some degree when two International Agricultural Research Centers – the International Center for Tropical Agriculture (CIAT) in Cali, Colombia and the International Institute for Tropical Agriculture (IITA) in Ibadan, Nigeria were founded in the late 1960s, and were mandated to work on cassava on a global scale (historically, CIAT in the Americas and Asia, and IITA in Africa). This investment motivated considerable parallel investment by national research programs around the world. Through the 1980s, most countries with significant cassava production had developed human and physical capital to improve the crop and the livelihoods of people who depended on it. However, by the 1990s, and especially in Latin America, many countries discontinued or severely cut funding to cassava research. A sort of *development fatigue* had set in for many traditional donors, and there was a rising belief that that the private sector would take on much of the responsibility for crop improvement research. In other crops, and especially maize and soybeans, research was more and more taken on by the private sector, which was able to get a return on investment through the sale of technologies such as seed and chemicals. But cassava provided much less opportunity for profit from the private sector, and therefore lagged in technology development. By early into the 21st century, some of the key donor agencies for agricultural development began to recognize the need and the potential to invest in cassava, especially the Bill and Melinda Gates Foundation. In addition, the CGIAR -- parent organization of CIAT and IITA -- renewed its interest in cassava and recognized the synergies that could be achieved by joint work on key starchy, vegetatively propagated crops. Thus, in 2011, the Root, Tuber and Banana CGIAR Research Program (RTB-CRP) began operation, involving four CGIAR centers and five key crops (cassava, potato, sweet potato, bananas/plantains and yams). At about the same time, some of first major private sector investment in cassava research was initiated to develop amylose-free (*waxy*) cassava varieties in Thailand and Colombia.

While cassava still receives far less research funding in proportion to its value, compared to other major food crops, there has been key interest in recent years from donors, governments, and a limited number of private sector organizations to better fund cassava research. Cassava has been reasonably well-positioned in the molecular revolution, in terms of development of breeding techniques and information. However, the practical applications remain largely a future hope, and impact to date from breeding has been though more conventional approaches.

3 Bringing together the latest research and development information, and expected outcomes of this book

There is no recent comprehensive review of cassava research and development. With the explosion of information in general, and specifically about crop research, scientists have a difficult challenge to stay broadly informed about any crop, even one in which they may be specialized. These two volumes bring together global experts across the spectrum of cassava production and utilization topics, to distill and analyze information toward the broader context of achieving sustainable cassava cultivation. Apart from the broad chapter contents, the reader is provided extensive reference lists for further consultation and in-depth learning. This book will enable single-source consultation of a wide range of topics relevant to cassava R&D well into the future.

4 Section and chapter overviews

These two volumes present a comprehensive review of the history, the current strategies and the future potential to further transform cassava value chains toward sustainable systems. Each volume divides into three complementary sections.

Volume 1: Part 1 describes the cassava plant and its uses. **Chapter 1** takes an ethno-biological journey through the crop's early history, informed especially by practices still carried out today in some of the crop's most traditional growing areas of South America's Amazon basin. **Chapter 2** describes the high diversity of food products from cassava, especially in Africa, where over 90% of cassava is destined for human food use. Many of the processes were brought from the crop's homeland in the America centuries ago, and expanded and modified to meet local needs. **Chapter 3** reviews the opposite end of the spectrum – cassava's diverse industrial uses, and especially starch. Much of SE Asia's production is destined for industrial uses, but Africa and Latin America are also advancing in these more sophisticated, value-added markets. **Chapter 4** shows the potential and some examples of the transformation of traditional products and markets into higher value markets in Africa, especially for high quality cassava flour (HQCF).

Part 2 of Volume 1 begins within a broad overview of cassava production, processing and use across Asia, Africa and Latin America, **Chapter 5, 6 and 7,** respectively. Although the crop was domesticated in the Americas, only about 18% of current production comes from this region, and over half from Africa. These chapters compare and contrast the commonalities and differences among the regions, with lessons from each that can support sustainable development goals. **Chapter 8** draws on the regional overviews to highlight some of the global challenges and opportunities for sustainable cassava development, and the drivers of research and policy for setting priorities. Clearly there is no single strategy that applies globally, and every strategy fits into the context of an evolving social, economic, agronomic and environmental environment, among others. **Chapter 9** discusses the need to target and involve small-holder farmers in the development of cassava technologies, with case studies from Africa. **Chapter 10** presents the Global Cassava Partnership for the 21st Century (GCP21) as a global support partnership for cassava research and development. GCP21 is a not-for-profit international alliance of 45 organizations, aiming to fill gaps in cassava R&D in order to unlock the potential of cassava for improving food security and income, especially for the poor.

Part 3 of Volume 1 includes a comprehensive coverage of production practices for sustainable intensification of cassava production. **Chapter 11** describes the full range of production practices that growers should take into account to improve yields, profitability and sustainability – land preparation, plant populations, cropping systems, weed management, pest and disease management, and harvesting. The chapter draws heavily on experiences from Asia, but with relevance around the world. **Chapter 12** reviews the critical management of *seed* in cassava – normally through the vegetative reproduction through stem pieces. Currently nearly all seed is managed under informal systems, but this is slowly changing. Practices and systems developed in India inform the needs, challenges and experiences globally to develop improved seed systems. **Chapters 13, 14 and 15** give extensive coverage to best practices for managing cassava nutrition to achieve high and sustainable yields, through fertility management and soil conservation. As the demands increase for higher yield through sustainable practices, effective long-term soil fertility maintenance is a core strategy. The chapters draw especially on very extensive research and production experience in Asia and Latin America, where fertilizer use is far more common than in Africa. The chapters cover the relationship between soil fertility and crop productivity, how to diagnose nutritional needs and disorders, and best practices to achieve sustainable productivity through nutrient application. **Chapter 16** describes rotation and intercropping in cassava cultivation, especially common in Africa. There is a growing body of evidence on the sustainability and income advantages of diversified crop systems. Because of cassava's long period in the field, and its slow early growth, there is a wide range of alternative options for managing multiple crops that are complementary to cassava's growth and development. Finally, **Chapter 17** of Volume 1 reviews the principles of mechanization for all aspects of cassava production, which until now has been limited, especially in Africa. The chapter uses a case study of the Cassava Mechanization and Agro-Processing Project of the Africa Agricultural Technology Foundation to illustrate principles, challenges and opportunities for cassava mechanization.

Volume 2 of *Achieving Sustainable Cassava Cultivation* covers genetic resources, breeding, and pests and diseases. **Part 1** focuses on genetic resources and breeding tools. **Chapter 1** reviews knowledge on cassava's unique growth and development features as a perennial crop managed as an annual, its vegetative propagation, and the fact that it has no phasic development as is the case for the grain and grain legume crops. This understanding is key to improving the crop through both management and breeding. **Chapter 2** describes the cassava's genetic resources, especially the *ex situ* collections managed by the International Agricultural Research Centers (CIAT and IITA), a fundamental resource for the crop's genetic improvement. **Chapter 3** delves into the genetic basis for cassava breeding, and provides novel and innovative strategies to move beyond the current plateau for improving cassava yield.In particular, the chapter describes in detail the rationale and the possible strategies to exploit heterosis and to make cassava breeding more efficient through the use of inbreeding. **Chapter 4** brings the reader up to date on the fast-moving repertoire of molecular knowledge that supports cassava breeding. While cassava lagged behind other major crops initially in the molecular revolution, it has been catching up fast in recent years and molecular techniques are poised to bring major benefits to the growers and consumers of this crop.

Chapter 5 provides the technical background and describes specific molecular tools for making cassava breeding more efficient. The use of marker assisted selection (MAS) and genomic selection (GS) are covered in some detail. **Chapter 6** reviews the current status and the potential for improving cassava through genetic engineering, or genetic modification through targeting the insertion of genes directly into the crop's genome to achieve novel traits. In spite of rapid scientific advances, the regulatory environment limits impact at the field level while governments and the public assess and absorb the potential risks and benefits.

Part 2 of Volume 2 looks at specific breeding goals for root yield and quality, and progress toward reaching them. **Chapter 7** uses a case study for cassava breeding programs in Thailand, which has had one of the world's most successful cassava improvement efforts through the collaboration of two local centers (The Department of Agriculture and Kasetsart University), with additional support from International Centers (CIAT) and the Thai cassava industry. The fact that breeding efforts were aimed exclusively at industrial markets allowed the Thai programs to focus on yield and starch content, and to gain valuable experience in breeding for these two traits. **Chapter 8** reviews a comprehensive program in West Africa to improve the nutritional content of cassava, for pro-vitamin A. The program supported by the HarvestPlus initiative demonstrates the many components of the value chain that need to be considered – from breeding through measuring nutritional impact at the household level. This case study is illustrative of the need for cassava improvement programs to be fully integrated with processors and consumers to develop successful products. **Chapter 9** reveals the importance and the complexity of breeding for the fine-tuned consumer preferences, especially in Africa where a plethora of different products and their regional variations present major challenges to breeders with regard to identifying priority traits and their prioritization for breeding. The chapter uses examples from Africa to sort through some of these complexities and guide breeders in their planning.

Part 3 of Volume 2 covers pest and disease management in cassava, including weeds. These are areas that have received insufficient attention in the past, due to a widespread belief that cassava is a rugged crop that will produce reasonably well without any need to manage pests and diseases. However, as production practices have intensified, and especially as insects, mites and pathogens have moved with ease around the world, it is now well-understood that sustainable production is possible only with good management of these problems. **Chapter 10** describes the major diseases affecting cassava and their integrated management. It is noteworthy that two of Africa's most devastating problems – cassava mosaic disease and cassava brown streak disease – have not been found in the Americas where the crop and nearly all other of its pests and disease have co-evolved. Asia, once nearly free of serious disease problems, is recently experiencing new challenges, for example with witches broom disease and Sri Lankan cassava mosaic disease. **Chapter 11** reviews the arthropod pest complex of cassava and the importance of integrated pest management strategies. Because of cassava's long growth cycle, along with other factors, pesticide applications are usually not economically or ecologically sustainable. Cassava entomologists have a long success in biological control of major pests, especially the cassava mealybug, once one of Africa's most devastating pest problems and now brought under control in most regions through a parasitic wasp introduced from cassava's homeland in the Americas. **Chapter 12** discusses integrated weed management for cassava, with a focus on Africa. Currently, the vast majority of weed control is done manually in Africa. However, the demands for more efficient and effective weed control are creating the need for research on new options, both mechanical and chemical, as well as refined crop management such as intercropping and supporting early-vigor and shading through fertilization, plant spacing and variety selection. In Africa, most weeding is done by women, and new options for better weed management can have broad gender implications for the continent.

Part 1

Cassava genetic resources and breeding tools

Chapter 1

Advances in understanding cassava growth and development

Virgílio Gavicho Uarrota, Deivid L. V. Stefen, Clovis Arruda de Souza and Cileide Maria Medeiros Coelho, University of the State of Santa Catarina (UDESC), Brazil; Rodolfo Moresco and Marcelo Maraschin, Federal University of Santa Catarina (UFSC), Brazil; Fernando David Sánchez-Mora, Technical University of Manabí (UTM), Ecuador; and Eduardo da Costa Nunes, Enilto de Oliveira Neubert and Luiz Augusto Martins Peruch, Santa Catarina State Agricultural Research and Rural Extension Agency (EPAGRI), Brazil

1 Introduction

The objective of this chapter is to provide insights and highlight advances in understanding cassava's fundamental physiological traits – essential to support the research of agronomists, plant breeders, plant protection specialists and others dedicated to sustainably improving cassava production, processing and use. The chapter explores in detail the growth and development of the plant's main structural components as well as the physiology and control of carbon partitioning between source (leaves) and sink (roots) for optimizing yields (e.g. maximizing financial returns, with sustainable practices). Cassava has several features that make it especially well-adapted to climate change, especially related to increasing

http://dx.doi.org/10.19103/AS.2016.0014.03

temperatures and less predictable rainfall. Special attention is also given to providing an understanding of the rapid post-harvest physiological deterioration (PPD) of cassava's roots, a feature that has profoundly influenced the ways that cassava is managed before and after harvest.

Cassava (*Manihot esculenta* Crantz) is a major food and industrial crop in tropical and subtropical Africa, Asia and the Americas, from about 30°S to 30°N latitudes (Cock, 1982, 2012) and below about 1500 masl. The plant is a perennial shrub grown mainly for its starchy roots. The crop and its wild relatives – some 100 species of the genus *Manihot* – existed only in the New World until trade routes opened between the Americas and Africa in the early sixteenth century (Pinto-Zevallos et al., 2016; FAO, 2013). Now, far more cassava is grown outside than within the centre of origin – about 53% in Africa and 35% in Asia (FAOSTAT, 2017). In Africa cassava is mainly used for human consumption, while in Asia, with some exceptions, it is mostly used for industrial starch and starch-based products, for animal feed and more recently for fuel ethanol (Stupak et al., 2006).

Cassava's tuberous starch storage roots provide over a quarter of the daily overall calorie consumption in some African countries, such as Angola and Mozambique, but for the rural poor it represents a much larger proportion of daily calories (Souza et al., 2016). In addition to the roots, which are the main harvested product, cassava leaves are consumed as a vegetable in at least 60% of the countries in sub-Saharan Africa, providing an important source of proteins, vitamins and micronutrients (USDA, 2016; Latif and Muller, 2015). Cassava leaves are also used as a protein supplement for livestock and are sold for further processing and use, providing important supplementary income for smallholders (Souza et al., 2016; Morgan and Choct, 2016).

Because of its tolerance to soil acidity, low soil fertility and drought, the crop is often grown by poor, smallholder farmers in marginal areas with poor soil and limited rainfall, on slopes where erosion is a high risk, and in forest margins. Good examples are the dry and infertile areas of northeast Brazil; northwest coastal regions of South America (Colombia and Venezuela); and the seasonally dry areas of southwest Nigeria, east and south of Lake Victoria in Tanzania, the northeast of Thailand and the eastern part of Java. It is highly flexible in harvest time, and mature roots can be left in the soil for months before harvesting (Okogbenin et al., 2013). This flexibility, along with vegetative propagation via stem cuttings and low input needs, makes cassava an excellent food security crop for smallholders (Stupak et al., 2006). The UN Food and Agriculture Organization (FAO) predicts that area planted to cassava will increase and production systems will intensify in the future, highlighting the need for developing strategies that improve the sustainability of production (Pinto-Zevallos et al., 2016).

Due to inherent tolerance of cassava to heat, water stress and poor soils, this crop is relatively adaptable to a warmer climate (El-Sharkawy, 2014). Jarvis et al. (2012) studied cassava's response to rising temperatures and to drought, relative to other major crops, as a possible answer for African climate change adaptation. They modelled the impacts that climate change will likely have on cassava itself, and on other important staple food crops for Africa including maize, millets, sorghum, banana and beans based on projections to 2030. Results indicate that cassava is actually positively impacted in many areas of Africa, with −3.7% to +17.5% changes in climate suitability across the continent. For other major food staples, they are all projected to experience negative impacts, with the greatest impacts for beans (−16% ± 8.8), potato (−14.7 ± 8.2), banana (−2.5% ± 4.9) and sorghum (−2.66% ± 6.45). In contrast, Sousa et al. (2016) concluded that, in the longer term, climate change by mid-century could depress aggregate cassava yield across sub-Saharan Africa by 18% in the absence of adaptation through breeding and management.

2 Overview of cassava growth and development

The characteristics of cassava growth and development of course share much in common with all crop plants, in that domestication and breeding have skewed photoassimilate partitioning towards products of human interest rather than towards optimum adaptation in natural environments. In the case of cassava, this means a high proportion of the plant's biomass is in the roots. At the same time, evolutionary forces and early farmer selection during and after domestication also modified growth and development characteristics related to traits such as pest and disease resistance, drought tolerance, root quality traits and adaptation to a range of cropping systems. Thus, there are key traits from cassava's evolutionary past, early domestication and further improvement that continue to feature strongly in the growth and development characteristics. This section introduces some of these key traits that distinguish cassava from other major crops, especially the cereal and grain legume crops where seeds are the main product of human interest.

One of the main features of cassava as a crop is that there is simultaneous development of the economically useful part (the roots) and the photosynthetic apparatus – the leaf *factory* – as compared to the phasic development of most determinant crops. This simultaneous development is vital to the understanding of cassava growth and development. Although *phases* of development have been identified for descriptive understanding (e.g. Matthews and Hunt (1994) and Alves (2002)), these are in fact somewhat artificial, and from a physiological perspective, there are no well-defined phases.

Cassava produces flowers, fruits and seeds, but, *per se*, these have little to no direct importance for growers. Flowering is important, however, to understanding plant growth and development, in that its initiation and the associated loss of apical dominance are the basis for the emergence of 3–4 lateral buds in the nodes just below the emerging flower apparatus, which will create the classical sympodial branching (or *forking*) characteristic of cassava (Fig. 1). Flowering times vary greatly among varieties and are also strongly influenced by environmental factors such as photoperiod, temperature and soil nutrient status. As a result, plant architecture, as influenced by number of branching levels, is also modified. If flowering occurs late or not at all in the growth cycle, plants will be very erect. Conversely, when flowering begins early, the plant will be highly branched and spreading. It should be noted that flowers and seeds are fundamental to the breeder for creating new genetic diversity leading to new varieties.

Cassava can be harvested whenever economic yield levels are reached. There is technically no defined period of maturity. Botanically, cassava is a perennial, though cultivated as an annual. While it is possible to harvest a few roots at a time and allow the plant to continue growing indefinitely, this is only done on a very small scale with garden-type production. In normal cultivation practice, entire plants are uprooted at harvest, thus ending the growth cycle. Roots typically reach desired size for consumption or processing at around nine months after planting, but the period of harvest can vary from 6 to 7 months (and even as early as three months, according to authors' experiences) to a few years. Industrial varieties that aim for starch production (as opposed to root quality for human consumption) can be harvested from 20 to 26 months after planting. In semi-arid regions of Northeast (NE) Brazil, cassava may remain in the ground for 4–5 years before harvest (Moore, 2016), as a food security crop.

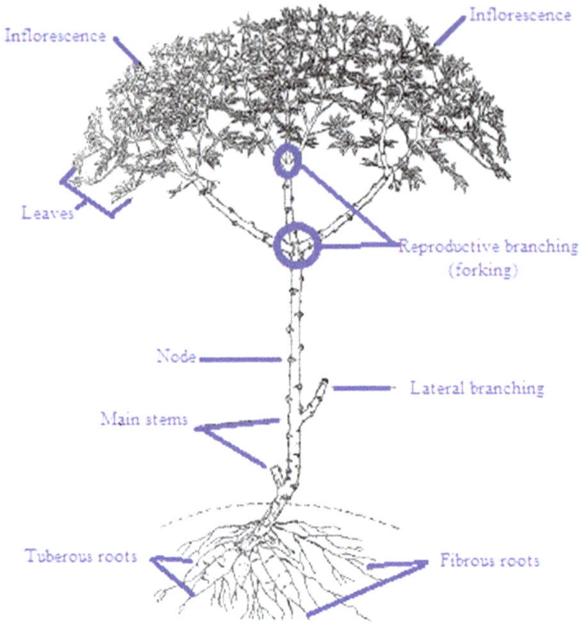

Figure 1 Diagram of cassava plant (adapted from FAO, 1999).

Cassava's geographic limitation is defined largely by the plant's sensitivity to both the low constant temperatures of the highland tropics (though the crop can be found at over 2000 masl in limited areas of the Andes of eastern South America and in the East African highlands) and its sensitivity to frost in temperate regions. Frost sensitivity determines the crop's northern and southern ranges, since it requires nearly a year to produce commercial-sized roots. Hence, even though cassava can grow well during the warm summer months of the temperate zone, it cannot produce a commercial yield prior to frost. In areas where the winter is mild and frost sporadic (e.g. southern Brazil, Paraguay and southern China) the cassava growth cycle can be extended to 18–24 months by cutting back stems prior to frost, and allowing regrowth in the spring.

From an evolutionary perspective, the development of starchy roots is most likely an adaptation of a perennial plant to periodically dry climates. These reserves allow the plant to tolerate extended periods of drought, mobilize these carbohydrate reserves when favourable conditions return, and continue growing. Like most domesticated plants, farmer-breeders selected for exaggerated expression of a naturally evolved plant feature, optimized for human use. Thus, while high levels of root starch may not have a survival function in most situations where cassava is currently grown, these reserves play a critical role in growth and development and in our understanding of how to increase yield and stability for specific production conditions.

The cassava plant can be grown from botanical seed or propagated vegetatively from stem cuttings. All commercial production is from stem cuttings, while botanical seeds are used mainly by plant breeders. Farmers, especially in very traditional systems, sometimes care for and select volunteer plants that have emerged from seeds that have fallen to the ground and germinated. In commercial production, cuttings are taken from mature, woody stems, another limitation to the minimum length of the required growing season. Stems of sufficient maturity to produce high-quality planting material typically are not available until about 6 months or more after planting. The planting piece, or *stake*, is normally 15–25 cm long and 1.5–2.0 cm in diameter. However, stems of varying maturities are capable of regenerating new plants, so long as they have viable buds. The quality of planting material (viability and ability to produce vigorous new plants) is strongly influenced by the physiological condition of the mother plant.

The fundamental structural unit of the cassava stem, or phytome, is the nodal unit consisting of a leaf, axillary bud (or buds) and the subtending node. Figure 1 shows the basic morphology of a mature cassava plant, with the main components to be discussed in the following sections: leaves; lignified stems, which develop a branched or forked pattern as a result of the initiation of flowering; fibrous roots for nutrient and water uptake; and starch storage or *tuberous* roots, which are the main economic product of cassava.

The developmental periods and main physiological events during the growth of cassava under favourable conditions according to days after planting (DAP) can be generalized as (Alves, 2002):

- *5–15 DAP:* Sprouting of the buds on the planting piece (or *stake*); first leaves begin to expand; first adventitious roots arise from the basal cut surface and from the nodes (base of buds) under the soil surface (Seesahai and Ousman, 2016).
- *15–90 DAP:* Leaf and root system development; by about 90 DAP, some roots predetermined as starch storage roots begin to swell with accumulated carbohydrates.
- *90–180 DAP:* The maximum growth of leaves and stems is achieved, light interception is maximum with a high rate of dry matter partitioning to leaves and roots, and rapid root bulking with increased translocation of assimilates to roots is achieved. Leaf senescence and leaf fall begin.
- *180–300 DAP:* The period of high carbohydrate translocation to roots; photoassimilate partitioning from leaves to roots is accelerated with highest rates of dry matter accumulation in the storage roots while leaf senescence increases; stems become lignified (Alves, 2002). In most cases the best time to harvest storage root for food is between 8 and 10 MAP (FAO, 2013).
- *300–360 DAP:* Leaf production rate decreases, translocation of sugars and conversion into starch in roots continues and the growth cycle is typically completed by harvest at around one year. Dormancy can be observed at this stage in subtropical environments; the rate of leaf production decreases, shoot growth is finished and only translocation of starch to roots continues (e.g. southern Brazil and southern China). The plant completes its 12-month cycle which can be followed by a new period of vegetative growth, dry matter accumulation in roots and dormancy again (Alves, 2002).
- *Post-harvest:* Typically within a few days after harvest, when roots are stored in ambient conditions, bluish-black spotting and streaking begin to appear throughout the root flesh, a process and condition known as PPD. As a result, roots need to be cooked or processed quickly after harvest, or managed in other ways to control this process. While both physiological and pathological processes are involved, this chapter includes a detailed review of the physiological aspects, an innate part of the plant's metabolic processes.

3 Leaf development, canopy formation and photosynthesis

The total biomass produced by any crop is influenced directly by the amount of the total solar radiation that is intercepted by the canopy and the efficiency with which it is used by the leaves in photosynthesis. Light interception in turn is closely related to leaf area (LA) index (LAI), the ratio of total one-sided LA per unit ground surface area (Chen and Black, 1992). The development and maintenance of LAI depends directly on the number of plants per unit land area, the number of leaves produced per plant, the size of those leaves and their longevity. The total number of leaves produced by the plant, their longevity and photosynthetic capacity are varietal characteristics, but which are also profoundly influenced by physical environmental conditions and by pests and diseases (Ospina and Ceballos, 2012). In general, LAI increases with crop age and then declines towards the end of the crop cycle. However, this pattern varies widely (Cock, 2011).

3.1 Leaf and canopy development

Leaf and canopy development are a means towards creating an efficient photosynthetic process. From a physiological standpoint, the outcome of shoot development and growth is the formation of crop canopy which in turn determines, in interaction with the environment, total biological productivity as well as economic yield (El-Sharkawy et al., 2004). During development, cassava plants develop typical canopy architecture by new leaf growth, branching and shedding older leaves.

The cassava canopy typically starts to develop at about 15 DAP the stem section propagules and reaches maximum light interception at 4–5 months (Alves, 2002; El-Sharkawy, 2004). This would appear slow compared with maize and soybean crops, which can achieve closed canopies within four weeks (Singer et al., 2011; Souza et al., 2016). Cassava stem segments should in theory represent a much larger reserve compared to seed, a factor which should in turn boost rapid development of leaves and canopy closure. This suggests that there has been little selection for accumulation of reserves in the stem, which could greatly improve the speed of canopy closure in cassava (Souza et al., 2016).

Leaves are the main photosynthetic organs, transforming radiant into chemical energy. A new cassava leaf is produced at each nodal unit, or phytome, except when the apex becomes reproductive. Leaves are palmate, with the number of lobes almost always being odd. Due to apical dominance and the indeterminate habit of the cassava plant, new leaves are formed continuously in a spiral manner but with rates that vary with the number of vegetative stems per plant and number of sympodial (reproductive) branches. This process is affected by the age of the plant, the nutritional status and general growth conditions (e.g. water stress, the vigour of the material, light, etc.). The rate of appearance of new leaves from a single apex is greatest when the plant is young, starting at a rate of about one new leaf per day and declining to one leaf per week in older plants, with little difference amongst most varieties (Cock, 2011). In the early growth stages, fully expanded leaves are small but they gradually increase in size, reaching their maximum size four to six months after planting (depending on the temperature), after which they become smaller.

Leaves are caducous, that is, with age they senesce and fall from the plant. This is an abscission process, with the formation of an abscission layer. During development, cassava plants develop typical canopy architecture by shedding older leaves. The life of a leaf ranges from about 60 to 180 days under normal conditions, increasing at later growth stages (Ceballos and de la Cruz, 2012). There are large varietal differences in leaf longevity, and pests and diseases can cause greatly reduced leaf life. Under cool temperatures leaf life can extend to about 200 days or more. The leaf life of cassava is short. Prolonging the life of leaves could improve the root yield and quality as well as permit more frequent foliage harvesting (Zhang et al., 2010).

Replacement of leaves requires a significant metabolic input and creates a competitive sink for photosynthetic assimilates, thus reducing root yields. In contrast to cereals, for example, in which reproductive development and seed formation follow vegetative development, photosynthesis in cassava has to be allocated for both leaf (source) and storage root (sink) growth simultaneously (Zhang et al., 2010). In cassava, it is generally assumed that production of storage roots is limited by the source tissue or LAI rather than by the sink strength of the roots. Experimental data suggest that root production is positively correlated with the lifespan of individual leaves (Tan and Cock, 1979). A more recent report also demonstrated that increased leaf retention could result in increased root yield under irrigated and stressed conditions (Lenis et al., 2006). Since cassava grows largely under marginal conditions, soil water deficiency is common. Leaf abscission in cassava helps the plant to reduce evaporation under water stress (Zhang et al., 2010). Leaf senescence can be induced by many different environmental and endogenous stimuli such as age, drought stress, nutrient deficiencies, fruit development, light, shading and hormone concentration (El-Sharkawy et al., 2007; Clark et al., 2004; Cock et al., 1979), higher temperatures, genotype, cropping system and management practices (Ekanayako et al., 1998).

According to Vile et al. (2005), leaf thickness (LT) also plays an important role in leaf and plant functioning and is related to species' strategies of resource acquisition and use. LT and mass in cassava are related to number of cells per leaf, cell expansion and proliferation. The amount of light absorbed by a leaf and the diffusion pathway of CO_2 through its tissues depend, at least partially, on its thickness. Negative relationships between LT and photosynthetic and growth rates have been observed. LT has therefore often been used as a tool to screen species and/or varieties for productivity or ecological performance (Vile et al., 2005). Under stress, leaves might become thicker, with smaller epidermal and mesophyll cells (Porto, 1983). Other adaptations include an increase in number and a reduction in the size of leaf hairs; abundance of coatings such as cutin, suberin and resins, leaf rolling or folding; thickening of outer epidermal walls and cuticle; and acceleration of the process of abscission (Porto, 1983). These mechanisms, together with changes in stomatal distribution and size, have the ability to reduce the amount of water lost by the plant and the effects of heat caused by excessive absorbed radiation (Porto, 1983).

According to a simulation model for substantially increased cassava yields with longer leaf life, the plant ideally has a leaf life of 15–20 weeks and an optimal LAI of 2.5–3.5 (Cock et al., 1979). In most cultivars, however, leaf life is much shorter, resulting in a suboptimal LAI, which in turn causes reduced root yields, especially in non-branching cultivars. For high levels of productivity it is obviously necessary to have sufficient LA to intercept most of the incoming radiation, but it has been suggested that many varieties of cassava do not have sufficient LA to achieve their potential productivity (Rosas et al., 1979). Replacement of leaves requires a significant metabolic input and creates a competitive sink for photosynthetic assimilates, thus reducing root yields.

Because of the small LA in the first three months and in the final 8–12 months, the seasonal average canopy area still limits the yield (El-Sharkawy et al., 2012; Ding et al., 2016). LA was found to increase by shading in cassava (Ding et al., 2016).

LAI is determined by leaf size, leaf formation rate and individual leaf life. According to Cock et al. (1979) leaf size reached a maximum within four months after planting and then decreased; the maximum was a varietal characteristic. Leaf life was reduced by shading, but in full daylight was determined by variety. Leaf formation rate per shoot apex showed little genetic variation and declined with time; large differences in leaf formation rate per plant were determined by differences in branching pattern. Top growth had preference over root growth, and root sink was not limiting when root number per plant was nine or more (Cock et al., 1979).

The branching behaviour of stems is a key determinant of canopy characteristics. Branching varies widely among varieties, ranging from zero to about four branching levels, rarely more. Height of first branching and subsequent branching levels are also strong factors in farmer preference, but the preferred branching habit varies across regions and farming systems. Varieties that branch earlier can have an advantage in faster canopy closure, and subsequent effects on weed control. Later branching types are typically more efficient in carbon partitioning, and provide easier access for various manual or mechanical management activities such as weeding. Overall, there is a shifting tendency for breeders to select for later branching types.

Mutual shading of the leaves will affect crop yield, and therefore optimized plant spacing is critical to achieving optimum LAI (Seesahai and Ousman, 2016; Alves, 2002). Streck et al. (2014), studying the effect of plant spacing on growth, development and yield of cassava, concluded that treatments with narrower spacing between plants (i.e. higher plant density) had higher LAI. On the other hand, the LA per plant, unlike the LAI, was higher in treatments with wider spacing whilst the final leaf size and the number of lateral shoots increase with increasing plant spacing.

Leaf harvesting, especially common in some West and Central African countries, can have a drastic effect on the cassava canopy through reduced LAI and total photosynthesis, to ultimately affect cassava growth and development and finally the root yield. Lockard et al. (1985) recommended that leaf collection should start 4 months after planting, and harvesting should be carried out at 2-month intervals in order to obtain the best yields of both leaves and storage roots.

3.2 Photosynthesis

Cassava photosynthesis has a high optimum temperature (35°C) and a wide plateau (25–35°C) corresponding to the temperature range under which cassava is largely cultivated. Leaf photosynthesis requires high saturation irradiance (1500 $\mu mol\ m^{-2}\ s^{-1}$) and the rates are greatly reduced by leaf air vapour pressure differences above 1.5 kPa, a reduction associated with stomatal closure. Field measurements of single-leaf photosynthesis among a wide range of varieties grown under rain-fed conditions showed that when light interception was not limiting, there were significant correlations between leaf photosynthesis, total biomass and root yield (El-Sharkawy, 2012; El-Sharkawy and Cock, 1990).

Most crop plants possess either the C3 or C4 photosynthetic cycle. The C4 plants tend not to light-saturate at high irradiance, have low photorespiration and high photosynthetic rates on a per unit LA, and are efficient in use of nitrogen and water. Cassava has normally been considered to be a typical C3 plant (as a dicotyledon with a lower photosynthetic rate). Under certain conditions, cassava exhibits an interesting photosynthetic C3–C4 intermediate behaviour which may have important implications for future plant selection (El-Sharkawy, 2006; El-Sharkawy and Tafur, 2010). Although it does not have the typical 'C4-Kranz' anatomy of C4 leaves, it does have low photorespiration, low CO_2 compensation point, high percentage of carbon fixation in C4 acids and a high PEP-carboxylase activity (15–35% of that in maize). Much evidence argues that cassava and some *Manihot* wild species represent an intermediate photosynthesis evolving from C3 to the C4 species (e.g. monocotyledon plants with high photosynthetic rates) (Cock, 2011). Parallel work feeding $^{14}CO_2$ to leaves of cassava showed an apparent intermediate pattern of initial C4 and C3 products (Cock et al., 1987). However, a subsequent detailed analysis of first products failed to show such an intermediate pattern, but rather a very typical C3 pattern (Sousa et al., 2016). According to El-Sharkawy et al. (2012), more research is needed to elucidate the photosynthetic behaviour of cassava, but they suggest that probably cassava is evolving biochemically towards a C4 photosynthetic pathway and therefore demonstrates C3–C4 intermediate photosynthetic behaviour. The overall findings indicate that cassava is a C3 species (Gleadow et al., 2009; Rosenthal et al., 2012).

4 Root development

The root system of cassava is composed of three root types (Seesahai and Ousman, 2016; El-Sharkawy et al., 2004):

1. Fine (fibrous) roots, which absorb water and nutrients. They are small, white and found all along the planting set and never become a tuberous root.
2. Thick roots that serve as anchorage. They also never become a tuberous root, but they do absorb water and nutrients (Knoth, 1993).
3. Starch storage roots (tuberous roots), which are specialized, underground, storage organs, and do not absorb water and nutrients, but only store food-starch molecules.

The elongation, growth angles from the main axis, lateral branching and longevity of all root classes determine the architecture of the root system. Each of these components is determined by a combination of genetic, physiological and environmental factors such as soil moisture, nutrients, temperature, pH and microbial communities (Khan et al., 2016). Root system architecture therefore refers to the spatial configuration of the root system or the explicit deployment of root axes.

Fibrous roots are formed initially within a few DAP. After about 2 months secondary thickening of some of the fibrous roots is visible and starch is deposited in the parenchyma (Cock, 2011). These roots initially form part of the fibrous system but, later, some (usually <10) begin thickening and become storage roots. The number of storage roots is determined, in most cases, by the conditions of a plant's early growth. The visible starch storage process begins at 6–8 weeks. Internal cell division gives rise to storage parenchyma and conversion of sugars to starches by the enzyme starch amylase (Seeshami and Ousman, 2016). The dry matter content of cassava roots (typically about 85% of dry matter is starch) is a varietal characteristic but also strongly influenced by environmental conditions; some varieties tend to always produce higher dry matter than others. Dry matter tends to increase as the plant reaches maturity.

In cassava, lateral roots have been shown to develop successively from the pericycle cells adjacent to the xylem pole (Villordon et al., 2014; Khan et al., 2016).

Factors that stimulate the change from fibrous root to storage root, and their chemical signals, are unclear. Chaweewan and Taylor (2015) showed that fibrous roots possess a distinct cortical region with a central stele, and develop from the vascular cambium of the stem. Only nodal roots undergo secondary thickening to develop as storage organs. Meristematic cells, from which the nodal roots are derived, may be initiated from remnants of the pericycle associated with protoxylem commonly described in relation to fibrous roots. A better understanding of the physiological origins of storage roots and the signals involved will be important for devising strategies to increase yields through advanced genetic tools and methods.

Yang et al. (2011) investigated differences in gene expression of three types of cassava roots: fibrous root (FR), developing storage root and mature storage root to gain better understanding of cassava root development at the transcriptional level. They proposed L-lactate dehydrogenase and aldehyde dehydrogenase as the rate-limiting enzymes of the glycolysis/gluconeogenesis pathway via enolase to be targets for improving starch yield in cassava. Sujikol et al. (2010), studying the cassava genome-wide expression at the onset of storage root formation, found sulphite reductase, calcium-dependent protein kinase (CDPK), ent-kaurene synthase (KS) and hexose transporter (HT) as root-specific expression patterns. They proposed that KS and HT may be involved in transient induction of CDPK expression, which may play an important role in the signalling pathway of storage root initiation. Sulphite reductase, on the other hand, may be involved in storage root development by facilitating sulphur-containing protein biosynthesis or detoxifying the cyanogenic glucoside content through aspartate biosynthesis.

Rooting response is also important for good crop establishment. Stakes capable of producing a good number of roots depends on stem age, size, thickness, number of nodes per stake and their orientation at planting (Legese et al., 2011). Experiments conducted by EPAGRI (Santa Catarina State Agricultural Research and Rural Extension Agency), Urussanga (Southern Brazil), showed that vertical planting resulted in greater weight of roots and more vigorous overall plant development (Neubert et al., 2015a,b).

It has been suggested that variability in root architecture in cassava in response to soil moisture can be used for breeding crops with stable yields in environments that are subject to soil moisture stress (Pardales et al., 2003).

5 Carbon partitioning in cassava

Plant yield and quality ultimately depend on composition of the storage products and biomass for each product, governed by an important metabolic process in higher plants, called carbon partitioning (Thelander, 2017). Plant carbon partitioning has been studied to some extent for cassava; however, there are still many issues that need to be investigated, especially regulatory mechanisms at the molecular level. For instance, which molecules play a role as a *switch* in orientating the carbon flux to a certain storage product? How does a regulatory molecule modulate the switch(es) in directing carbon partitioning? How do environmental conditions affect carbon partitioning at the molecular level? (Thelander, 2017).

The balance between top growth and root growth is key to understanding the productivity of cassava, especially in terms of managing the crop for maximum yield and for resilience under stress, for example, drought or pests (Cock, 2011). In contrast to cereals, in which reproductive development and seed formation follow vegetative development, photosynthesis in cassava has to be partitioned for both leaf and associated stems (source) and storage root (sink) growth simultaneously throughout the plant's life. In cassava, it is generally assumed that production of storage roots is limited by the source tissue or LAI rather than by the sink strength of the roots (Zhang et al., 2010; Porto, 1983). Dry matter production was found to be more reduced when water stress was imposed early in the plant growth cycle. The stressed plants delayed the allocation of dry matter to the storage roots (Porto, 1983).

The dynamic partitioning of biomass into shoot and storage roots has been discussed by Gray (2000). According to the model, the amount of biomass partitioned into storage root production during cassava crop growth can be conceptualized in terms of the mass balance Equation 1.

$$\left[\frac{dR_s}{dt} = A_d - R - G_1 - G_{rf} - \frac{dC}{dt}\right], \quad (1)$$

where A_d is canopy photosynthetic rate, R_s storage roots, G_1 gross growth rate of leaves and petioles, R total respiratory substrate demand, G_{sb} gross growth rate of stems and branches, G_{rf} gross growth rate of fibrous roots and C reserve substrate carbon pool.

According to this model the growth rate of storage roots will increase if any or all of the terms following a *minus* sign are reduced or if A_d is increased. Based on the equation, biomass allocation to storage root growth equals the amount of substrate carbon that remains once the growth demands of the other cassava plant components have been fulfilled (Gray, 2000). The two major sinks for assimilate during cassava crop growth are the woody component of the shoot and the storage roots. The balance of assimilate allocation between these two sinks determines both harvest index and yield for the cassava crop (Cock et al., 1979).

According to a model described by Matthews and Hunt (1993), potential dry matter production is modified by the effects of light, temperature, water stress and vapour pressure. Leaf and stem growth are assumed to be the dominant sinks for assimilate, with fibrous roots receiving a fraction of that allocated to the shoot, decreasing as the crop ages (Matthews and Hunt, 1993).

In a study by Siriwat and co-workers (2012) that integrated gene expression data by Yang and co-workers (2011) with metabolic pathways (cell wall biosynthesis, starch biosynthesis and respiration pathway) it was found that the genes involved in starch biosynthesis are more upregulated, in contrast to the expression of genes in the cell wall biosynthesis and respiration. The results may indicate that the carbon source from α-D-glucose-1-phosphate (G1P) might be mostly used for starch biosynthesis rather than cell wall biosynthesis and respiration pathways during cassava root development.

However, under conditions of increased canopy photosynthesis, it appears that the excess is allocated to the tuberous roots. Increasing crop photosynthesis may therefore result in a larger than expected increase in yield, given that a 30% increase in photosynthesis at the leaf level results in a more than three-fold increase in the harvested yield of tuberous roots (Souza et al., 2016; Rosenthal et al., 2012; Rosenthal and Ort, 2012; El-Sharkawy et al., 1990; Cock, 1976). Root and shoot yields and total biomass in all cultivars were also significantly correlated with seasonal average LAI (El-Sharkawy and Tafur, 2010). Nitrogen supply also increased shoot dry mass to a greater extent than root dry mass, thus leading to an increased shoot to root ratio (Cruz et al., 2003). Photosynthetic rates decreased with increasing N-deficiency. Such a decrease was circumstantially associated with a limited capacity for carbon export from source leaves under N-limitation (Cruz et al., 2003). The sink capacity of branches was enhanced greatly by N fertilization; at the high N rates the proportion of dry matter allocated to branches increased almost linearly over time (Manrique, 2008).

Sink limitations in the roots might influence total dry matter production by feedback that reduces the photosynthetic rate of the leaves (Williams, 1974). Small reductions in root number did not affect total dry matter production or root yield (Cock et al., 1979). Tan and Cock (1979) observed the relations between top and root, the sink capacity of the roots and the effects of changing the branching habit on the yield. They concluded that apices and roots were competing sinks; roots accepted excess carbohydrates beyond what was needed to sustain shoot growth and development, in normal situations. Reduced sink did not limit total growth. Yield was increased by 75% by branching control. Good yields can be obtained by either late-branching types or continuous branching types with two branches at each branch level (Tan and Cock, 1979).

Building on earlier work by El-Sharkawy et al. (1990), which showed significant correlation in leaf photosynthesis, total biomass and root yield, Izumi et al. (1999) found that the dry weight of both shoots and roots increased rapidly with the LA.

Understanding of sink-source relations is key to the concept of developing *early* varieties of cassava. Selection of early yielding genotypes has become important in the national cassava breeding programmes in Africa and Asia where mounting pressure on land for urban and industrial development is compelling farmers to intensify production, and also in semi-arid regions where early yielding genotypes can be harvested after only one cycle of rain. Strategies for achieving earlier yields than standard varieties include:

1 An earlier initiation of root bulking followed by the same bulking rate;
2 The same time of initiation of bulking followed by more rapid bulking rate;
3 A combination of both earlier bulking and higher bulking rate.

Identification of such traits is of great importance to understanding and improving cassava development, growth and productivity.

Chipeta and co-workers (2016), studying the traits associated with early bulking in cassava, reported that harvest index, storage root number, storage root diameter and storage root length can be used as the selection criteria to achieve high fresh storage root yield and dry mass yield. The study suggested that both assimilation (source) and storage root (sink) capacities are important for determining early yield. Tumuhimbise and co-workers (2014) found significant genotype effects for early bulking. Also, GxE interactions underline the need to conduct multi-locational trials in order to identify both generally and specifically adapted genotypes with early yield.

6 Cassava response to environmental conditions: solar radiation and temperature

Sunlight, water, temperature, humidity, adequate soil moisture and access to nutrients are all required for growth and development. Various authors (El-Sharkawy, 2012; Edwards and

Kang, 1978; Irikura et al., 1979; Keating and Evenson, 1979; Connor et al., 1981; Connor and Cock, 1981; Keating et al., 1982a,b; Fukai and Hammer, 1987; Iglesias et al., 1994; Santisopasri et al., 2001) have studied the response of cassava to variations in a range of environmental factors. This section discusses the effects of solar radiation and temperature. Section 7 discusses the effects of water availability, and Section 8 addresses salinity and factors involved in climate change which are expected to have great impact on crop productivity, namely increased atmospheric gases such as carbon dioxide (CO_2), methane (CH_4), nitrous oxide (N_2O) and ozone (O_3) (Huber, 2011). While mineral nutrition is a key to cassava growth and development, the topic is discussed extensively in other chapters.

6.1 Solar radiation

Daily solar radiation reaching individual cassava leaves is affected by latitude (primarily by effect on day length), cloud cover and shading by intra- or intercrop effects. The study of the shade response of cassava is very important for both intercropping management and genetic improvement for conditions where solar radiation may be reduced, including high cloud cover environments (Ding et al., 2016).

As reported by Fukai et al. (1984) low radiation led to leaves with high specific LA, particularly in young plants. However, because of reduced leaf dry matter increase, LAI was slightly reduced. In the canopies with full ground cover, more than 90% of the incident radiation was intercepted by the top half of the plant. Light extinction coefficients were higher in the canopies under heavy shade, indicating an adaptation in canopy structure to reduced light environment.

For younger plants, reduction in solar input to 78% of maximum had significant effects on both root and leaf growth (reduction of 86% and 47%, respectively), while it had very little effect on plants in the late bulking stage. Reduction in solar input to 32% of maximum reduced crop growth rate (CGR) to about half that of the control regardless of the plant age. Reduced radiation affected dry matter growth of roots and hence reduced harvest index of plants of all ages. Storage root number was significantly reduced by shading during the root bulking initiation stage, but not during the other stages (Fukai et al., 1984).

Fibrous root length has also been reported to be affected by solar radiation (Aresta and Fukai, 1984; Irikura et al., 1979). Shading the plants for 35 days reduced the rate of increase in root length density and hence decreased total root length. Reduction in solar input to 32% resulted in a 50% reduction in elongation rate of fibrous roots. Reduction to 78% resulted in a similar reduction in elongation rate when plants were at root bulking stage, but much less reduction when plants were at the bulking initiation stage. The results indicate that under limited carbohydrate supply, shoots were a stronger sink than both fibrous roots and storage roots. At bulking initiation stage, storage roots were a weaker sink than fibrous roots, but during bulking both root types were about equal (Aresta and Fukai, 1984).

According to Okoli and Wilson (1986), plant height increased while LAI decreased as percentage of shade increased. Stem and leaf dry weight also decreased with increasing degree of shade. All levels of shade delayed tuber bulking. Compared to the 0% shade (control), the 20, 40, 50, 60 and 70% shade levels reduced cassava yield by 43, 56, 59, 69 and 80% respectively.

Under a leaf canopy (shade condition), both light quality and intensity could be changed – relatively higher far-red (FRE) but lower red (R) and blue light. Both R:FRE ratio and blue light, as well as the total light intensity, are signals that can induce shade-avoidance syndrome that causes the elongation of stems and petioles. Plants grow tall and spindly for sunlight competition at the expense of leaf development, and finally the yield was reduced.

Ding and co-workers (2016), studying cassava growth under both full sunlight and natural shade in intercropping systems, found that gene expression profiles of these two conditions exhibited similar developmental transitions. Genes related to cell wall and basic cellular metabolism were highly expressed in immature leaves, genes involved in lipid metabolism and tetrapyrrole synthesis were highly expressed during the transition stages, and genes related to photosynthesis and carbohydrates metabolism were highly expressed in mature leaves. Compared with the control, shade significantly induced the expression of genes involved in light reaction of photosynthesis, light signalling and DNA synthesis/chromatin structure; however, the genes related to anthocyanins biosynthesis, heat shock, Calvin cycle, glycolysis, tricarboxylic acid cycle, mitochondrial electron transport, and starch and sucrose metabolisms were dramatically depressed.

When cassava is intercropped with grain crops such as maize, rice or grain legumes (beans, cowpeas or groundnuts) that mature earlier, shading occurs in the early stages of development, reducing cassava's productivity. The study of the shade response of cassava is very important for both intercropping management and genetic improvement (Ding et al., 2016).

Intercropping cassava with plantation crops, in their early stages of development, is common, especially in India and Indonesia. Intercropping experiments in coconut stands revealed that cassava had no adverse effect on the main crop of coconut provided the same intercrop was not grown in the same land every year and both the intercrop and main crop were manured separately and adequately. Five to fifteen per cent increase in yield of coconut palms was obtained when cassava, elephant foot yam, greater yam, lesser yam and taro were grown in rotation (Nayar and Suja, 2007). The experimental evidence in general indicated that intercropping in areca nut gardens was not harmful to the main crop (Nayar and Suja, 2007). Its association with banana or coconut was found to be beneficial; however, when grown with eucalyptus or leucaena, the root yields were reduced by 60–80% (George et al., 2000). Eucalyptus grew vigorously in association with cassava but had the greatest shade effect on cassava (Ghosh et al., 1987).

6.2 Temperature

Temperature affects sprouting, leaf size, leaf formation, storage root development and consequently general plant growth (Alves, 2002). Sprouting and emergence are delayed by low soil temperature and this creates potential problems of weed control and increased chance of plant losses due to pathogen attack (Keating and Evenson, 1979). CGR declined to zero or near zero for all planting dates in late winter in experiments conducted in Queensland, Australia (Keating et al., 1982). Sprouting rates of cassava stem cuttings have been shown to be significantly reduced at higher altitudes, presumably because of lower temperatures. Covering the soil with plastic improves the soil environment by increasing temperature and lowering water loss (Neubert et al., 2015a,b). In contrast, temperatures above 30°C usually reduce leaf longevity from 200 to 120 days, decreasing LAI and biomass partitioning into tuberous roots (Alves, 2002). Temperatures between 18 and 16°C delay leaf appearance and expansion, which consequently reduce total plant biomass. For these lower temperatures, there is a delay in, and reduction of, biomass partitioning into tuberous roots, which can decrease yield.

Irikura et al. (1979) found a significant genotype by temperature interaction for yield in cassava. The results showed that under different temperature regimes, the optimum mean LAI for high rate of root bulking between 8 and 16 months remained similar (about 3), but the genotype which expressed the optimum LAI differed among environments. Therefore, selection of highest yielding varieties is highly dependent on the varietal adaptation to a given temperature. There was a varietal temperature interaction in the number of branches at each branch point. Leaf formation rate per apex decreased with time and was always less at lower temperatures (Irikura et al., 1979).

In a study by Keating et al. (1982), higher CGRs were associated with high LAI given high solar radiation or high mean air temperatures (>22°C). Keating et al. (1982b) confirmed that higher temperatures, long photoperiods and high LA indices are associated with reduced proportion of total dry matter assimilation partitioned to storage organs over a given time period.

The effects of soil temperature on the sprouting and sprout elongation of cassava stem cuttings are another factor that influences cassava growth. Sprouting rates of cassava stem cuttings have been shown to be significantly reduced at higher altitudes, presumably because of lower temperatures (Keating and Evenson, 1979). The minimum temperature for sprouting and sprout elongation was found to lie between 12 and 17°C, and the maximum was between 36 and 40°C. Whilst the rate of sprout elongation versus temperature response curve was shown to be quadratic over a wide temperature range, it was linear over the range of temperatures normally experienced in the field (Keating and Evenson, 1979).

Brown and co-workers (2016) studied the implications of the interactive effects of temperature and drought on cassava growth and root toxicity, and showed that both temperature and drought affected cassava physiology and chemistry. While temperature alone drove differences in plant height and above-ground biomass, drought and temperature × drought interactions mostly affected root yield, as well as leaf and root chemistry, including C:N, nitrogen and cyanogenic potential (CNp: total cyanide released from cyanogenic glucosides). The magnitude of drought effects on root yield and toxicity were greater at high temperature. Thus, increases in root CNp were not merely a consequence of reduced root biomass (Brown et al., 2016). Those findings confirmed that cassava is adaptable to forecast temperature increases, particularly in areas of adequate or increasing rainfall; however, in regions forecast for increased incidence of drought, the effects of on both food quality (root toxicity) and yield present a greater threat to future food security and indicate an increasing necessity for processing of cassava to reduce toxicity.

7 Cassava response to environmental conditions: water availability

Soil water availability is one of the strongest influencing factors on cassava growth and development and photosynthetic capacity (Wang et al., 2012). Water stress in cassava highly affects fresh root yield and dry matter content (Bakayoko et al., 2009). Aina et al. (2008) reported a decline in plant height, stem girth, number of roots and yield under water stress. Water stress has been found to influence root quality functional properties such as starch granule size, lower swelling power, lower peak viscosity, increased pasting and onset temperatures, and cyanide content (Santisopasri et al., 2001). Leaf size and LAI are reduced in response to water shortage as is overall biomass production (Okogbenin et al., 2013; El-Sharkawy, 2012; El-Sharkawy et al., 2007; Alves and Setter, 2000; Connor et al., 1981; Connor and Cock, 1978). The reduction of CGR during water stress was found to be more pronounced in the shoots than in the roots, particularly in varieties with vigorous vegetative growth. Upon recovery from water stress, cassava rapidly regenerates new leaves and the LAI of previously stressed cassava plants becomes higher than that of non-stressed plants.

Plants have evolved complex mechanisms to sense water availability in their environment. They may reprogramme their metabolism and growth in response to drought stress, leading to a variety of physiological, biochemical and morphological changes at molecular and whole-plant levels. These include stomatal closure, deepened root systems, reduced shoot growth, leaf senescence and dormancy (Zhao et al., 2014). Studying different strategies in two cassava cultivars, Zhao and co-workers (2014) found that under moderate drought, plants adapted a survival mode by early stomatal closure, wilting symptoms and reduced photosynthesis. The result was growth arrest in one cultivar but the other plants responded to moderate drought stress by senescence and shedding of old leaves, while the young leaves remained green and turgid and had a higher photosynthesis capacity (Zhao et al., 2014).

Many varieties develop maximum dry matter by end of the rainy season, and maintain relatively stable dry matter content, at least through the early phase of the dry season. This is one reason that the dry season is the most common period for farmers to harvest the crop. However, when rains begin, carbohydrate reserves in the roots are mobilized to supply new leaf growth, and dry matter can drop drastically, by up to 20% points or more (e.g. from 40% at beginning of dry season to 20% after beginning of next rainy season) (C. Hershey, pers. comm.).

Mechanisms associated with cassava's adaptation to drought combine dehydration avoidance and tolerance. Several traits have been identified and associated with these mechanisms (Okogbenin et al., 2013).

Cassava can extract deeply stored water in soils in dry and semi-arid environments (El-Sharkawy et al., 2007). The fine root system, while sparse, can extend to more than two metres into the soil to exploit water that would be unavailable to many annual crops.

A tight stomatal control over leaf gas exchange reduces water loss when plants are subjected to soil water deficits as well as to high atmospheric evaporative demands, thus protecting the plant from severe dehydration. Unlike the root growth in search of water, which is limited because cassava has sparse root system or leaf loss and osmotic adjustment (OA) that involve the sacrifice of growth reserves previously assimilated (yield penalty), stomata closure reduces the current carbon assimilation capacity of plant under stress with a view to increasing the water-use efficiency (WUE) to maintain the existing yield and/or increase it (Oguntunde, 2005).

When water is available, cassava maintains a high stomatal conductance with high internal CO_2 concentration, but when water becomes limiting, the stomata are closed in response to even small decreases in soil water potential. Alves and Setter (2004) showed that mature and expanding cassava leaves decreased their osmotic potential to a relatively limited extent in response to a 6-day water deficit. In contrast, the youngest (folded) cassava leaves decreased their osmotic potential to a greater extent than mature leaves (Alves and Setter, 2004). Total sugars (sucrose, glucose and fructose) did not contribute to OA in either mature or expanding leaves. Although the concentration of proline in mature leaves increased significantly in response to water stress, its contribution to the total change in osmotic potential was found to be insignificant. K-salts were the major contributors to total leaf osmolyte concentration. K-salts were considered as primary osmolytes in cassava (Alves and Setter, 2004).

Abscisic acid (ABA) levels in mature leaves were found to be lower than those in younger ones in both control and stressed treatments. Higher levels of ABA in young leaves were attributed to a lower capacity to metabolize (break down) ABA leading to higher net ABA accumulation, and of a translocation of ABA from mature source leaves to the young sink leaves. In addition to the well-established role of ABA in closing stomata, there is evidence that ABA has a role in regulating solute accumulation and thus OA (Alves and Setter, 2004). Although ABA increased substantially in all cassava genotypes, only in certain genotypes was it accompanied by modest OA due to accumulation of K-salts. Thus, the lack of greater OA was apparently due to limited upregulation of membrane transport systems, not due to a lack of ABA signal or sensing of water potential decline. Another role of ABA is to stimulate the accumulation of a set of specific proteins that are thought to confer osmoprotective function during water stress and in tissues undergoing desiccation. One subset of these proteins are the dehydrins (Alves and Setter, 2004).

During prolonged water deficits, cassava reduces its canopy by shedding older leaves and forming smaller new leaves leading to less light interception, another adaptive trait to drought (El-Sharkawy et al., 2007). Other studies have shown a reduction in photosynthesis in response to water shortage (Palta, 1984). The crop can recover when water becomes available by rapidly forming new leaves with higher photosynthetic rates compared to unstressed crops, thus compensating for yield losses with final yields approaching those in well-watered crops (Ike and Thurtell, 1981; Okogbenin et al., 2013). This suggests that the ability to regulate numerous plant processes to rapidly change course is the key for cassava's success under unfavourable weather. Stem, petiole and total non-structural carbohydrates serve as a substantial carbohydrate buffer that may contribute to cassava's ability to withstand extended droughts during which photoassimilate production is stopped (Alves et al., 2006).

Early stomatal closure, reduction of various photosynthetic proteins in leaves, early growth quiescence, and activation of drought proteins such as aquaporins, myo-inositol 1-phosphate synthases and proteins involved in antioxidant systems and secondary metabolism were also found by Zhao and co-workers (2015) studying cassava strategies to drought. Signalling proteins such as protein kinases, RNA-binding proteins, transcription factors and histone deacetylases were found (Zhao et al., 2015). Linamarin was found to play a role in nitrogen reallocation in cassava under drought (Zhao et al., 2015).

Studies conducted on 10 cassava clones evaluated for drought tolerance at the International Institute of Tropical Agriculture showed that, while most clones studied increased stomatal conductance throughout the day, clones TMS91934 and TMS84751 diminished their stomatal conductance, giving a better yield because they depleted the soil water more slowly (Okogbenin et al., 2013). Results also suggested a strong influence of the water table on growth (as reflected in height and leaf formation).

Under drought, changes in the biosynthesis, content and distribution of plant growth regulators such as ABA within plant organs and tissues, particularly in roots, leaves and buds, may play an important role in sensing changes in both soil water and atmospheric humidity, and in controlling stomatal movements, leaf formation and extension, root growth and bud dormancy. They may also be involved in other biological functions such as the expression of dehydrin and other proteins that are thought to stabilize macromolecular structure. Studies in cassava show that cassava varieties rapidly accumulate large amounts of ABA, linked to the cessation of leaf expansion growth and transpiration. The high ABA readings are almost completely reversed once water levels are recovered. This rapid return to control levels corresponds with a rapid recovery of LA growth rates (Alves and Setter, 2000).

One means of increasing drought tolerance is by accumulation of osmotically active solutes, so that turgor and turgor-dependent processes may be maintained during episodes of dry down. In some plant species, OA allows cell enlargement and plant growth under high water stress, and allows stomata to remain partially open and CO_2 assimilation to continue at low water potentials that would otherwise be inhibitory (Okogbenin et al., 2013). However, the extent of OA has been found to be quite limited in cassava's tolerance to drought. In studies with potted greenhouse-grown cassava, Alves (1998) found that the largest increases in solutes after a few days of water deficit occurred in the youngest and folded (expanding) leaves, with only about 0.14 MPa OA increase in mature leaves, pointing to a limited effect of OA in the latter leaves (Okogbenin et al., 2013).

Studying water deficit in cassava, Alves and Setter (2003) also concluded that K-salts were the major contributors to total osmolyte concentration in both mature and expanding leaves, accounting for approximately 60% of osmotic potential. The concentration of K-salts increased in response to water stress and was positively correlated with the extent of OA. In contrast, total sugars (sucrose, glucose, fructose) decreased during water deficit, showing a negative correlation with OA.

Pot experiments by Calatayud et al. (2000) showed that after 45 days of limited water supply, cassava exhibited pronounced reduction in shoot growth, high leaf fall and decreased stomatal conductance. In the field however, what cassava does is to reduce leaf production (rate and size) and maintain its green leaves. In cassava-growing environments, it is common to see green patches of cassava in an otherwise barren landscape at the height of the dry season. An amplified heliotropic response and drooping which minimizes radiant energy interception were also observed, suggesting that leaves are sensitive to high irradiance. In well-irrigated plants CO_2-saturated oxygen evolution and net photosynthesis in air were markedly higher in young (expanding) leaves than in mature leaves. The mature leaves of drought-adapted plants displayed a residual rate of net photosynthesis and deteriorated photosystem II (PSII). In young leaves at moderate irradiance, net photosynthesis (Pn) was depressed in stressed plants. Moreover, photochemical quenching of chlorophyll a (Chl a) fluorescence and the quantum efficiency of PSII in young leaves were comparable in both control and stressed plants. In contrast, at high irradiance, Pn was almost null. The heliotropic response and drooping displayed by young leaves under water limitation is thus an important strategy in cassava tolerance to drought.

8 Cassava response to environmental conditions: salinity, atmospheric CO_2 and other greenhouse gases

8.1 Salinity

Rising sea levels and salinization of groundwater are becoming increasing problems in many parts of the world, reducing the area of land available for agricultural crops. Globally, one-fifth of irrigated land is salt-affected, resulting in huge losses in productivity (Gleadow et al., 2016). Cassava has the ability to cope with a wide range of environmental stresses and continues to produce under certain poor growing conditions, such as low nutrients and drought, but it is generally considered to be highly sensitive to salinity. In a study of pre-tuberous plants, large effects were found on growth as a result of salinity and only 30% of plants survived at the highest salt concentration (Carretero et al., 2008). Gleadow et al. (2016) reported that cassava plants in more advanced growth stages were able to tolerate 100 mM NaCl but in younger plants 40 mM was sufficient to retard plant growth severely. Nutrient analysis showed that plants were only able to exclude sodium at low concentrations. The foliar cyanogenic glucoside concentration in young plants increased under moderate salinity stress but was lower in plants grown at high salt. Ion leakage at severe stress can also be observed. Cassava stores nitrogen in leaves for shoot recovery or in storage roots for re-sprouting upon alleviation of salinity.

Changes in cassava tissue chemistry in response to salinity were attributed to a combination of factors, including differences in the balance in trade-offs in resource allocation between growth and defence in plants at different stages of development and the possible reduction in oxygen radicals. Cassava does not tolerate salinity well, especially at early stages of development, and is not suitable for planting in regions contaminated with even relatively low levels of salt. Impacts may be minimized by irrigating with less saline water, or during periods of high rainfall to allow time for plants to become established before they are exposed to higher concentrations of salt.

8.2 Atmospheric CO_2 and other greenhouse gases

The top four greenhouse gases are carbon dioxide (CO_2), methane (CH_4), nitrous oxide (N_2O) and ozone (O_3). Greenhouse gases have a major indirect impact on crop growth and development through their effect on climate change (discussed in previous sections) but also may have direct effects on plant growth and development.

The current atmospheric CO_2 concentration is approximately 400 ppm and is expected to reach between 730 and 1020 ppm near the end of this century unless effective measures of reversal are implemented. These higher levels are having direct effects on plant growth and development, but also indirect effects such as rising temperatures, which in turn have their own secondary effects such as less predictable rainfall. This section will discuss the direct effects only, since temperature and water availability – which can change as an indirect result of rising greenhouse gases – have been discussed in previous sections.

The interaction between CO_2 concentration and plant-soil water is a particularly important issue since water deficit constitutes one of the main factors limiting food production in various parts of the world. In a study conducted by Cruz and co-workers (2016) in cassava, it was found that elevated CO_2 delayed the onset of water deficit in cassava by reducing the stomatal conductance and decreasing the transpiration rate. Elevated CO_2 was also found to increase the net photosynthesis and instantaneous transpiration efficiency and, as a result, biomass production was greater for plants grown under elevated CO_2 levels during water deficit. Increase in the atmospheric CO_2 level will increase WUE of cassava and hence the crop performance under water-deficit conditions. A study by Rosenthal et al. (2012) researching the effect of higher CO_2 levels on cassava growth also found higher above-ground biomass and cassava storage root dry mass. High photosynthetic rates and photosynthetic stimulation by elevated CO_2, larger canopies and a large sink capacity all contributed to cassava's growth and yield stimulation. Large gaps remain in our understanding of other global change impacts on cassava productivity (Rosenthal et al., 2012). Average global temperatures are predicted to increase between 1 and 6°C during this century (IPCC, 2007). The proportional enhancement of C_3 photosynthesis by elevated CO_2 is expected to increase with temperature because both the specificity of Rubisco for CO_2 over O_2 and the ratio of solubilities of CO_2 versus O_2 decline with increasing temperature (Long, 1991). This prediction appears to be borne out in cassava. Increasing day/night temperatures from 28°C/21°C to 33°C/26°C for chamber-grown cassava at 700 ppm CO_2 caused a reported 232% yield stimulation compared to 150% in elevated CO_2 without the increase in temperature (Imai et al., 1984), consistent with the notion of greater stimulation by CO_2 at higher growth temperatures. The photosynthetic temperature optimum of cassava is predicted to increase from 28°C at current CO_2 to 31°C at 585 ppm (Rosenthal and Ort, 2012). Thus, increases in temperature and CO_2 may have a synergistic effect on cassava productivity (Long, 1991), but this important prediction remains to be confirmed under field conditions. Finally, the effect of tropospheric ozone on cassava productivity is completely unexplored (Rosenthal et al., 2012). Elevating CO_2 stimulates C_3 photosynthesis by increasing the substrate for carboxylation, CO_2, and by reducing photorespiration (Rosenthal and Ort, 2012).

It has also been suggested that mid- and high-latitude regions might reap the benefits of warming and CO_2 effects via increasing total production and yield of cassava, coupled with greater WUEs (Noerwijati and Budiono, 2015).

In the last two decades, a large number of studies have focused on the effects of NO_3 and NH_4^+ fertilization on plant photosynthetic physiology. While NH_4^+ is known to be detrimental to photosynthesis and accumulation of dry mass in cassava plants, it may be important in maintaining photosynthetic activity under CO_2-enriched conditions to the extent that lower nitrate assimilation is responsible for photosynthetic acclimation (Cruz et al., 2013). However, the positive effect of NH_4^+ fertilization on cassava photosynthetic CO_2 response eventually can lead to a toxicity problem that reduces biomass production. The challenge in future studies must be to determine how to manage NH_4^+ fertilization so that the photosynthetic benefit observed in the initial growth phase of cassava may persist throughout the crop cycle.

Domestic livestock produce large amounts of CH_4 as part of their normal digestive process. Also, when animals' manure is stored or managed in lagoons or holding tanks, CH_4 is produced. When livestock and manure emissions are combined, the agriculture sector is the primary source of CH_4 emissions (EPA, 2017). Nitrous oxide is emitted when people add nitrogen to the soil through the use of synthetic fertilizers. Agricultural soil management is the source of N_2O emissions (EPA, 2017).

An often overlooked aspect of global atmospheric change is that surface ozone (O_3) concentrations are rising. Although there are few measurements, it appears that pre-industrial levels were <10 ppb. According to some research, concentrations of trophospheric ozone have risen by 30% since the Industrial Revolution (EPA, 2017). Ozone is one of the most powerful oxidants known and is phytotoxic. It enters the leaf via the stomata and, while too reactive to penetrate beyond the mesophyll cell walls, causes the production of a range of oxidizing radicals and a signal cascade. While extreme concentrations of ozone will cause leaf necrosis, a more common effect is loss of photosynthetic capacity and accelerated leaf senescence. Over the past 30 years, ozone was found to have caused a loss of 10% of maize yield and 5% of soybean yield. Little is known about the impacts of ozone on cassava, but assuming similar levels of damage to temperate root crops at a loss of 10% could be expected, possibly rising to 20% by mid-century (Souza et al., 2016).

9 Post-harvest physiological deterioration

Cassava roots are highly perishable when compared with roots of other temperate and tropical root and tuber crops. This may be associated, in evolutionary and physiological terms, with the fact that they do not have endogenous dormancy, and have no function in propagation and no primordial buds where sprouting can occur (Coursey and Booth, 1977; Passam and Noon, 1977). In general, due to physiological (see Fig. 2) and microbial deterioration, cassava roots cannot be stored under ambient conditions for more than about three days, without some pre- or post-harvest treatment or management (Wenham, 1995).

Early research on PPD relied on limited information about the underlying nature of the process (Rickard and Coursey, 1981). On the microbial side of deterioration, the number of fungal and bacterial species isolated from roots in various storage conditions shows that the deterioration is a complex process, often involving several species of microorganisms (Ekundayo and Daniel, 1973; Wegmann, 1970; Burton, 1970). Research also suggested that initial vascular discolouration had a physiological basis (Booth, 1976; Noon and Booth, 1977; Rickard et al., 1979). On the basis of his observations, Booth (1976) made a clear distinction of the PPD types in cassava roots, dividing them into primary decay (a physiological process) (Fig. 3a) and secondary deterioration (Fig. 3b), which is commonly derived from microbial activity. Primary decay due to physiological deterioration is the principal cause of loss of acceptability of the root. Secondary decay occurs when pathogens enter through wounds or injuries caused during harvest or handling (Wenham, 1995).

Traditionally, the problem of deterioration has been overcome by postponing harvest until roots can be consumed or processed. The disadvantages of this system are:

1 Agricultural land is occupied for long periods, preventing use for other crops;
2 Starch content often (but not always) declines;
3 Palatability decreases as the root becomes more fibrous (Rickard and Coursey, 1981);
4 Cooking time increases (Wheatley and Gomez, 1985).

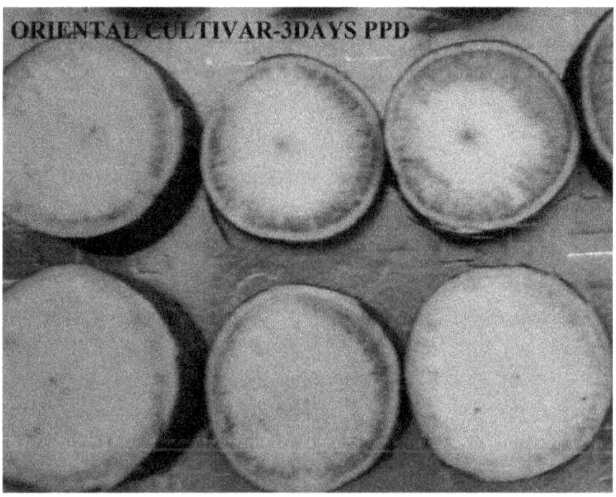

Figure 2 Photo from the author illustrating physiological deterioration of sliced cassava roots (*Branco* variety) (3 days after harvest).

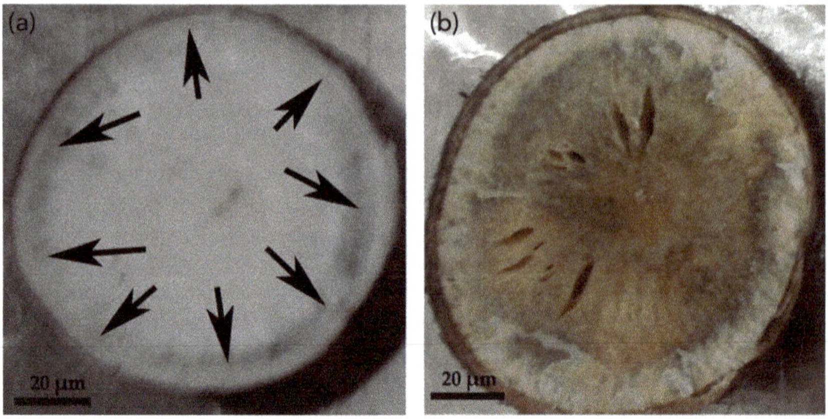

Figure 3 Illustrative images of the types of post-harvest deterioration of cassava roots. (a) Primary or physiological deterioration. Arrows indicate bluish-black areas, features associated with PPD after 3 days of storage. (b) Typical image of secondary deterioration, 5 days of storage (photos from the author).

The recognized perishability in cassava roots creates difficulties for the post-harvest storage of large volumes of fresh roots and restricts the managed supply of raw materials to multiple end uses. This is one of the reasons that in many countries, especially in Latin America, urban consumers have replaced cassava with other sources of carbohydrates such as rice and wheat.

The problem of deterioration also results in crop losses. Studies carried out by CIAT in Asia, in collaboration with local partners, indicate losses ranging from 10 to 12% in India and 5.3% in Indonesia (CIAT, 1987). FAO's latest estimates for China point to values on the order of 3% and of 15–30% and 27% for Ghana and Ivory Coast, respectively. In a recent evaluation of global cassava production, FAO suggests that losses in tuberous roots range from 30 to 60%. In the case of cassava in Africa, the losses in 2002 were estimated at 19 million t out of a total of 101 million t produced in the continent (NRI, 2014). Financial losses can also be significant, taking into account loss of quality and corresponding value as the crop deteriorates (Westby et al., 2005). In Ghana, for example, the combined effect of the economic impact on the losses of cassava in monetary terms was estimated at up to US$500 million per year (NRI, 2014). Research suggests that, whilst there are variations between countries, the principal losses in the value chain are at the processing, distribution and retail stages (Hall, 1998; Wenham, 1995; Westby et al., 2002; Naziri et al., 2014). This is illustrated in Fig. 4.

The first in-depth studies on the deterioration of fresh cassava were conducted in 1970s by the International Center for Tropical Agriculture (CIAT) of Colombia and the Natural Resources Institute (NRI) of the United Kingdom. Research led to the development of simple systems based on packing in polyethylene bags (to retain high moisture), and chemical treatment such as thiabendazole to prevent the onset of microbial deterioration (Wheatley, 1989).

In many cases, physiological decay develops from the sites of tissue damage during harvest, handling or transportation. Initial signs are dark pigmentation of the vascular tissues, which is referred to as 'vascular streaking' (Fig. 3a). This symptom is followed by the

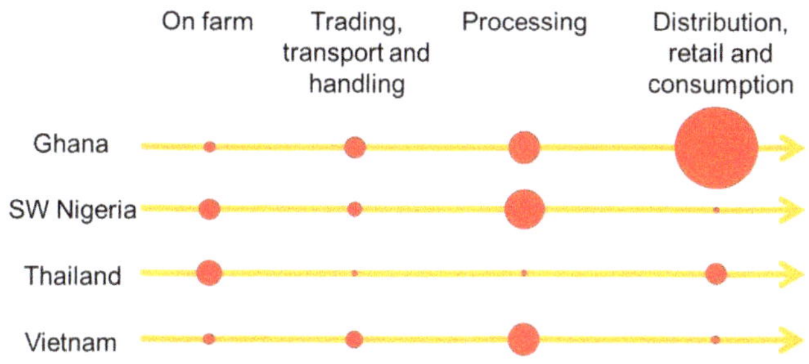

Figure 4 The extent of physical losses of cassava roots throughout the stages of the value chain in Ghana, Nigeria, Thailand and Vietnam. The size of the circle indicates proportionately the level of loss in the stage referenced. Image adapted from Natural Resources Institute (NRI, 2014).

generalized discolouration (colour change) of storage parenchyma. According to Bowles (1990), the tissue injury results in a cascade of defence responses and subsequent occlusion of the injured area, generating a protective barrier characterized by the formation of periderm. Biochemical responses usually involve enzymatic activation (e.g. glucanase and quitinases), the synthesis of inhibitors of proteases and the production of hydroxyproline-rich glycoproteins. Enzymes associated with the phenylpropanoid pathway, for example, phenylalanine ammonia lyase and chalcone synthase, are produced, resulting in the synthesis of phenolic compounds which act directly in the process of physiological stress defence. The major phenolic compounds associated with the development of physiological deterioration were characterized, and included scopoletin, scopolin, esculin, proanthocyanidins, catechin and gallocatechin (Rickard, 1981, 1982, 1985; Tanaka et al., 1983; Wheatley, 1982; Wheatley and Schwabe, 1985; Lorenzo et al., 1987; Fritzemeier et al., 1987; Hahlbrock and Scheel, 1989).

Studies conducted at the Federal University of Santa Catarina (Southern Brazil) by our research group during 2011–15 in collaboration with CIAT tested different methods of induction of cassava physiological deterioration with the aim of understanding the biochemical, histological and metabolomic changes related to cassava post-harvest deterioration (PHD) (Fig. 5) (Uarrota et al., 2014, 2016; Uarrota and Maraschin, 2015a,b). Treatments include (1) removing the distal and proximal parts of the root and covering with parafilm, (2) leaving the entire root and (3) slicing the roots . The results brought new understanding of the process. Removing the distal and proximal parts and covering the distal part with parafilm mimic small wounds that can happen during harvest. Leaving the entire root had an advantage as it mimics how cassava roots are harvested and PPD develops at low levels. Slicing the entire root accelerated the PPD process, since the entire cassava section comes into contact with oxygen and PPD is activated rapidly.

The field of metabolomics has received growing interest by the scientific community, given its extensive range of applications in functional genomics and, more globally, in the characterization of biological systems (Costa et al., 2016; Mozzi et al., 2012; Nielsen and Jewetti, 2007; Villas-Boas et al., 2007; Schauer and Fernie, 2006). Metabolomics can be defined as the identification and quantification of all intracellular and extracellular metabolites with low molecular mass (Costa et al., 2016; Mozzi et al., 2012; Nielsen and Jewetti, 2007; Villas-Boas et al., 2007; Schauer and Fernie, 2006). Applications include study of metabolic systems, measuring biochemical phenotypes, understanding and

Figure 5 Three different methods of cassava physiological deterioration analysis; from top to bottom: distal and proximal ends removed, and covered with parafilm; whole; sliced (photos from the author).

reconstructing networks, discriminating between samples, identifying biomarkers of disease, analysing food and beverages, studying plant physiology or fostering drug discovery (Costa et al., 2016; Mozzi et al., 2012; Nielsen and Jewetti, 2007; Villas-Boas et al., 2007; Schauer and Fernie, 2006). Metabolomics approaches have recently been used to assess the natural variance in metabolite content between individual plants, an approach with great potential for the improvement of the compositional quality of crops (Schauer and Fernie, 2006).

Metabolomic analyses in cassava PPD studies showed increases in carotenoids, flavonoids, anthocyanins, phenolics, reactive scavenging species and enzymes (superoxide dismutase family, hydrogen peroxide and catalase) until 3–5 days post-harvest (Uarrota et al., 2014). Increases in ascorbate peroxidase (APX) and guaiacol peroxidase (GPX) activity, as well as total protein contents occurred from 3 to 5 days of storage and were correlated with the delay of PPD. Cassava samples stained with periodic acid-Schiff (PAS) highlighted the presence of starch and cellulose. Degradation of starch granules during

PPD was also detected. Slight metachromatic reaction with toluidine blue was found to be indicative of increasing acidic polysaccharides and may play an important role in PPD delay (Uarrota et al., 2016; Uarrota and Maraschin 2015a,b,c). Besides the results found by our research group, future studies must be performed in order to better understand the genes involved and metabolic pathways that lead to physiological deterioration.

10 Summary and future trends

Understanding of crop growth and development is fundamental to improving cassava productivity and sustainability through both breeding and management, especially for the dramatic adaptations that will be needed to confront climate change. Cassava has key features which distinguish it from the principal cereal and grain legume crops. In cassava there are no distinct phases of development, but instead continuous and competing development of both source (photosynthesizing leaves and the stem support system) and sink (the starch storage roots). One of the ongoing challenges for cassava physiologists and others on crop improvement teams is that cassava is very often grown under challenging conditions of drought, suboptimal soil nutrition, and a range of pest and disease stresses. Thus, while it is critical to understand crop models for maximizing yields under ideal conditions, it is even more important to understand the interactions of the plant with factors that will limit yield, and to develop physiologically based strategies for crop improvement and crop management that account for these yield-limiting externalities. For example, plants that would develop too vigorous a canopy (and low harvest index) under good growing conditions may be optimal for yield under drought stress. Generally, crop scientists should avoid developing crop phenotypes for adaptation to poor management, but on the other hand, it must be recognized that cassava will likely be produced under suboptimal conditions in most parts of the world, well into the future.

Modelling predicts a higher suitability of cassava in a globally warming climate relative to most other food crops, confirming its high level of tolerance to drought and high temperatures. Cassava has high optimum temperature for photosynthesis and growth, and responds positively to elevated CO_2, which point to its potential as a food, feed and energy crop in tropical and subtropical zones adversely affected by climate changes (El-Sharkawy, 2012).

Cassava exhibits a photosynthetic behaviour intermediate in some respects to classical C3 or C4 species. Although it does not have the typical 'C4-Kranz' anatomy of C4 leaves, it does have low photorespiration, a low CO_2 compensation point, a high percentage of carbon fixation in C4 acids and high PEP-carboxylase activity. Good correlations between single-leaf photosynthetic rates and yield among a wide genetic base of materials tested give promise of the possibility to improve yield potential through selection for higher photosynthetic rates.

While photoassimilate partitioning has been well studied under a wide range of genotypes and environmental conditions, the triggers that stimulate the change from fibrous root to starch storage root are still poorly understood. New studies underway through metabolomics and through more detailed root morphological and physiological studies should shed light on possible means to manipulate the triggers for starch storage to achieve earlier and/or higher yields.

Cassava is unusual among root and tuber crops in its rapid PPD, initially a physiological process but often progressing as well to phytopathological issues. The process of PPD has challenged physiologists for decades. Whilst much has been learned about some of the metabolomics differences between more resistant and more susceptible genotypes, there has been little progress in improving this trait through breeding. One of the principal challenges has been the high influence of environmental factors, and the high intra-plant and even intra-root variability, which give low confidence in any evidence for genetic differences. As cassava becomes more and more an industrial crop requiring transport and storage, this trait will need new emphasis in research. Basic research is required on the understanding of the pathways, genes and physiological mechanisms that lead to rapid deterioration of cassava roots.

Physiology research is advancing rapidly. Sensor technologies and analytical tools such as ground-penetrating radar, soil sensors, photosynthesis analysers, chromatography and spectroscopic techniques are being improved to monitor cassava growth more precisely. A new era that integrates genomic, proteomic, transcriptomic, metabolomic and big data analysis, along with plant and crop physiology and agronomy, for better understanding of the factors involved in cassava growth will be of crucial importance and will accelerate advances in cassava research towards sustainable production.

11 Where to look for further information

Further information on cassava growth and development can be found at the International Centre of Tropical Agriculture (CIAT-Colombia), the International Institute of Tropical Agriculture (IITA-Nigeria), Natural Resources Institute (NRI-United Kingdom) University of Greenwich, Santa Catarina State Rural Extension Agency (EPAGRI), Experimental Station of Urussanga, the Federal University of Santa Catarina (Brazil) and the Brazilian Agricultural Research Corporation (EMBRAPA).

12 References

Aina, O. O., Dixon, A. G. O. and Akirinde, E. A. (2007). Effect of soil moisture stress on growth and yield of cassava in Nigeria. *Pakistan Journal of Biological Sciences*, 10(18), pp. 3085–90.

Alves, A. (2002). Cassava botany and physiology. In R. Hillocks, J. Thresh and B. Anthony (Eds) *Cassava: Biology, Production and Utilization*, 1st ed. UK: CABI publishing, pp. 1–23.

Alves, A. and Setter, T. (2000). Response of cassava to water deficit. *Crop Science*, 40(1), p. 131.

Alves, A. A. C. and Setter, T. L. (2004). Response of cassava leaf area expansion to Water deficit: Cell proliferation, cell expansion and delayed development. *Annals of Botany*, 94, pp. 605–13.

Alves, A. A. C., Fregene, M. A., Ceballos, H., Setter, T. L., Duque, L., Silva, A. F., Ferguson, M. E., Mkamilo, G. and Osei, C. (2006). Identifying the physiological and genetic traits related to drought tolerance in cassava [poster][online]. Empresa Brasileira de Pesquisa Agropecuária (EMBRAPA), Embrapa cassava & Tropical Fruits, São Paulo, BR, 1p. Available at: http://hdl.handle.net/10568/57966.

Aresta, R. B. and Fukai, S. (1984). Effects of solar radiation on growth of cassava (Manihot esculenta Crantz). II. Fibrous root length. *Fields Crops Research*, 9, pp. 361–71.

Bakayoko, S., Tschannen, A., Nindjin, C., Dao, D., Girardin, O. and Assa, A. (2009). Impact of water stress on fresh tuber yield and dry matter content of cassava (Manihot esculenta Crantz) in Côte d'Ivoire. *African Journal of Agricultural Research*, 4, pp. 21–7.

Booth, R. (1976). Storage of fresh cassava (*Manihot esculenta*). I. Post-harvest deterioration and its control. *Experimental Agriculture*, 12, p. 103.
Bowles, D. (1990). Defense-related proteins in higher plants. *Annual Review of Biochemistry*, 59, pp. 873–907.
Brown, A., Cavagnaro, T., Gleadow, R. and Miller, R. (2016). Interactive effects of temperature and drought on cassava growth and toxicity: Implications for food security? *Global Change Biology*, 22(10), pp. 3461–73.
Burton, C. (1970). Diseases of tropical vegetables in the Chicago market. *Tropical Agriculture (Trinidad)*, 47: 303–13.
Calatayud, P. A., Llovera, E., Bois, J. F. and Lamaze, T. (2000). Photosynthesis In Drought-adapted cassava. *Photosynthetica*, 38, pp. 97–104
Carretero, C. L., Cantos, M., García, J. L., Azcón, R. and Troncoso, A. (2008). Arbuscular-mycorrhizal contributes to alleviation of salt damage in cassava clones. *Journal of Plant Nutrition*, 31, pp. 959–71.
Ceballos, H., Okogbenin, E., Pérez, J., López-Valle, L. and Debouck, D. (2010). Cassava. In J. Bradshaw (Ed.) *Handbook of Plant Breeding. Root and Tuber Crops*, 1st ed. UK: Springer, pp. 53–96.
Ceballos, H. and de la Cruz, G. (2012). Cassava taxonomy and morphology. In B. Ospina and H. Ceballos (Eds) *Cassava in the Third Millennium. Modern Production, Processing, Use and Marketing System*. Cali, Colombia: CIAT publication, pp. 15–28.
Chaweewan, Y. and Taylor, N. (2015). Anatomical Assessment of Root Formation and Tuberization in Cassava (Manihot esculenta Crantz). *Tropical Plant Biology*, 8(1–2), pp. 1–8.
Chen, J. and Black, T. (1992). Defining leaf area index for non-flat leaves. *Plant, Cell and Environment*, 15(4), pp. 421–9.
Chipeta, M., Shanahan, P., Melis, R., Sibiya, J. and Benesi, I. (2016). Early storage root bulking index and agronomic traits associated with early bulking in cassava. *Field Crops Research*, 198, pp. 171–8.
CIAT. CENTRO INTERNACIONAL DE AGRICULTURA TROPICAL. (1987). Global cassava research and development. The cassava economy of Asia: Adapting to economic change. Documentation of cassava demand studies conducted in Asia in collaboration with national programs. Draft version prepared for the Technical Advisory Committee for the Nairobi meeting. CALI-COLOMBIA.
Clark, D., Dervinis, C. and Barrett, J. (2004). Drought-induced Leaf Senescence and Horticultural Performance of Transgenic PSAG12IPT Petunias. *Journal of the American Society for Horticultural Science*, 129, pp. 93–9.
Cock, J. (1976). Characteristics of high yielding cassava varieties. *Experimental Agriculture*, 12(02), p. 135.
Cock, J. (1982). Cassava. A basic energy source in the tropics. *Science*, 218(4574), pp. 755–62.
Cock, J., Franklin, D., Sandoval, G. and Juri, P. (1979). The ideal cassava plant for maximum yield. *Crop Science*, 19(2), p. 271.
Cock, J., Porto, M. and El-Sharkawy, M. (1985). Water use efficiency of cassava. III. Influence of air humidity and water stress on gas exchange of field grown cassava. *Crop Science*, 25(2), p. 265.
Cock, J. H. (2012). Cassava: A basic energy source in the tropics. In R. H. Howeler (Ed.) *The Cassava Handbook. A Reference Manual based on the Asian Regional Cassava Training Course*. Cali, Colombia: Thail and Centro Internacional de Agricultura Tropical (CIAT), pp. 23–38.
Cock, J. H., Riaño, N. M., El-Sharkawy, M. A., Lopéz, F. Y. and Bastidas, G. (1987). C3-C4 intermediate photosynthetic characteristics of cassava (*Manihot esculenta* Crantz). II. Initial products of $^{14}CO_2$ fixation. *Photosynthetic Research*, 12, pp. 237–41.
Connor, D. J. and Cock, J. H. (1981). Response of cassava to water shortage II. Canopy dynamics. *Field Crops Research*, 4, pp. 285–96.
Connor, D. J., Cock, J. H. and Parra, G. E. (1981). Response of cassava to water shortage. I. Growth and Yield. *Field Crops Research*, 4, pp. 181–200.
Coursey, D. and Booth, R. (1977). Post-harvest problems of non-grain staples. *Acta Horticulturae*, 53, pp. 23–33.

Costa, C., Maraschin, M. and Rocha, M. (2016). An R package for the integrated analysis of metabolomics and spectral data. *Computer Methods and Programs in Biomedicine*, 129, pp. 117–24.
Cruz, J., Alves, A., LeCain, D., Ellis, D. and Morgan, J. (2013). Effect of elevated CO_2 concentration and nitrate: Ammonium ratios on gas exchange and growth of cassava (Manihot esculenta Crantz). *Plant and Soil*, 374(1–2), pp. 33–43.
Cruz, J., Alves, A., LeCain, D., Ellis, D. and Morgan, J. (2016). Elevated CO_2 concentrations alleviate the inhibitory effect of drought on physiology and growth of cassava plants. *Scientia Horticulturae*, 210, pp. 122–9.
Ding, Z., Zhang, Y., Xiao, Y., Liu, F., Wang, M., Zhu, X., Liu, P., Sun, Q., Wang, W., Peng, M., Brutnell, T. and Li, P. (2016). Transcriptome response of cassava leaves under natural shade. *Scientific Reports*, 6, p. 31673.
Edwards, D. and Kang, B. (1978). Tolerance of cassava (Manihot esculenta crantz) to high soil acidity. *Field Crops Research*, 1, pp. 337–46.
Ekanayake, I. J., Osiru, D. S. O. and Porto, M. C. M. (1998). *Physiology of Cassava*. Edited by Oyetunde, A., Intrnational Institute of Tropical Agriculture, Nigeria, pp. 1–35.
Ekundayo, J. and Daniel, T. (1973). Cassava rot and its control. *Transactions of the British Mycological Society*, 61, pp. 27–32.
El-Sharkawy, M. (2012). Stress-tolerant cassava: The role of integrative ecophysiology-breeding research in crop improvement. *Open Journal of Soil Science*, 2(2), pp. 162–86.
El-Sharkawy, M., Tafur, S. and Lopez, Y. (2012). Cassava productivity, photosynthesis, ecophysiology, and response to environmental stresses in the tropics: A multidisciplinary approach to crop improvement and sustainable production. In B. Ospina and H. Ceballos (Eds), *Cassava in the Third Millennium. Modern Production, Processing, Use and Marketing System*, 1st ed. Cali, Colombia, pp. 29–88. Centro Internacional de Agricultura Tropopical (CIAT); Latin American and Caribbean Consortium to Support Cassava Research and Development (CLAYUCA) and Technical Centre for Agricultural and Rural Cooperation (CTA). CIAT publication n. 377, p574. ISBN: 978-958-694-112-9
El-Sharkawy, M. (2004). Cassava biology and physiology. *Plant Molecular Biology*, 56(4), pp. 481–501.
El-Sharkawy, M. (2006). International research on cassava photosynthesis, productivity, eco-physiology, and responses to environmental stresses in the tropics. *Photosynthetica*, 44(4), pp. 481–512.
El-Sharkawy, M. (2014). Global warming: Causes and impacts on agroecosystems productivity and food security with emphasis on cassava comparative advantage in the tropics/subtropics. *Photosynthetica*, 52(2), pp. 161–78.
El-Sharkawy, M. and Cock, J. (1990). Photosynthesis of Cassava (Manihot esculenta). *Experimental Agriculture*, 26(03), p. 325.
El-Sharkawy, M. and De Tafur, S. (2010). Comparative photosynthesis, growth, productivity, and nutrient use efficiency among tall- and short-stemmed rain-fed cassava varieties. *Photosynthetica*, 48(2), pp. 173–88.
El-Sharkawy, M., Cock, J., Lynam, J., del Pilar Hernàndez, A. and Cadavid, L. (1990). Relationships between biomass, root-yield and single-leaf photosynthesis in field-grown cassava. *Field Crops Research*, 25(3–4), pp. 183–201.
El-Sharkawy, M. A. (2007). Physiological characteristics of cassava tolerance to prolonged drought in the tropics: Implications for breeding cultivars adapted to seasonally dry and semiarid environments. *Brazilian Journal of Plant Physiology*, 19(4), pp. 257–86.
EPA. Environmental Protection Agency. (2017). Available at: https://search.epa.gov/epasearch/epasearch?querytext=Greenhouse+Gases&areaname=&areacontacts=&areasearchurl=&typeofsearch=epa&result_template=2col.ftl, accessed on 28 May 2017.
Fageria, N. (2012). Effects of mineral nutrition on root growth of crop plants. In N. Fageria (Ed.) *The Role of Plant Roots in Crop Porduction*, 1st ed. CRC Press, Francis & Taylor Group, Boca Raton and New York, pp. 299–349. DOI:10.1201/b12365-8, ISBN: 978-1-4398-6737-2.
FAO. (1999). *CASSAVA Post-Harvest Operations*. International Institute of Tropical Agriculture (IITA). Authored by Bokanga, M. and Edited by Mejia, D. and Lewis, B. Ibadan, Nigeria, pp. 1–37.

FAO. (2013). *Save and Grow Cassava: A Guide to Sustainable Production Intensification*. Rome: Food and Agriculture Organization of the United Nations.
Freitas, J., da Silva Santos, V. and de Oliveira, E. (2016). Inbreeding depression in cassava for productive traits. *Euphytica*, 209(1), pp. 137–45.
Fritzemeier, K. H., Cretin, C., Kombrink, E., Rohwer, F., Taylor, J., Scheel, D. and Hahlbrock, K. (1987). Transient induction of phenylalanine ammonia-lyase and 4coumarate: CoA ligase mRNAs in potato leaves infected with virulent or avirulent races of Phytophthora infestans. *Plant Physiology*, 85, pp. 34–41.
Fukai, S. and Hammer, G. (1987). A simulation model of the growth of the cassava crop and its use to estimate cassava productivity in northern Australia. *Agricultural Systems*, 23(4), pp. 237–57.
Fukai, S., Alcoy, A., Llamelo, A. and Patterson, R. (1984). Effects of solar radiation on growth of cassava (Manihot esculenta crantz.). I. Canopy development and dry matter growth. *Field Crops Research*, 9, pp. 347–60.
George, J., Mohankuman, C. R., Nayr, G. M. and Ravindran, C. S. (2000). Cassava agronomy research and adoption of improved practices in India-Major achievements during the past 30 years. India: CTCRI, Central Tuber Crops Research Institute, CIAT Publication.
Ghosh, S. P., Nair, G. M., Pillai, N. G., Ramanujam, T., Mohankumar, B. and Lakshmi, K. R. 1987. Growth, productivity and nutrient uptake by cassava in association with four perennial species. *Tropical Agriculture*, 64, pp. 233–6.
Gleadow, R., Pegg, A. and Blomstedt, C. (2016). Resilience of cassava (Manihot esculentaCrantz) to salinity: Implications for food security in low-lying regions. *Journal of Experimental Botany*, 67(18), pp. 5403–13.
Gleadow, R. M., Evans, J. R., McCaffery, S. and Cavagnaro, T. R. (2009). Growth and nutritive value of cassava (Manihot esculenta Cranz.) are reduced when grown in elevated CO_2. *Plant Biology*, 11, pp. 76–82.
Gray, V. (2000). A comparison of two approaches for modelling cassava (Manihot esculenta Crantz.) crop growth. *Annals of Botany*, 85(1), pp. 77–90.
Hahlbrock, K. and scheel, D. (1989). Physiology and molecular biology of phenylpropanoid metabolism. *Annual Review of Plant Physiology and Plant Molecular Biology*, 40, pp. 347–69.
Howeler, R. H.; Oates, C. G.; Allem, A. C.; Chuzel, G., Henry, G., Hershey, C. H., Müller-Sämann, K. M.; Okogun, A., Sanginga, N., Souza, L. S., Sriroth, K. and Westby, A. (2001). Strategic environmental assessment: An assessment of the impact of cassava production and processing on the environment and biodiversity. In: *Validation Forum on the Global Cassava Development Strategy* (2001, Rome, Italy). Proceedings. Food and Agriculture Organization of the United Nations (FAO); International Fund for Agricultural Development (IFAD), Rome, IT. v. 5, pp. 1–137.
Huber, S. (2011). Grand challenges in plant physiology: The underpinning of translational research. *Frontiers in Plant Science*, 2, pp. 1-5. Doi: 10.3389/fpls.2011.00048.
Iglesias, C., Calle, F., Hershey, C., Jaramillo, G. and Mesa, E. (1994). Sensitivity of cassava (Manihot esculenta Crantz) clones to environmental changes. *Field Crops Research*, 36(3), pp. 213–20.
Ike, I. F and Thurtell, G. W. (1981). Response of indoor-grown cassava to water deficits and recovery of leaf water potential and stomatal activity after water stress. *Journal of Experimental Botany*, 32, pp. 1029–34.
Imai, K., Coleman, D. F. and Yanagisawa, T. (1984) Elevated atmospheric partial pressure of carbon-dioxide and dry matter production of cassava (Manihot esculenta Crantz). *Japanese Journal of Crop Science*, 53, pp. 479–85.
Irikura, Y., Cock, J. and Kawano, K. (1979). The physiological basis of genotype – temperature interactions in cassava. *Field Crops Research*, 2, pp. 227–39.
Izumi, Y., Yuliadi, E., Sunyoto, S. and Iijima, M. (1999). Root system development including root branching in cuttings of cassava with reference to shoot growth and tuber bulking. *Plant Production Science*, 2(4), pp. 267–72.
IPCC. (2007). Summary for policy makers. In Solomon, S., Quin, D., Manning, M., Chen, Z., Marquis, M., Averyt, K. B., Tignor, M., Miller, H. L. (Eds) *Climate Change 2007: The Physical Science Basis. Contribution of Working Group 1 to the Fourth Assesment Report of the Intergovernmental*

Panel on Climate Change. Cambridge, UK and New York, NY: Cambridge University Press, pp. 1–18.

Jarvis, A., Ramirez-Villegas, J., Herrera Campo, B. and Navarro-Racines, C. (2012). Is cassava the answer to African climate change adaptation? *Tropical Plant Biology*, 5(1), pp. 9–29.

Keating, B. and Evenson, J. (1979). Effect of soil temperature on sprouting and sprout elongation of stem cuttings of cassava (Manihot esculenta crantz.). *Field Crops Research*, 2, pp. 241–51.

Keating, B., Evenson, J. and Fukai, S. (1982a). Environmental effects on growth and development of cassava (Manihot esculenta Crantz.) I. Crop development. *Field Crops Research*, 5, pp. 271–81.

Keating, B., Evenson, J. and Fukai, S. (1982b). Environmental effects on growth and development of cassava (Manihot esculenta Crantz.) III. Assimilate distribution and storage organ yield. *Field Crops Research*, 5, pp. 293–303.

Keating, B. A., Evenson, J. P. and Fukai, S. (1982c). Environmental effects on growth and development of cassava (Manihot esculenta Crantz). II. Crop growth rate and biomass yield. *Field Crops Research*, 5, pp. 283–92.

Khan, M., Gemenet, D. and Villordon, A. (2016). Root system architecture and abiotic stress tolerance: Current knowledge in root and tuber crops. *Frontiers in Plant Science*, 7, pp. 1–13. Doi: 10.3389/fpls.2016.01584.

Knoth, J. (1993). *Traditional Storage of Yams and Cassava and its Improvement*. 1st ed. Eschborn: GTZ.

Latif, S. and Müller, J. (2015). Potential of cassava leaves in human nutrition: A review. *Trends in Food Science & Technology*, 44(2), pp. 147–58. https://doi.org/10.1016/j.tifs.2015.04.006

Lebot, V. (2009). *Tropical Root and Tuber Crops: Cassava, Sweet Potato, Yams and Aroids*. Crop production science in horticulture (17). Wallingford, UK: CAB books, CABI.

Legese, H., Gobeze, L., Shegro, A. and Geleta, N. (2011). Impact of planting position and planting material on root yield of cassava (Manihot esculenta Crantz). *Journal of Agricultural Science and Technology*, [online] 5(4), pp. 448–54. Available at: http://www.davidpublishing.com/davidpublishing/Upfile/8/23/2011/2011082310716468.pdf [Accessed 19 December 2016].

Lenis, J., Calle, F., Jaramillo, G., Perez, J., Ceballos, H. and Cock, J. (2006). Leaf retention and cassava productivity. *Field Crops Research*, 95(2–3), pp. 126–34.

Lockard, R. G., Saqui, M. A. and Wounuah, D. D. (1985). Effects of time and frequency of leaf harvest on growth and yield of cassava (*Manihot esculenta* Crantz) in Liberia. *Field Crops Research*, 12, pp. 175–80.

Long, S. P. (1991) Modification of the response of photosynthetic productivity to rising temperature by atmospheric CO_2 concentrations has its importance been underestimated? *Plant Cell and Environment*, 14, pp. 729–39.

Lorenzo, G. De, Ranucci, A., Bellinncamp, D., Salvi, G. and Cervone, F. (1987). Elicitation of phenylalanine ammonia-lyase in Daucus curota by oligogalacturonides released from sodium polypectate from homogeneous polygalacturonase. *Plant Science*, 51, pp. 147–50.

Magalhães, J. E. S., Ferreira, E. A., Oliveira, M. C., Pereira, G. A. M., Silva, D. V. and Santos, J. B. (2016). Effect of plant-biostimulant on cassava initial growth. *Revista Ceres*, 63(2), pp. 208–13. DOI:https://dx.doi.org/10.1590/0034-737X201663020012.

Manrique, L. (1990). Effects of nitrogen fertilization on growth and yield of cassava in Hawaii: II dry matter production and efficiency of partitioning. *Communications in Soil Science and Plant Analysis*, 21(13–16), pp. 1817–30.

Matthews, R. and Hunt, L. (1994). GUMCAS: A model describing the growth of cassava (Manihot esculenta L. Crantz). *Field Crops Research*, 36(1), pp. 69–84.

Moore, S. (2016). Growth period of the cassava plant. Available at: http://homeguides.sfgate.com/growth-period-cassava-plant-64334.html, Accessed on 29 November 2016.

Morgan, N. and Choct, M. (2016). Cassava: Nutrient composition and nutritive value in poultry diets. *Animal Nutrition*, 2(4), pp. 253–61.

Mozzi, F., Ortiz, M. E., Bleckwedel, J., Vuyst, L. D. and Pescuma, M. (2012). Metabolomics as a tool for the comprehensive understanding of fermented and functional foods with lactic acid bacteria. *Food Research International*, 54(1), pp. 1152–61.

Nayr, T. C. R. and Suja, G. (2007). Production potential of root and tubers in association with plantation crops. In *Proceedings of the 13th Triennial Symposium of the International Society for Tropical Root Crops (ISTRC)*: tropical root and tuber crops: opportunities for poverty alleviation and sustainable livelihoods in developing countries.. Arusha, Tanzânia, pp. 186–94.

Naziri, D., Quaye, W., Siwoku, B., Wanlapatit, S., Tu, V. P. and Bennett, B. (2014). Not all those who wander are lost: A comparative analysis of postharvest losses in cassava value chains in Ghana, Nigeria, Thailand and Vietnam. Paper submitted for the 14th Congress of the European Association of Agricultural Economists 'Agri-food and Rural Innovations for Healthier Societies', Ljubljana, Slovenia.

Neubert, E. O. and Back, A. J. (2015a). Avaliação do comportamento do genótipo IAC 576/70 cultivado em canteiros revestidos com lona plástica e sob duas formas de disposição das manivas. 16 congresso Brasileiro de Mandioca. 09–13 November, Foz de Iguaçu, Paraná, Brasil.

Neubert, E. O., Moreto, A. L., Peruch, L. A. M., Soares, A. J. and Beber, J. (2015b). Avaliação do tempo de cozimento de genótipos de mandioca de mesa (aipim) em diferentes épocas de colheita. 16 congresso Brasileiro de Mandioca. 09–13 de November, Foz de Iguaçu, Paraná, Brasil.

Nielsen, J. and Jewett, M. C. (2007). The role of metabolomics in system biology, Topics in Current Genetics 18. In *Metabolomics: A Powerful Tool in Systems Biology*, pp. 1–8. Edited by Nielsen, J. and Jewett, M. C., Springer-Verlag, Berlin and Heidelberg 284p. ISBN: 978-3-540-74718-5.

Noerwijati, K. and Budiono, R. (2015). Yield and yield components evaluation of cassava (Manihot esculenta Crantz) clones in different altitudes. *Energy Procedia*, 65, pp. 155–61.

Noon, R. and Booth, R. (1977). Nature of post-harvest deterioration of cassava roots. *Transactions of the British Mycological Society*, 69, pp. 287–90.

NRI. (2014). Postharvest Loss Reduction Centre. Roots and Tubers. *Postharvest.nri.org*. Available on: http://postharvest.nri.org/scenarios/roots-and-tubers, Accessed on 4 December.

Oguntunde, P. (2005). Whole-plant water use and canopy conductance of cassava under limited available soil water and varying evaporative demand. *Plant and Soil*, 278(1–2), pp. 371–83.

Okogbenin, E., Setter, T., Ferguson, M., Mutegi, R., Ceballos, H., Olasanmi, B. and Fregene, M. (2013). Phenotypic approaches to drought in cassava: Review. *Frontiers in Physiology*, 4, pp. 1–15. Doi: 10.3389/fphys.2013.00093

Okoli, P. and Wilson, G. (1986). Response of cassava (Manihot esculenta Crantz) to shade under field conditions. *Field Crops Research*, 14, pp. 349–59.

Ospina, B. and Ceballos, H. (2012). *Cassava in the Third Millennium*. Cali, Colombia: CIAT, p. 574.

Palta, J. A. (1984). Influence of water deficits on gas-exchange and the leaf area development of cassava cultivars. *Journal of Experimental Botany*, 35(10), pp. 1411–9.

Pardales, J. R. and Yamauchi, A. (2003) Regulation of root development in sweetpotato and cassava by soil moisture during their establishment period. *Plant Soil*, 255, pp. 201–8.

Passam, H. and Noon, R. (1977). Deterioration of yams and cassava during storage. *Annals of Applied Biology*, 85, pp. 436–40.

Pinto-Zevallos, D., Pareja, M. and Ambrogi, B. (2016). Current knowledge and future research perspectives on cassava (Manihot esculenta Crantz) chemical defenses: An agroecological view. *Phytochemistry*, 130, pp. 10–21.

Porto, M. C. M (1983). Physiological Mechanisms of Drought Tolerance in Cassava (Manihot Esculenta Crantz). PhD thesis. University of Arizona, p. 137.

Rickard, J., Marriott, J. and Gahan, P. (1979). Occlusions in cassava xylem vessels associated with vascular discoloration. *Annals of Botany*, 43, pp. 523–7.

Rickard, J. (1981). Biochemical changes involved in the post harvest deterioration of cassava roots. *Tropical Science*, 23, pp. 235–7.

Rickard, J. (1985). Physiological deterioration of cassava roots. *Journal of the Science of Food and Agriculture*, 36, pp. 167–76.

Rickard, J. E. and Coursey, D. G. (1981). Cassava storage. Part 1: Storage of fresh cassava roots. *Tropical Science*, 23, pp. 1–32.

Rosas, C., Cock, J. and Sandoval, G. (1976). Leaf fall in cassava. *Experimental Agriculture*, 12(04), p. 395.

Rosenthal, D., Slattery, R., Miller, R., Grennan, A., Cavagnaro, T., Fauquet, C., Gleadow, R. and Ort, D. (2012). Cassava about-FACE: Greater than expected yield stimulation of cassava (Manihot esculenta) by future CO 2 levels. *Global Change Biology*, 18(8), pp. 2661–75.

Rosenthal, D. M. and Ort, D. R. (2012) Examining cassava's potential to enhance food security under climate change. *Tropical Plant Biology*, 5, pp. 30–8.

Santisopasri, V., Kurotjanawong, K., Chotineeranat, S., Piyachomkwan, K., Sriroth, K. and Oates, C. G. (2001). Impact of water stress on yield and quality of cassava starch. *Industrial Crops and Products*, 13, pp. 115–29.

Schauer, N. and Fernie, A. R. (2006). Plant metabolomics: Towards biological function and mechanism. *Trends in Plant Science*, 11, pp. 508–16.

Seesahai, A. and Ousman, M. (2016). *Understanding the Cassava Plant- Its Growth, Development, Physiology and Tuberization Process*, Centre for Ageing Research and Development in Ireland (CARDI) publishing, Ireland, pp. 1–6.

Siriwat, W., Kalapanulak, S., Suksangpanomrung, M., Netrphan, S., Meechai, A. and Saithong, T. (2012). Transcriptomic data integration inferring the dominance of starch biosynthesis in carbon utilization of developing cassava roots. *Procedia Computer Science*, 11, pp. 96–106.

Sojikul, P., Kongsawadworakul, P., Viboonjun, U., Thaiprasit, J., Intawong, B., Narangajavana, J. and Svasti, M. (2010). AFLP-based transcript profiling for cassava genome-wide expression analysis in the onset of storage root formation. *Physiologia Plantarum*, 140(2), pp. 189–298.

Souza, A., Massenburg, L., Jaiswal, D., Cheng, S., Shekar, R. and Long, S. (2016). Rooting for cassava: Insights into photosynthesis and associated physiology as a route to improve yield potential. *New Phytologist*, 213(1), pp. 50–65.

Streck, N., Pinheiro, D., Zanon, A., Gabriel, L., Rocha, T., Souza, A. and Silva, M. (2014). Effect of plant spacing on growth, development and yield of cassava in a subtropical environment. *Brangatia*, 73(4), pp. 407–15.

Stupak, M., Vanderschuren, H., Gruissem, W. and Zhang, P. (2006). Biotechnological approaches to cassava protein improvement. *Trends in Food Science & Technology*, 17(12), pp. 634–41.

Tan, S. L. and Cock, J. (1979). Branching habit as a yield determinant in cassava. *Field Crops Research*, 2, pp. 281–9.

Tanaka, Y., Data, E. S., Hirose, S., Taniguchi, T. and Uritani, I. (1983). Biochemical changes in secondary metabolites in wounded and deteriorated cassava roots. *Agricultural and Biological Chemistry*, 47, pp. 693–700.

Thelander, M. (2017). Towards understanding carbon partitioning between starch and other storage carbon compounds. Available at: http://lcpu.se/?page_id=440, accessed on 25 May 2017.

Tumuhimbise, R., Melis, R., Shanahan, P. and Kawuki, R. (2014). Genotype × environment interaction effects on early fresh storage root yield and related traits in cassava. *The Crop Journal*, 2(5), pp. 329–37.

Uarrota, V. and Maraschin, M. (2015a). Metabolomic, enzymatic, and histochemical analyzes of cassava roots during postharvest physiological deterioration. *BMC Research Notes*, 8(1), pp. 1–15. DOI: 10.1186/s13104-015-1580-3.

Uarrota, V. and Maraschin, M. (2015b). *Postharvest Physiological Deterioration of Cassava Roots: New Insights*. LAP Lambert Academic Publishing. Germany, 156p. ISBN-13: 978-3-659-71778-9.

Uarrota, V., Moresco, R., Coelho, B., Nunes, E., Peruch, L., Neubert, E., Rocha, M. and Maraschin, M. (2014). Metabolomics combined with chemometric tools (PCA, HCA, PLS-DA and SVM) for screening cassava (Manihot esculenta Crantz) roots during postharvest physiological deterioration. *Food Chemistry*, 161, pp. 67–78.

Uarrota, V., Moresco, R., Schmidt, E., Bouzon, Z., da Costa Nunes, E., de Oliveira Neubert, E., Peruch, L., Rocha, M. and Maraschin, M. (2016). The role of ascorbate peroxidase, guaiacol peroxidase, and polysaccharides in cassava (Manihot esculenta Crantz) roots under postharvest physiological deterioration. *Food Chemistry*, 197, pp. 737–46.

Uarrota, V., Nunes, E., Peruch, L., Neubert, E., Coelho, B., Moresco, R., Domínguez, M., Sánchez, T., Meléndez, J., Dufour, D., Ceballos, H., Becerra Lopez-Lavalle, L., Hershey, C., Rocha, M. and

Maraschin, M. (2015c). Toward better understanding of postharvest deterioration: Biochemical changes in stored cassava (Manihot esculenta Crantz) roots. *Food Sci Nutr*, 4(3), pp. 409–22.

USDA (2016). *National Nutrient Database for Standard Reference*. Cassava, raw. Nutrient values and weights are for edible portion. Release 28 slightly revised May, 2016

Vile, D. (2005). Specific leaf area and dry matter content estimate thickness in laminar leaves. *Annals of Botany*, 96(6), pp. 1129–36.

Villas-Boas, S., Roessner, U., Hansen, M. A. E., Smedsgaard, J., Nielsen, J. (2007). *Metabolome Analysis: An Introduction*. Wiley-Interscience, A John Wiley & Sons, Inc., Publication, Hoboken, NJ, USA, 312p. ISBN: 9780470105511, DOI: 10.1002/0470105518.

Villordon, A., Ginzberg, I. and Firon, N. (2014). Root architecture and root and tuber crop productivity. *Trends in Plant Science*, 19(7), pp. 419–25.

Wang, M., Shi, S., Lin, F., Hao, Z., Jiang, P. and Dai, G. (2012). Effects of soil water and nitrogen on growth and photosynthetic response of Manchurian ash (Fraxinus mandshurica) seedlings in Northeastern China. *PLoS ONE*, 7(2), p. e30754.

Weerarathne, L., Marambe, B. and Chauhan, B. (2017). Does intercropping play a role in alleviating weeds in cassava as a non-chemical tool of weed management? – A review. *Crop Protection*. 95, pp. 81–8.

Wegmann, K. (1970). Untersuchungen uber die ursachen des schnellen verderbens von Maniokwurzeln und due wirksamkeit von Konservierrungstoffen in Maniokmehl. *Brot und Geback*, 24, pp. 175–8.

Wenham, J. (1995). *Post-harvest Deterioration of Cassava*. Rome: Food and Agriculture Organization of the United Nations.

Westby, A., Lankford, B., Coulter, J., Orchard, J. and Morton, J. (2005) Rural infrastructure to contribute to African agricultural development: The cases of irrigation and post-harvest. Background paper for the Commission for Africa.

Westby, A. (2002). Cassava utilization, storage and small-scale processing. In R. Hillocks, J. Thresh and A. Bellotti (Eds), *Cassava: Biology, Production and Utilization*, 1st ed. Wallingford, UK: CABI International, pp. 281–300.

Wheatley, C. (1982). Studies on cassava (*Manihot esculenta* Crantz) root postharvest physiological deterioration. Ph.D. Thesis. University of London.

Wheatley, C. C. (1989). Studies on *cassava* (Manihot esculenta Crantz) root post- harvest physiological deterioration. University of London (Ph.D. Thesis).

Wheatley, C; Gomez, G. (1985). Evaluation of some quality characteristics in cassava storage roots. *Qualitas Plantarum Plant Foods for Human Nutrition*, 35, pp. 121–9.

Wheatley, C. and Schwabe, W. (1985). Scopoletin Involvement in Post-Harvest Physiological Deterioration of Cassava Root (*Manihot esculenta* Crantz). *Journal of Experimental Botany*, 36, pp. 783–91.

Williams, C. (1974). Growth and Productivity of Tapioca (Manihot utilissima) IV. Devlopment and Yield of Tubers. *Experimental Agriculture*, 10(01), p. 9.

Yang J., An, D. and Zhang, P. (2011). Expression profiling of cassava storage roots reveals an active process of glycolysis/gluconeogenesis. *Journal of Integrative Plant Biology*, 53, pp. 193–211.

Zhang, P., Wang, W., Zhang, G., Kaminek, M., Dobrev, P., Xu, J. and Gruissem, W. (2010). Senescence-inducible expression of isopentenyl transferase extends leaf life, increases drought stress resistance and alters cytokinin metabolism in cassava. *Journal of Integrative Plant Biology*, 52(7), pp. 653–69.

Zhao, P., Liu, P., Shao, J., Li, C., Wang, B., Guo, X., Yan, B., Xia, Y. and Peng, M. (2014). Analysis of different strategies adapted by two cassava cultivars in response to drought stress: Ensuring survival or continuing growth. *Journal of Experimental Botany*, 66(5), pp. 1477–88.

Chapter 2

Conservation and distribution of cassava genetic resources

Michael Abberton, Badara Gueye, Tchamba Marimagne and Folarin Soyode, International Institute of Tropical Agriculture (IITA), Nigeria

1 Introduction

The conservation of genetic diversity is an essential activity for any crop and underpins current and future crop improvement programmes. Cassava *Manihot esculenta* Crantz is a vitally important crop in many parts of the world. Significant germplasm collections have been established in international and regional genebanks and have been used extensively by plant breeders. Conservation of genetic resources falls into two main categories: *ex situ* and *in situ*. *Ex situ* includes all germplasm removed from its situation in the wild or on

http://dx.doi.org/10.19103/AS.2016.0014.14

farmers' fields. Generally this refers to genebanks at the international, regional, national or local levels. Cultivated cassava is a clonally propagated crop and this is a major factor in how it is conserved in genebanks. As shown in detail below, approaches to conservation are only rarely through true (botanical) seed but rather as a field collection which is regenerated every two years, as *in vitro* plantlets or, in the future, cryopreservation. This situation is different for wild species of *Manihot* which are typically propagated through seed and conserved in that form. It seems likely that a significant proportion, at least 50%, of all cassava clonal genotypes is conserved in genebanks. However, the situation with respect to the wild relatives of cassava is very different, with many species represented only very sparsely in genebanks and an unknown but likely small proportion of the diversity of these species being conserved. There is a large potential for *in situ* conservation approaches with respect to these wild relatives which is hitherto unrealised.

2 Origins and genetic diversity of cassava

Cassava belongs to the family Euphorbiaceae; the possible origin of the genus *Manihot* is in Mexico/Central America and the cultivated species may have been domesticated from a wild species progenitor possibly along the south-western edge of the Amazon rainforest (Allem, 2002). The highest genetic diversity is seen in Brazil and that within Central America is also high. Cassava is monoecious, with separate male and female reproductive organs on the same plant and this is reflected in the high level of outcrossing observed.

The genus *Manihot* has the following attributes (from Hershey, 2008):

- Perennial
- Sporadic – do not dominate vegetation in which they occur
- Shade intolerant
- Weak competitors
- Frost sensitive
- Adaptations for drought avoidance and tolerance – include storage roots with starch
- All have linamarin which when cells are damaged produces HCN
- High morphological variability within species

All *Manihot* species have $2n = 2x = 36$ chromosomes. The classification of Rogers and Appan (1973) has been generally accepted with the genus divided into 98 species in 19 sections (Rogers and Appan, 1973; Rogers and Fleming, 1973; Allem, 2002). A major feature is that Northern and Southern hemisphere species groupings are very distinct (Allem, 1987, 1994, 2002). Allem discovered *M. esculenta* subspecies *flabellifolia* in Brazil and subsequent molecular analysis shows *M. esculenta* lies within the range of variation for this species. The ssp. *flabellifolia* is a likely progenitor of *M. esculenta*, possibly with landrace 'intermediaries' (Allem, 2002). All the wild species of the genus seem to be mainly seed propagated and the process of domestication is observed mainly in increased root thickening, swollen leaf scars, shortened internodes and thicker stems, serving as carbohydrate reserves. The cultivated species is mainly vegetatively propagated and this has very important implications for its conservation and breeding. Deletre et al. (2016) showed that unconscious selection of seedlings from sexual reproduction, with morphology similar to that of the crop being cultivated, can act to maintain distinct landraces.

3 *Ex situ* conservation of cassava genetic material

Major international collections of cassava are held at CIAT and IITA, two institutes of the CGIAR. CIAT in Colombia has the most extensive cassava collection. It holds more than 6000 cassava accessions with the great majority of these coming from South America including around 2400 from Colombia and 1500 from Brazil (http://isa.ciat.cgiar.org/urg/cassavacollection.do). IITA conserves over 4000 accessions of the crop mainly from countries of sub-Saharan Africa. More than 90% of this collection originates in West Africa (http://my.iita.org/accession2). The cassava collection at EMBRAPA-Brazil is about 4000 accessions. Other important genebanks are those in CTCRI-India, INIA-Peru, NRCRI-Nigeria, IAN-Paraguay, SRCV-Benin, NARO-Uganda, D.R. Congo and PGRC/CRI-Ghana.

Genebank standards have been published by FAO and updated in 2014 (FAO 2014). They outline the principles and practices of conservation of *ex situ* conservation.

Standard operating procedures (SOPs) should be in place for all genebank activities from acquisition to distribution and including all aspects of data collection and management. These are key components of quality management systems.

At IITA, germplasm is maintained in the field and *in vitro* and cryobanking of cassava has recently commenced. It also has a small field collection of wild *Manihot*. Both IITA and CIAT engage in 'black box' safety duplication of *in vitro* plantlets: at CIP (Lima, Peru) for CIAT and the IITA/Africa Rice facility at Cotonou, Republic of Benin, for IITA. The term 'black box' means that germplasm is simply maintained *as it is* without, for example characterisation or distribution. However, this term is somewhat misleading when applied to *in vitro* duplication of clonal crops since considerable time and effort are required to maintain viable contamination-free plantlets. In the next few years cryobanking at IITA, with germplasm also cryopreserved at another site, will represent a sustainable long-term approach to safety duplication.

4 Field conservation of cassava genetic material

Conservation of genetic resources in the field is a common *ex situ* approach in cassava. It has the advantage that germplasm can be phenotypically characterised and evaluated and that accessions are maintained in ways that avoid the possibility of genetic or epigenetic changes introduced through tissue culture. The identity of accession, through continuous field propagation from the original, should be maintained if high standards of field procedures and sample tracking are applied. This can be very important in establishing whether accessions (e.g. those held *in vitro*) are 'true to type'. However, field conservation also has disadvantages, particularly the resource (labour, land, time) requirements and the potential loss of accessions from pests, diseases, abiotic stress and so on. Cassava performance (and related to risk of loss in the case of field genebanks) has been linked to the type of soil where it is planted; selection of a well-aerated, loose and light sandy loam soil is preferred. Other factors to be considered during land selection for a field genebank include risk of flooding, prevalence of pests and diseases, and unsolicited human intervention. Before the planting of cassava in the field, there are a number of activities that are carried out in land preparation. These include mowing, ploughing, harrowing and ridging. Except where irrigation facilities are available, the planting of cassava is carried out at the outset of the rainy season which is around April in the derived savannah and some parts of the humid forest regions of Nigeria. The right selection of healthy mother plants for the cassava stem cuttings is key for success in cassava sprout rate. Disease-free stem cuttings with enough energy reserves ensure better sprout rates and these are usually located in the mid-section of healthy and mature cassava plants. A spacing of 1 m by 1 m between plants in ridges and an alleyway of about 1 m to 1.5 m is usually required for maintenance and research purposes and this is done by the methodological marking of the field. Weeds are either removed manually with the use of hoes and cutlass or through the application of herbicides. Manual weeding is laborious while herbicide application may be observed to be expensive in the short term. In most cases, a combination of both approaches is employed for the combating of weeds in cassava fields. The frequency and need for weeding may be different in cassava fields depending on soil and canopy type. A typical approach is the application of pre-emergence herbicide shortly after planting before sprouting of the cassava stems. Post-emergence herbicide application comes any time from the period the planted cassava stems have sprouted to the time they are harvested and this can be carried out from three to four times in a yearly cycle.

Cassava plants are prone to a number of pests and diseases in the field. These include rodents, although generally good weed management is enough to ward them off. Hoppers and whiteflies are also threats and insecticide is usually sufficient for acceptable levels of protection. There are a number of important diseases of cassava fields, especially cassava mosaic disease, cassava bacterial blight, cassava brown streak disease and cassava anthracnose disease. These diseases are best prevented by the use of healthy plant materials that are free from their causative pathogens and the maintenance of high hygienic standards in field cultural practices. Frequent field monitoring is an essential component of cassava field management practices.

Field conservation has the distinct advantage that it allows the collection of a wide range of information about accession description and performance. Most field genebank managers will want to take advantage of this opportunity, not available for *in vitro* genebanks. It is important that accessions have not only passport data but information on a range of morphological descriptors, that is, characterisation data. Phenotypic characterisation takes place in the field and for cassava at IITA it spans one year. Data are taken up to four times at an interval of three months and is collected from both the aerial and root parts. Data collected are both quantitative (e.g. number of roots) or qualitative, for example, leaf colour. Measurements made on the aerial parts include colour of unexpanded apical leaf, colour of first expanded apical leaf, petiole colour, petiole length, absence/presence of stipules, anthocyanin pigmentation, incidence and severity of diseases. Data taken from the underground parts include storage root form, absence/presence of root peduncle, root surface colour, root surface texture, starch content, harvest index and poundability.

5 Core collections of cassava genetic material

In genebanks which have large numbers of germplasm accessions for a particular crop the concept of a 'core collection' has often been applied. A core collection is essentially a fraction of the total number – often around 10% – which captures most of the genetic diversity as shown commonly by morphological descriptors or a combination of morphological characters and molecular markers. A number of such core collections have been developed in cassava. They have proved useful as a focus for evaluation of germplasm for important agronomic traits and as a starting point for molecular characterisation of the diversity in the collection. At CIAT a core collection of 630 accessions was developed and the diversity and redundancy levels of this core were initially assessed in 1999 using microsatellite, AFLP and isozyme markers (Chavarriaga-Aguirre et al., 1999). This work suggested that morphological characterisation was effective at ensuring 'genetically representative non-redundant samples'. However, core collections have their limitations, particularly where a subgroup enriched for accessions with a particular trait, for example drought tolerance, is required. Core collections may not adequately capture useful rare alleles of a species. In recent years a number of approaches to developing traits-specific subsets have been described, including the Focused Identification of Germplasm Strategy (FIGS) that draws upon knowledge of the geographic origin and evolutionary history of accessions (Khazaei et al., 2013). It seems likely that such approaches, integrated with the use of genomic data, will become an important part of ensuring greater use of the wealth of cassava germplasm held in genebanks.

6 *In vitro* conservation of cassava genetic material

The limitations of field conservation mean that the two major international collections at CIAT and IITA have *in vitro* conservation at the centre of their strategy and operations for cassava germplasm conservation and exchange. Slow growth conservation has many advantages for clonal crops such as cassava. It constitutes a viable alternative to complement and reduce the large land size required for field banks. Plant tissue culture is a powerful tool for a safer and faster way to multiply large quantities of material for distribution, duplication in other genebanks and international exchange (easier plant material transport); and also for breeding purposes. Slow growth storage is, however, for short- to medium-term conservation, after which the plantlets are subcultured when signs of deterioration/necrosis are visible (Gueye et al., 2012). *In vitro* conservation offers a means of maintaining valuable gene combinations in a small space, protected against pest and disease attack, soil problems and climatic changes; gives possibility of sanitation and with high multiplication potential; and opens the door for further research possibilities (e.g. somatic embryogenesis and protoplast fusion). They are generally more economical and less risky in a long-term perspective, as compared to field collections. Therefore, effort and research have been concentrated on improving the *in vitro* regeneration (subculture duration), which is the major leverage point for cost-effectiveness.

The conservation of cassava needs a small quantity of material and allows longer duration between two regenerations or subcultures, using slow growth storage. The principle is to place the *in vitro* plantlets under slow growth conditions, through adaptation of key growth factors (light, temperature, culture medium and growth retardants). A number of cassava genebanks around the world have *in vitro* tissue culture facilities as either the principal or a complementary conservation system, with the possibility to clean the germplasm from diseases and pests via meristem culture and/or other sanitation methods such as thermo- or chemotherapy. *In vitro* slow growth conservation methods require technical expertise, facilities and operating budget. According to Hershey (2008), there are about 70 cassava genebanks worldwide but only 12 countries (Ng and Ng, 2002) have *in vitro* facilities for conservation. The CIAT genebank conserves its collection exclusively in tissue culture. A few genebanks (mainly EMBRAPA, Brazil and CIAT, Colombia) have seed banks to conserve seeds of wild species or breeding material. A few, including CIAT with its extensive collection from the centre of origin for cassava, are also initiating DNA banks.

Many of these laboratories combine *in vitro* techniques targeted for multiple purposes, for example, for pathogen cleaning, rapid multiplication, gene bank conservation and exchange. *In vitro* conservation of cassava is still far less common than field conservation. The largest national *in vitro* collections are held in Brazil and Argentina, while both Thailand and Vietnam are building their *in vitro* genebank inventories. There appears to be very few *in vitro* cassava genebanks in Africa. Apart from the international collections held at CIAT and at IITA (the two CGIAR centres sharing the international mandate for cassava research, breeding and improvement) all other *in vitro* genebanks have national or regional focus (e.g. National Genebank of China for vegetatively propagated crop species, CTCRI-India and EMBRAPA in Brazil). The *in vitro* conservation procedures are available in IITA and CIAT manuals/handbooks (Dumet et al., 2007; Mafla et al., 2009), also accessible on those CG centres' websites (www.iita.org, https://ciat.cgiar.org/).

7 Cryopreservation of cassava genetic material

Cryopreservation, with respect to the conservation of clonal crops, is the maintenance of plant material at ultra-low temperature, normally in liquid nitrogen at −196°C, using cryogenic techniques. At such low temperature, biological activities and metabolism are stopped, eliminating the need to regularly rejuvenate or regenerate the plant. It is currently a supplementary tool to improve conservation of germplasm in a long-term perspective. For species where the techniques are well-developed and reliable, cryopreservation is the most reliable technique for long-term storage of plant genetic resources (Popov et al., 2005). It also ensures the safe long-term conservation of the physiological state of the genetic resources, thus reducing genetic changes through multiple regeneration cycles. Cryopreservation helps to overcome many of the *in vitro* maintenance disadvantages such as labour-intensive subcultures, potential for fungal and bacterial contaminants and tissue ageing (Benson, 2008). Many studies confirmed that it is economically competitive with other conservation systems (Harvengt et al., 2004; Reed et al., 2004; Keller et al., 2008). Thus, cryoconservation techniques have been increasingly used for the long-term storage. In the last 25 years, several cryogenic techniques have been developed, especially those based on vitrification methods (the transition of water directly from the liquid phase into an amorphous or 'glassy' phase, whilst avoiding the formation of crystalline ice) such as encapsulation dehydration, pre-culture dehydration and encapsulation/vitrification. The main requirement for using the cryopreservation method is that it should be simple, economical, reproducible and should allow a relatively high regrowth rate (Leunufna and Keller, 2003).

Work on cassava cryoconservation was initiated in the mid-1982s at CIAT and many researchers have improved on the techniques since then (Escobar and Roca, 1997; Escobar et al., 1997). To date, cryogenic techniques used for cassava long-term conservation have met with different levels of success.

Specific cassava protocols were developed testing and optimising a range of different cryogenic methods. The CIAT method is based on rapid freezing of highly proliferating cassava meristem/shoot tips as explant. Cryobanking of cassava shoot tips has been started at the CIAT cassava genebank after adaptation of protocols from other crops in different laboratories (Escobar et al., 1997; Gonzalez-Arnao et al., 2007). IITA tested various cryogenic techniques according to local conditions (Dumet et al., 2013). Working on numerous taxa, cryopreservation via droplet vitrification showed high efficiency with 48 accessions of cassava maintained at IITA. However, work remains to be done on achieving an adequate recovery level for a majority of the cassava collection. There appear to be a certain number of recalcitrant accessions, on which research should continue on improving recovery, before committing to large-scale cryopreservation of any genebank.

Cryopreservation is also used for other purposes, such as the cryopreservation of embryogenic tissues for genetic transformation and for botanic seeds.

8 Conservation of cassava genetic material as true seed

Because all known cultivars of cassava, including all landrace varieties in farmers' fields and in genebanks, are highly heterozygous, the plants that result from cassava true seed segregate for many traits. As a result, none of the progeny from a given plant, even if self-pollinated, will be genetically identical to the parent plant. This is in contrast to the normal vegetative propagation of cassava, where indefinite generations of propagation will yield identical genotypes. In terms of genebanks for cassava, this means that cloned offspring from any single individual of a genetically uniform variety can faithfully represent the genotype of that variety in a genebank. This is the principal under which field and *in vitro* genebanks are managed.

On the other hand, seeds can preserve the genes of the parents, if a large enough population of seeds is maintained. There are, however, several challenges to seed conservation for genebanks. First, cassava tends to produce few seeds, and some varieties are not known to flower at all (or produce seeds). Producing adequate-sized populations for purposes of gene conservation would be a challenge for some landrace varieties. Secondly, as mentioned above, seed conservation would not allow reproducing the original variety from which they were derived. Since it is often the specific gene combinations that give special value to farmers' varieties, this value would not be recaptured without extensive breeding. Thirdly, in order to retain just the set of genes existing in a particular variety, without *contamination* with genes from other varieties, it is necessary to do self-pollination of each variety or accession of interest. Since cassava suffers inbreeding depression in the S_1 generation, this reduced vigour can create additional challenges for maintaining and regenerating subsequent generations. However, it would also be possible to inter-cross S_1 plants from a given clonal variety, to recover some vigour in subsequent generations while not expanding the gene population beyond what was present in the original variety.

In summary, seed conservation is an option for cassava genebanks, but, given the complexity of management, will not likely be done by more than a few major genebanks, if done at all on a systematic basis.

From the point of view of seed viability, cassava should not present any difficulties. Seed of cultivated cassava can be stored under the same conditions as for many seed-propagated crops. Ellis et al. (1981a) stated that, 'there can now be no doubt that cassava seeds are orthodox' and 'no problems are envisaged with long-term cassava seed storage' (Ellis et al., 1981a,b). However, conservation of cultivated cassava as seed is rare and at a low level.

9 Data collection and management in genebanks

Different and distinct stages are involved in managing genetic resources. Each stage generates different types of data, including information related to the identity of the accession and where collected (passport data), and the agro-morphological characterisation of the plants (characterisation and evaluation data). Full genebank data should be available and publicly accessible through well-managed databases. Genebanks information should be integrated into the global portal for plant genetic resources implemented for national and international genebanks.

Passport data are used as one of genebanks' ways to manage diversity. This has helped genebank curators and users to group the materials conserved in a more logical and comprehensive order. The minimum descriptors used can serve as a linkage between genebank data and external data sources such as breeders' evaluation data. One can link geographical passport descriptors to spatial data sources and relate to climatic data. IITA cassava passport data are available in both its public accessions database (http://my.iita.org/accession2/) and in Genesys (https://www.genesys-pgr.org/welcome).

Similar to a person's passport information page, accession passport data constitutes the basic information used for the general management of accessions: its entry into the genebank collections and description parameters related to where the accessions were originally collected. Materials enter the collection through collection expeditions, from breeding programmes, or acquisition from other institutes. In each case, the material will already have some identifier assigned by the collector, breeder or other institute. Accession name is the vernacular name of the material and is commonly captured by the collector or assigned by the breeder. Genebank accessions obtained through collecting missions should maintain data about the site and dates of the collecting and collector information. Lines developed by breeding programmes of the institute may be included in the collection. Information provided by the breeders should include the pedigree, ancestral information of the material, along with names and identifiers used by the breeding programme and the codes and names of institutes that developed the material. Materials coming from other institutes and genebanks must be accompanied by accession passport data as documented in the source genebank.

Accession documentation should capture any identifiers provided by the source institute. This data allows for validation and curation of passport data between the genebanks and allows researchers to obtain material from either collection. The Multi-crop Passport Descriptors (MCPD V.2) (http://www.bioversityinternational.org/e-library/publications/detail/faobioversity-multi-crop-passportdescriptors-v2-mcpd-v2/) are a revision of the original FAO/IPGRI publication released in 2001, expanded to accommodate emerging needs, such as the broader use of GPS tools, and the implementation of the International Treaty on Plant Genetic Resources for Food and Agriculture (ITPGRFA) (http://www.planttreaty.org) Multilateral System for access and benefit sharing.

The Crop Trust (https://www.croptrust.org/) manages an online data portal for accession-level data management-GENESYS. GENESYS is a global portal to information about Plant Genetic Resources for Food and Agriculture. It is a gateway from which information about germplasm accessions and genebanks around the world can be easily found and accessions ordered (https://www.genesys-pgr.org/welcome).

10 Germplasm distribution

International exchange of germplasm is a vital part of the system of genetic resource conservation and utilisation. A key aspect of this is arrangements for phytosanitary testing and import/export approvals. In cassava this has historically been a constraint and challenge to safe germplasm movement and this is still the case. The second major aspect of germplasm distribution is the system of international law and regulation governing access and transfer. Here a major component is the ITPGRFA. A key element of this Treaty is the development of a multilateral system for the conservation and distribution of a number of major crops listed in Annex 1 of the Treaty. These Annex 1 crops can be held by genebanks and distributed to those requesting accessions (for purposes of research, breeding and training for agriculture) using a Standard Material Transfer Agreement. Effectively this multilateral system obviates the need for bilateral negotiations for germplasm access and distribution under the Convention for Biological Diversity. Cassava is listed under Annex 1 of the Treaty but the related wild species are excluded. In addition, the genebanks of the CGIAR (e.g. CIAT and IITA) have a special status under the Treaty (Article 15) in which accessions are designated as held 'in trust' for future generations.

11 *In situ* conservation of cassava genetic material

Whilst *ex situ* conservation of cassava genetic resources is clearly key for their continued survival and utilisation, it may not be possible to conserve all landraces in this way and it is unlikely that a significant proportion of the wild relatives will be conserved *ex situ* in the near future. So it is important that conservation in the wild or in farmers' fields – that is, *in situ* conservation – is developed as an alternative and ideally complementary approach. This also has the advantage of allowing continued evolution of germplasm in response to changing environment, including climate, and abiotic and biotic stress.

Hershey (2008) estimated that about 27 000 landraces of cassava exist *in situ* and that less than half of these are in genebanks (the precise number is difficult to ascertain since the extent of duplication of accessions within genebanks has not yet been fully evaluated).

There are a number of significant threats to cassava landraces *in situ*. These include loss of habitat, pests and diseases, and perhaps most significantly, replacement of landraces with improved varieties that may have a more narrow genetic base. In some cases the best option may be to secure these landraces in genebanks. However, in other cases the conservation *in situ* can be equally or more practical. Maxted and Kell (2009) considered the case of cassava wild relatives and outlined the priorities for their *in situ* conservation in terms of species and geographical area. The wild relatives are particularly important in the light of evidence that a number of them can cross easily with the cultivated species and that interspecific hybridisation has been important in the evolution and breeding of cassava (Bredeson et al., 2016).

Currently there is very little ongoing effort to systematically conserve cassava landraces *in situ*, and there is a strong case for more research and more investment to be deployed, including the use of molecular tools to survey diversity existing both *in situ* and in genebanks.

12 Molecular genetic studies of cassava diversity

As with many crops the tools of molecular genetics are being deployed with increasing power and scope in cassava and they have many potential applications in diversity analyses and genebank management as well as pre-breeding and breeding.

In recent years considerable progress has been made with respect to use of single-nucleotide polymorphism (SNPs) to study diversity in cassava and application of next-generation sequencing to both marker development and identification of quantitative trait loci or the position of important single genes (Ferguson et al., 2012).

Wang et al. (2014) sequenced two genotypes: W14 *M. esculenta* ssp. *flabellifolia* and KU50 a bred variety common in Southeast Asia with higher root yield and starch content. They confirmed that the 'level of heterozygosity in cassava is among the highest found in sequenced plant genomes'. They also found a greatly reduced linamarin and lotaustralin content in KU50 relative to W14 and speculated that the domestication of cassava around 7000–12 000 years involved changes in the pattern of carbon flux.

Sequencing studies also reveal synteny with other sequenced Euphorbiaceae genomes, for example, *Ricinus communis* and *Jatropha curcas.* Bredeson et al. (2016) produced a high-quality assembly of the cassava genome sequence and compared it to wild accessions and related species. They found evidence of significant interspecific hybridisation in the evolutionary history of cassava.

Peña-Venegas et al. (2014) studied diversity amongst cassava cultivated by different ethnic groups in the Amazon and Moura et al. (2016) used microsatellites to distinguish between two groups of sugary cassava accessions held in Brazilian genebanks. This work helps demonstrate the increasingly important role of molecular genetic approaches for genebank management. Rabbi et al. (2015) used genome-wide SNP markers derived from genotyping by sequencing methods to track variety ancestry and adoption. Again, such approaches offer a powerful tool for the identification of duplicates in collections and for variety identification. Pariyo et al. (2013) used an earlier generation of molecular markers, simple sequence repeats (SSRs), to relate genetic diversity in cassava from South, East and Central Africa to resistance to cassava brown streak disease.

In recent years there have also been efforts to improve genetic mapping in cassava and Soto et al. (2015) described a high-density map related to a physical map and the whole-genome sequence.

13 Where to look for further information

Further information can be found in the References section below and from the websites of IITA and CIAT in particular.

14 Acknowledgements

The authors are grateful for the financial support offered to IITA GRC from the Crop Trust and the CGIAR funders.

15 References

Allem, A. C. 1987. *Manihot esculenta* is a native of the neotropics. *Plant Genet. Resour. Newsl.*, 71: 22–4. IBPGR/FAO.

Allem, A. C. 1994. The origin of *Manihot esculenta* Crantz (*Euphorbiaceae*). *Genet. Resourc. Crop Evol.*, 41: 133–50. Netherlands.

Allem, A. C. 2002. The origins and taxonomy of cassava. In Hillocks, R. J., Thresh, J. M. and Bellotti, A. C. (Eds), *Cassava: Biology, Production and Utilization*. CABI Publishing: Wallingford, UK, pp. 1–16.

Benson Erica, E. (2008). Cryopreservation of phytodiversity: A critical appraisal of theory and practice. *Crit. Rev. Plant Sci.*, 27(3): 141–219.

Bredeson, J. V., Lyons, J. B., Prochnik, S. E. and et al. (2016). Sequencing wild and cultivated cassava and related species reveals extensive interspecific hybridization and genetic diversity. *Nat. Biotechnol.* 34, 562–70. doi:10.1038/nbt.3535.

Chavarriaga-Aguirre, P., Maya, M. M., Tohme, J. Duque, M. C., Iglesias, C., Bonierbale, M. W., Kresovich, S. and Kochert, G. (1999). Using microsatellites, isozymes and AFLPs to evaluate genetic diversity and redundancy in the cassava core collection and to assess the usefulness of DNA-based markers to maintain germplasm collections. *Mol. Breed.*, 5: 263.

Deletre, M., Hodkinson, T. R. and McKey, D. (2016). Perceptual selection and the unconscious selection of 'volunteer' seedlings in clonally propagated crops: An example with African cassava (Manihot esculenta Crantz) using ethnobotany and population genetics. *Genet. Resour. Crop Evol.*, 64, 665–80.

Dumet, D., Diebiru, E., Adeyemi, A., Akinyemi, O., Gueye, B. and Franco, J. (2013). Cryopreservation for the in perpetuity conservation of yam and cassava genetic resources. *CryoLett.*, 34(2): 107–18.

Dumet, D., Ojuderie, O. and Adeyemi, A. (2007). *IITA Genebank Manual Series, Cassava in vitro genebanking*. International Institute of Tropical Agriculture: Ibadan, Nigeria.

Ellis, R. H., Hong, T. D. and Roberts, E. H. (1981a). The influence of dessication on cassava seed germination and longevity. *Ann. Bot.*, 47: 173–5.

Ellis, R. H., Hong, T. D. and Roberts, E. H. (1981b). An investigation of the influence of constant and alternating temperature on the germination of cassava seed using a two dimensional temperature gradient plate. *Ann. Bot.*, 48: 241–6.

Escobar, R. H., Mafla, G. and Roca, W. M. (1997). A methodology for recovering cassava plants from shoot tips maintained in liquid nitrogen. *Plant Cell Rep.*, 16: 474–8.

Escobar, R. and Roca, W. M. (1997). Cryopreservation of cassava shoot tips though rapid freezing. *Afr. J. Root Tuber Crops*, 2: 214–15.

FAO (2014). Genebank Standards for Plant Genetic Resources for Food and Agriculture. http://www.fao.org/3/a-i3704e.pdf.

Ferguson, M. E., Hearne, S. J., Close, T. J., Wanamaker, S., Moskal, W. A., Town, C. D., de Young, J., Marri, P. R., Rabbi, I. Y. and de Villiers, E. P. (2012). Identification, validation and high-throughput genotyping of transcribed gene SNPs in cassava. *Theor. Appl. Genet.*, 124: 685–95.

Gonzalez-Arnao, M. T., Panta, A., Roca, W. M., Escobar, R. H. and Engelmann, F. (2007). Development and large scale application of cryopreservation techniques for shoot and somatic embryo cultures of tropical crops. (http://www.springerlink.com/content/y42117245j364361/fulltext.pdf).

Gueye, B., Adeyemi, A. and Debiru, M. (2012). Standard operation procedures (SOP) for IITA *in vitro* genebank. IITA Annual Report. p. 59.

Harvengt, L., Meir-Dinkel, A., Dumas, E. and Collin, E. (2004). Establishment of a cryopreserved genebank of European elms. *Can. J. For. Res.*, 34: 43–55.

Hershey, C. (2008). A global conservation strategy for cassava (*Manihot esculenta*) and wild Manihot species. Summary of stakeholder deliberations and recommendations prepared for the Global Crop Diversity Trust. Available from: http://isa.ciat.cgiar.org/urg/urgweb_folder/files/unitfiles/A%20Global%20Conservation%20Strategy%20for%20Manihot%20%20August%202008–2.pdf.

Khazaei, H., Street, K., Bari, A., Mackay, M. and Stoddard, F. L. (2013). The FIGS (focused identification of germplasm strategy) approach identifies traits related to drought adaptation in Vicia faba genetic resources. *PLoS One*, 8(5): e63107. doi:10.1371/journal.pone.0063107. Print 2013.

Keller, E. R. J., Kaczmarczyk, A. and Senula, A. (2008). Cryopreservation for plant genebanks: A matter between high expectations and cautious reservations. *Cryolett.*, 29: 53–62.

Mafla, G., Roa, J. C., Aranzales, E. and Debouck, D. 2009. *Handbook of Procedures for in vitro Germplasm Conservation of the Genus Manihot*. CIAT: Cali, Colombia, p. 56.
Maxted, N. and Kell, S. (2009). Establishment of a global network for the *in situ* conservation of crop wild relatives: Status and needs. Commission on Genetic Resources for Food and Agriculture. Background Study Paper No 39.
Moura, E. F., Sousa, N. R., Moura, M. F., Costa dias, M., Souza, E. D., de Farias Neto, J. T. and Sampaio, J. E. (2016). Molecular characterization of accessions of a rare genetic resource: Sugary cassava (*Manihot esculenta* Crantz) from Brazilian Amazon. *Genet. Resour. Crop Evol.*, 63(4): 583–93.
Ng, N. Q. and Ng, S. Y. C. (2002). Genetic resources and conservation. In Hillocks, R. J., Thresh, J. M. and Bellotti, A. (Eds), *Cassava: Biology, Production and Utilization*. CABI Publishing, pp. 167–77.
Pariyo, P., Tukamuhabwa, P., Baguma, Y., Kawuki, R. S., Alicai, T., Gibson, P., Kanju, E., Wanjala, B. W., Harvey, J., Nzuki, I., Rabbi, I. Y. and Ferguson, M. (2013). Simple sequence repeats (SSR) diversity of cassava in South, East and Central Africa in relation to resistance to cassava brown streak disease. *Afr. J. Biotechnol.*, 12(28): 4453–64.
Peña-Venegas, C. P., Stomph, T. J., Verschoor, G., Becerra Lopez-Lavalle, L. A. and Struik, P. C. (2014). Differences in Manioc Diversity Among Five Ethnic Groups of the Colombian Amazon. *Diversity*, 6: 792–826. doi:10.3390/d6040792.
Pilatti, F. K., Aguiar, T., Simões, T., Benson, E. E. and Viana, A. M. (2011). In vitro and cryogenic preservation of plant biodiversity in Brazil. *In Vitro Cell. Dev. Biol. Plant*, 47(1): 82–98.
Popov, A. S., Popova, E. V., Nikishina, O. N. and Vysotskaya, O. N. (2005). Cryobank of plant genetic resources in Russian Academy of Sciences. *Int. J. Refrig.*, 28(6): 1–8.
Rabbi, I. Y., Kulakow, P. A., Manu-Aduening, J. A., Dankyi, A. A., Asibuo, J. Y., Parkes, E. Y., Abdoulaye, T., Girma, G., Gedil, M. A., Ramu, P., Reyes, B. and Maredia, M. Y. (2015). Tracking crop varieties using genotyping by-sequencing markers: A case study using cassava (*Manihot esculenta* Crantz). *BMC Genetics* 16: 115.
Reed, B. M., Engelmann, F., Dulloo, M. E. and Engels, J. M. M. (2004). Technical guidelines for the Management of Field and in vitro Germplasm Collections. IPGRI Handbooks for Genebank. No 7. Rome, IT.
Rogers, D. J. and Appan, S. G. (1973). 'Flora neotropica monograph.' Manihot and Manihotoides (Euphorbiaceae): A computer-assisted study. *Fl. Neotrop. Monogr.*, 13.
Rogers, D. J. and Fleming, H. S. (1973). A monograph of *Manihot Esculenta* with an explanation of the taximetrics methods used. *Econ. Bot.*, 27(1): 1–113.
Soto, J. C., Ortiz, J. F., Perlaza-Jiménez, L., Ximena Vásquez, A., Becerra Lopez-Lavalle, L. A., Mathew, B., Léon, J., Bernal, A. J., Ballvora, A. and López C. E. (2015). A genetic map of cassava (Manihot esculenta Crantz) with integrated physical mapping of immunity-related genes. *BMC Genomics*, 16: 190.
Wang, W., Feng, B., Xiao, J. and et al. (2014). Cassava genome from a wild ancestor to cultivated varieties. *Nat. Comm.* DOI:10.1038/ncomms 6110.

Chapter 3

Developing new cassava varieties: tools, techniques and strategies

Hernán Ceballos, Nelson Morante, Fernando Calle, Jorge Lenis and Sandra Salazar, International Center for Tropical Agriculture (CIAT), Colombia

1 Introduction

Cassava (*Manihot esculenta* Crantz) is a crop of significant economic and social relevance. Along with rice, maize, sugarcane and banana/plantain, it is among the most important sources of energy in the diet in many tropical countries (Burns et al., 2010). Cassava is a key food security crop in sub-Saharan Africa and in several Asian and American countries, such as Indonesia, Brazil, Paraguay and Haiti. In addition, cassava is the second most important source of starch worldwide (Stapleton, 2012); thus, it is also a key commodity for agro-industrial processes, including the production of carburant ethanol and cooking fuel (Anyanwu et al., 2015; Chen et al., 2016).

There are almost 100 *Manihot* species described (Rogers and Appan 1973), all of Neotropical origin. Duputié and co-workers (2011) suggested that the genus originated in Mesoamerica, where it first diversified, before colonizing South America. The origin of cultivated cassava is still unclear. The prevailing hypothesis is that it originated in South America (Allem, 2002; Olsen and Schaal, 2001; Nassar and Ortiz 2008). The crop, which

http://dx.doi.org/10.19103/AS.2016.0014.15

might have been domesticated more than once, most likely in Brazil, spread quickly to other regions. More than 2000 years ago the Mochica culture in coastal Peru had already produced ceramic pieces representing cassava roots and stems (Fig. 1). Early European sailors soon recognized the advantages of the crop and carried it to Africa. Cassava was introduced into Asia slightly later: by the Spanish from Mesoamerica (through the Philippines) and by traders from Africa (Byrne, 1984).

All parts of this remarkable crop can be exploited. Its most common product is the starchy root. Its foliage has an excellent nutritional quality for animal and human consumption and offers great potential for further exploitation in food and feed. The stems are used for commercial multiplication through cuttings, along with minor applications such as mushroom culture.

According to Gonçalvez Fukuda et al. (2002), one of the first reports on cassava variety assessment and selection was published in Bahia, Brazil, in 1899 (Zehntner, 1919). Modern breeding programmes were initiated during the first half of the twentieth century (Byrne, 1984; Jennings and Iglesias, 2002) in Brazil (Graner, 1935, Lozano et al., 1978; Normanha, 1970), Ghana (Hahn et al., 1979), India (Abraham, 1957), Indonesia (Bolhuis, 1953), Madagascar (Cours, 1951) and Tanzania (Nichols, 1947; Jennings, 1957). The latter was the East African Agriculture and Forestry Research Organization, based at Amani (then Tanganyika). This breeding programme generated the first source of resistance to cassava mosaic disease (CMD), which is still widely used. CMD is a devastating disease in Africa, India and Sri Lanka. However, with the exception of Brazil, most of these early efforts were discontinued as a result of the dismantling of the colonial system. The creation of cassava improvement programmes at the International Institute of Tropical Agriculture (IITA based in Ibadan, Nigeria) and Centro Internacional de Agricultura Tropical (CIAT in Cali, Colombia) broke the previous isolation that breeding efforts had had in the past (Byrne, 1984) and contributed to the establishment of vigorous and successful programmes in the most important cassava-growing countries.

Cassava is frequently considered as a polyploid species (Westwood, 1990). Magoon and co-workers suggested, in 1969, that certain portions of the genome may be duplicated and, therefore, that cassava may be a segmental allotetraploid. However, analyses conducted during diakinesis and metaphase I consistently indicate the presence of 18 small and similar bivalents, although some irregularities have been observed (Hahn et al., 1990; Wang et al., 2011). Cassava is, therefore, a functional diploid ($2n = 2x = 36$) (Westwood, 1990; De Carvalho et al., 2002; Nassar and Ortiz, 2008).

Farmers grow clonally propagated hybrids that can be either landraces or improved clones. Cassava is an outcrossing species that shows strong inbreeding (Kawano et al., 1978; Rojas-C. et al.,2009; Kawuki et al., 2011; de Freitas et al., 2016) and heterotic responses (Bueno, 1991; Cach et al., 2006; Calle et al., 2005; Ceballos et al., 2015; Chipeta et al., 2013; Easwari Amma et al., 1995; Jaramillo et al., 2005; Kulembeka et al., 2012; Lokko et al., 2006; Njenga et al., 2014; Owolade et al., 2006; Parkes et al., 2013; Were et al., 2012; Zacarias and Labuschagne, 2010). The progenitors used by cassava breeders are usually elite hybrids and, therefore, heterozygous. In this regard, cassava is in a unique situation among the major staple crops: it is the only diploid species whose breeding is not based on inbred progenitors (Table 1). All major cereal and legume crops are bred using homozygous progenitors. Other root and tuber crops are polyploids, with the exception of *Solanum phureja* (one of the diploid potato species cultivated in the Andean Region of South America, Grüneberg et al., 2009). The diploid and heterozygous nature of cassava

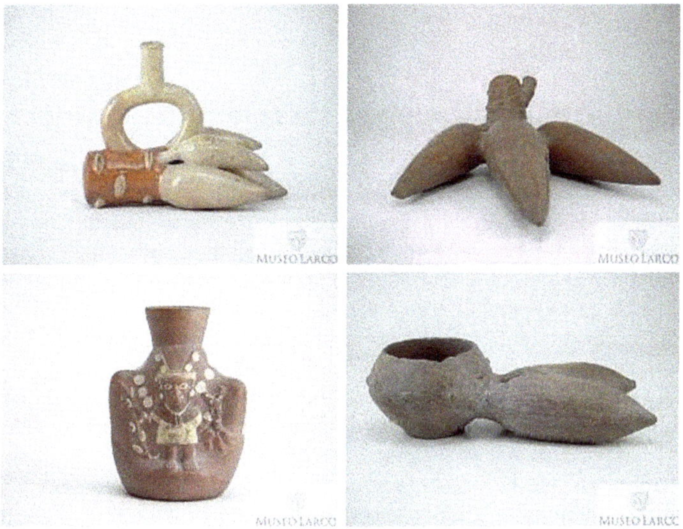

Figure 1 Illustration of pottery from the Mochica culture in coastal Peru. As long as 2500 years ago, cassava had already spread across South America. Source: Larco Museum, Lima, Peru.

Table 1 Characteristics of the main crops grown worldwide. Cassava is the only diploid crop whose breeding is based on the use of heterozygous progenitors (adapted from Grüneberg et al., 2009)

Crop	Ploidy level	Multiplication organ	Type of progenitor
		Grain crops	
Maize	2x	Seed	Inbred lines
Wheat	6x	Seed	
Rice	2x	Seed	
Sorghum	2x	Seed	
Soya bean	2x	Seed	
Sunflower	2x	Seed	
		Vegetatively propagated crops	
Cassava	2x	Stem cuttings	Heterozygous clones
Bananas	3x	Corms	
Potato	4x (2x to 5x)	Tubers	
Sweet potato	6x	Stem cuttings	
Sugarcane	8x	Stem cuttings	
Yam	3x–10x	Tubers	
Taro	4x	Corms	

offers unique opportunities to make its genetic enhancement more efficient as will be illustrated in this chapter.

1.1 Commercial multiplication of cassava through stem cuttings

Cassava is a perennial species cultivated usually as an annual crop. A common pattern is to plant stem cuttings soon after the arrival of the rains (after the dry season in sub-humid environments) or with increasing temperatures during spring (in subtropical regions). Plants are usually grown for an average of 10–12 months and then harvested. However, harvest may range from 6 months to a few years in some situations. In tropical areas farmers harvest their fields at the end of the dry season. At this time, root quality is optimum (e.g. ideal dry matter content) and the stems would need to be stored for a short period until the planting season begins with the arrival of the rains.

Before harvesting the roots, farmers collect the main stems which are then tied together in bunches (≈20–50 stems). Depending on the variety and growing conditions stems will vary from 0.5 to 2.0 m in length. Any part of the cassava stem can be used for propagation purposes. However, depending on the age of the plant certain sections of the stem would be preferable, that is, those sections that are properly lignified and have good carbohydrate reserves. In 10- to 12-month-old cassava, these traits will typically correspond to the middle half section of the stems (Ceballos and Calle, 2010). Farmers cut the stems into planting *stakes* or *cuttings*. Mechanical planters that cut the stems into stakes have been developed. Ideal stakes should have from 5 to 7 nodes and a length of about 20 cm. To prevent dehydration during storage, it is recommended that, just prior to their planting, the stems be cut into planting stakes. There is no dormancy period and stakes can be planted immediately after harvest.

A key constraint for the breeding scheme in cassava is the low multiplication rate of stem cuttings. In general, it is accepted that a maximum of eight (at best ten) cuttings can be expected from each plant. Particular genotypes and under optimum conditions this figure can increase considerably. When hundreds or thousands of genotypes need to be screened in uniform conditions, eight cuttings is a reasonable number of cuttings per plant that breeders can expect to have in each and every genotype. This low reproductive rate implies that several years are required until enough planting material is available for multilocation trials.

The physiological and nutritional status of stems may not be uniform across all genotypes evaluated, which complicates matters further. This variation affects randomly the performance of the germplasm, thus increasing the experimental error of the evaluation trials. In addition to physiological and nutritional factors controlling the quality of planting material (which depends on the status of the harvested field and the conditions and length of the storage period), there is a genetic component that occasionally affects sprouting capacity of the cuttings of certain clones. Farmers require a vigorous, uniform and quick sprouting of cuttings from the varieties they adopt and grow. This creates a dilemma for the breeder because planting materials of the many genotypes screened are usually handled with extreme caution and care so its quality is uniform across the many genotypes to evaluate (e.g. to minimize experimental error in trials). In doing so, however, breeders often cannot adequately assess the genetic potential related to sprouting capacity of the planting material. Figure 2b shows an excellent hybrid that reached the last stage of the evaluation and selection process. Unfortunately, after several years of evaluation, it was only at that final stage that this hybrid showed a poor sprouting capacity when the stems were stored for more than 3–4 weeks and had to be rejected after many years of evaluation.

It has been suggested that perhaps one of the most important impacts of climate change will be on the sprouting capacity of the stems (Ceballos et al., 2011). Weather patterns are changing and becoming increasingly erratic. The time the rainy season begins and ends is no longer as predictable as it used to be in some areas. Stored stems, therefore, may have to wait for more than 1–2 months until they can be cut into stakes and planted. Extended storage of stems drastically reduces their sprouting capacity. Erect plant types may be stored for a longer period before they dehydrate and loose sprouting capacity (Fig. 2). An erect plant architecture, therefore, can be a key adaptive trait in response to the impact of climate change.

Another important issue in relation to stems and plant architecture is the possibility of mechanization. More advanced mechanical planters have been designed to cut the stems into stakes. Erect plant architecture produces long stems which are better suited for

Figure 2 Illustration of plant types and its impact on the quality of planting material. (a) Contrast between branching and erect phenotypes. (b) A plot in a multilocation trial with poor sprouting. (c) and (d) Length and amount of planting material harvested from branching and erect plant types, respectively.

mechanical planting. It is envisioned that labour invested in cultivating cassava will have to be minimized as a way to reduce production costs, even in Africa. Mechanization will become increasingly important and, along with it, the justification for developing cultivars with an erect plant architecture.

1.2 Flowering biology of cassava

Cassava is a diclinous and monoecious species: either female (pistillate) or male (staminate) flowers are produced in inflorescences (racemes or panicles) within the same plant (Perera et al., 2012). Pistillate flowers occupy the lower portion of the inflorescence and open 10–14 days before the male flowers which are located towards the apex on the same inflorescence. Cassava naturally outcrosses (mostly by bees). Self-pollination can occur if male and female flowers on different branches or on different plants of the same genotypes open simultaneously (Jennings and Iglesias, 2002).

Inflorescences always develop at the apex of the stem. Sprouting of the buds below the inflorescence allows further growth of the plant. Every flowering event, therefore, results in branching. Some genotypes flower early and several times (starting three months after planting (MAP) and up to five times during a growth cycle) and others flower little or late (e.g. once at eight MAP). Farmers frequently prefer erect, non-branching types because this plant architecture facilitates cultural practices and results in good production of vegetative planting material; also, its transport and storage is easier. Longer stems can withstand longer storage period and may be a key trait to deal with climate change (Ceballos et al., 2011). However, the advantages for farmers to grow erect clones result in another dilemma for the breeder because the production of botanical seed from these genotypes is slow and, ultimately, more expensive.

There are indeed many cassava clones with profuse, early and frequent flowering. These materials would facilitate greatly the production of segregating seed. However, as stated above, this flowering behaviour is closely related to a low branching, which is generally undesirable. Since the two characteristics are under genetic control, the progeny from early flowering types tend to inherit the trait and, along with it, the undesirable branching plant architecture. Perhaps, one of the most important areas of research to accelerate and improve the impact of breeding would be to induce flowering in cassava. Conventional breeding would benefit by reducing the costs and time currently required to make crosses and obtain seeds. Induction of flowering would also facilitate the development of inbred progenitors through successive self-pollinations. The benefits of the introduction of inbreeding have been mentioned (Ceballos et al., 2015, 2016a) and will be described further in this chapter. Accelerated and synchronized flowering in cassava crossing blocks would also be useful for the implementation of genomic selection as well.

Synchronization of flowering for planned crosses can be a challenge because some elite clones flower relatively early, whereas others flower late and perhaps only once, or possibly not at all. The scarcity of flowers in erect, non-branching types only complicates matters further. Because of this and the time required for the seed to mature, it takes generally more than a year to obtain seeds from planned crosses. On average, between one and two seeds (out of the three possibly formed in the trilocular fruit) per pollination are obtained. This low and variable fertility has been known for decades. Jennings (1963) speculated that the reduced seed set in cassava may be the result of the species evolution towards vegetative reproduction. Some of the irregularities observed during meiosis can also explain the low fertility rate in cassava. There is no evidence of any type of incompatibility in cassava (Westwood, 1990). Environmental conditions also influence flowering and seed set. Several publications illustrate the procedures for controlled pollinations in cassava (Byrne, 1984; Gonçalvez Fukuda et al., 2002; Jennings and Iglesias, 2002; Kawano, 1980).

Male sterility is a known phenomenon in cassava. It has been used to measure pollen flow and to create barriers to pollen movement among pollination nurseries designed for open pollination, such as polycross blocks. Directed crosses, using and registering a known source of pollen, result in full-sib families (Kawano, 1980). In open pollinations, the origin of pollen is not known and results in half-sib families with individuals sharing the identity only of the female progenitor.

2 Cassava breeding objectives

The most common cassava product is the starchy root. Cassava is the third most important source of dietary energy after maize and rice in the tropical and subtropical regions where it is grown. It is therefore a key food security staple crop for millions of people (Caccamisi, 2010). Roots are important for starch production as well. As stated earlier, cassava is the second most important source of starch worldwide (Stapleton, 2012), which places the crop as a relevant commodity for agro-industrial processes as well. Dried root chips have traditionally been used for animal feed and, more recently, for the production of ethanol (Nguyen et al., 2007; Sriroth et al., 2010, and this volume). Cassava foliage has an excellent nutritional quality for animal and human consumption and offers great untapped potential (Babu and Chatterjee, 1999; Benesi et al., 2010).

The role of cassava for food security – either through direct consumption by both rural and urban poor, or as a source of income for producers, which in turn contributes greater ability to purchase food – cannot be overemphasized. Cassava is a very rustic crop that grows well under marginal conditions where few other crops could survive (Burns et al., 2010; El-Sharkawy, 2016). A large proportion of varieties is drought tolerant and can produce reasonably well in degraded soils relative to more demanding crops. Varieties are typically resistant to the most important diseases and pests in their target environments, so that farmers normally do not apply pesticides. The crop is naturally tolerant to acidic soils. Cassava is a perennial shrub and it does not have a clearly defined harvesting time (e.g. roots do not reach a specific maturity stage). This offers a convenient flexibility that benefits farmers and contributes to the recognized resilience of cassava, but also results in some undesirable variation in root quality traits.

The most direct way for human consumption of roots is by boiling them. There is a large variation in cooking time. Roots from some varieties soften after only 10–20 minutes of boiling, whereas those from other genotypes remain 'hard' after 60 minutes. In addition to the quantitative measure related to cooking time, there is a qualitative '*mealiness*' component that is difficult to predict as it may be influenced by both genetics and environmental conditions. There are also many different regional and ethnic-related alternative ways to consume cassava roots. In Africa there are root-processing techniques (that may or may not require fermentation) such as gari, fufu, baton and high-quality flour. In the Americas farinha and casabe are very common in certain regions. Indonesia (one of the few Asian countries where direct human consumption is generalized) has several unique products from cassava roots. In many regions of the world the starting point for processing is fermentation of peeled roots by submerging them in water. Roots from certain varieties soften as the fermentation occurs, while others remain hard, even after several days of fermentation. Softening of the root tissue after fermentation is a key requirement for those areas were mechanical grinding is not practised. Many of these regional/ethnic uses of cassava have specific quality requirements that have not been properly defined from the physico-chemical standpoint. Better understanding of the factors affecting *fermentability*, *poundability* and/or *mealiness* would facilitate the selection process in breeding. Starch, high-quality flour, ethanol and dried chips for animal feeding, on the other hand, only require a competitive price and acceptable levels of dry matter (e.g. starch) content. In the case of starch production, roots with white parenchyma and peel without pigments are desirable.

As the conditions for cassava production, processing and use evolved, there was a need to adapt cassava breeding to an increasingly wider set of socio-economic and environmental conditions. Participatory breeding approaches (Ashby et al., 1987; Gonçalvez Fukuda et al., 2001, 2002; Grüneberg et al., 2009; Manu-Aduening et al., 2006) are ideally suited for conditions where cassava plays a more important role in food security and on-farm consumption, in low-input conditions and/or marginal environments. However, for more commercial, market-oriented production of cassava, centralized, on-station breeding efforts, highly attuned to production and market needs, have proven to be extremely successful. Results largely explain the high productivity of cassava in Asia (Hershey et al., 2001; Kawano, 2003). Cassava production in Asia grew rapidly in the last quarter of the twentieth century in response to demand for starch and dried root chips and, more recently, for bioethanol production. Breeding for these end uses allows concentrating on few traits (e.g. fresh root production, dry matter content and ideal plant type architecture). The cassava breeder, therefore, must have a clear understanding of the target farming conditions as well as the main end uses for the varieties that he or she will develop. Subsistence farming and production for diverse traditional markets especially benefit from the inputs provided by farmers through participatory breeding approaches, whereas breeding for commercial production and for industrial markets can be more efficiently made through more centralized breeding efforts that include on-station and on-farm selection.

Cassava has very special characteristics related to its uses which, in turn, have important implications in relation to breeding objectives. Below is a brief description of the most relevant ones.

2.1 Cyanogenic glucosides

Cyanogenic glucosides (CG) such as linamarin are found in every tissue except in seeds. The CG, synthesized in the leaves and transported to the roots, is broken down by the enzyme linamarase to produce the volatile poison, hydrogen cyanide (Andersen et al., 2000; Du et al., 1995; McMahon et al., 1995; Mkumbira et al., 2003; Wheatley and Chuzel, 1995). Linamarin and linamarase accumulate in different parts of the cell, thus preventing their contact and the subsequent formation of free cyanide. CG products can be toxic to humans, causing paralysing diseases such as Konzo, and even death if roots or leaves are not properly processed (Vandegeer et al., 2013).

Most processing methods disrupt the tissues, allowing the enzyme to act on the glucoside for a rapid release of cyanide. CG accumulation varies with genotypes, environment, agronomic practices, age of the plant and plant tissue (e.g. leaves and peel of the root tend to have higher values than the root parenchyma) (Cock, 1985). Low cyanogenic potential (CNP) has frequently been an important breeding objective (Dixon et al., 1994). Cultivars with less than 100 mg CG kg^{-1} fresh weight in the roots are considered *'sweet'*. Above this level cassava roots are considered *'bitter'*. Nonetheless, the relationship between bitterness and CNP is not very exact, and there may be other components of the root that confer bitterness or sweetness, such as tannins and sugars. Depending on the processing methodologies, ultimate uses and cultural preferences, clones with sweet or bitter (even very bitter) roots may be preferred (Benesi et al., 2010). Drought conditions result in higher CNP of the leaves and roots (Vandegeer et al., 2013). This is unfortunate because it is under drought that cassava emerges as the recognized food security crop it is. As people rely more heavily on cassava roots with higher CNP, it is not surprising, therefore, to observe increases in the incidence of Konzo in drought years.

2.2 Short shelf life of stored roots

Cassava roots are not tubers (i.e. modified stems, such as potato) and, therefore, cannot be used for reproductive purpose. Tubers can be stored for longer periods compared with cassava roots, which have a short shelf life (Beeching et al., 1993; Reilly et al., 2007; Uarrota et al., 2015). Within one or two days after harvest there is a rapid initiation of post-harvest physiological deterioration (PPD). Until recently, little useful genetic variation to delay or reduce PPD had been found and the solution to this problem remains as one of the most important goals for cassava research. Roots can be protected from PPD by cultural practices such as pruning about two weeks before harvesting (van Oirschot et al., 2000) or applying wax or paraffin to their surface. Recent reports suggest that there may be genetic variation for tolerance to PPD (Morante et al., 2010). Transgenic approaches to extend shelf life of cassava roots after harvest have been attempted (Zidenga et al., 2012). However, this was accompanied with the reduction of root yields in the field.

In addition to PPD, roots stored in ambient conditions lose an estimated 1% of starch per day as a result of conversion into sugars and respiration of root tissue (Sánchez et al., 2013). Currently, long-term storage of cassava roots can only be achieved through freezing.

2.3 Fluctuation in root quality traits

Cassava roots do not reach a *'physiological maturity'* and their properties (particularly dry matter content and CNP) vary considerably depending on the age of the plant, environmental conditions and management (Benesi et al., 2004, 2008; Maziya-Dixon et al., 2007; Sanni et al., 2008; Sriroth et al., 1999; Vandegeer et al., 2013). Variation in root quality traits creates considerable problems both for roots used directly for food and for different industrial processes.

Typically, farmers plant cassava at the beginning of the rainy season and harvest it at the end of the dry season (just before the return of the rains). A key factor in this scheme is that DMC of the roots, at least in the varieties grown by farmers, should reach a maximum at harvest time. This is important because when cassava is used for processing (dried chips and pellets for animal feeding or the starch, flour or ethanol industries), its price is directly linked to DMC of the roots. The efficiency in the production of ethnic products such as gari, fufu, casabe and so on also depends on DMC. A low DMC means a longer period in the drying patios (for the production of dried chips) or higher extraction cost per unit of derived product, and amount of effluents (for the starch industry, for example). In other words, low DMC implies higher processing costs for a range of cassava products. This is the reason why cassava breeders have paid so much attention to DMC in the new varieties released for industrial purposes (Kawano, 2003; Kawano and Cock, 2005).

When the rains return the plant hydrolyses the starch and mobilizes other nutritional components stored in the roots to re-initiate growth. This means that, within 2–3 weeks after the arrival of the rains, DMC will be reduced drastically (Bakayoko et al., 2009). As an example of this phenomenon, we cite results from a trial of 1350 genotypes by CIAT in the sub-humid environment of the northern coast of Colombia. The harvest of these clones was split into two dates. The first took place during the dry season, and the average DMC was 32.4%. The second harvest was made after the rains had arrived and the average DMC dropped to 26.7% (CIAT, 2001). This illustrates the drastic impact that the environment (rainfall) has on a key component of the overall cassava productivity and commercial value. In high latitude regions of South America, cassava can be harvested just before the onset of winter (8–10 MAP), or left in the field through the winter, and harvested after a second vegetative growth cycle (e.g. 18–20 MAP). In this system, productivity in late harvests increases considerably and DMC can be recovered to acceptable levels (Kvitschal et al., 2009; Sagrilo et al., 2008).

Developing clones with high and more stable DMC will become an even more relevant objective in the years to come because of climate change, which is already altering the normal patterns of rainfall. Fluctuations in the arrival and end of the rainy season may not only affect sprouting of cuttings (if starting of rainy season is delayed) but also the quality of roots harvested by farmers (e.g. if rainy season begins early and roots are harvested after DMC has been reduced) and create logistic problems to different industries that rely on the availability of raw materials for their operations.

2.4 Starch, nutritional and peel characteristics in the root

In addition to the quantitative DMC trait, other quality characteristics in the root have been addressed by cassava breeding projects. In recent years, two commercially relevant starch mutations have been identified in cassava. The first one was a spontaneous mutation of amylose-free or '*waxy*' starch which was reported in 2007 (Ceballos et al., 2007a). Several studies describing the special functional properties and commercial applications of amylose-free starch were published thereafter (Morante et al., 2016; Rolland-Sabaté et al., 2012, 2013; Sánchez et al., 2010), as well as the potential of transgenic cassava with this trait (Koehorst-van Putten et al., 2012; Rolland-Sabaté et al., 2013; Zhao et al., 2011). Important breeding efforts are currently ongoing to develop commercial cassava varieties with amylose-free starch. In that regard, basic information regarding the relationship between this trait and DMC has been published (Karlström et al., 2016), as well as the potential of a molecular marker that can be used for breeding purposes (Aiemnaka et al., 2012).

The second starch mutation was induced through the gamma ray irradiation of seeds and produces a small-granule starch phenotype (Ceballos et al., 2008). This particular mutation offers interesting advantages to the ethanol and sweeteners industries because the starch is more easily hydrolysed into simple sugars.

Regarding nutritional value of cassava roots, for many years breeding has addressed the issue of CG as described in the previous section. Cassava roots have low levels of proteins and efforts to identify high-protein cassava roots have a long history (Bolhuis, 1953). In spite of recent efforts (Akinbo et al., 2012; Ceballos et al., 2006) little progress has been achieved. A major problem is that the standard protocol for the practical quantification of protein is based on the indirect measurement of N, which has proven to be inappropriate (Dufour et al., 2012). Significant gains have been achieved, on the other hand, to increase the levels of carotenoids content based on the available variability reported several years ago (Iglesias et al., 1997). During the past few years, important progress has been made in improving the protocol for carotenoids quantification (Belalcazar et al., 2016; Ceballos et al., 2012a; Chávez et al., 2005; Davrieux et al., 2016; Ortiz et al., 2011; Sánchez et al., 2014), which has increased the heritability of the trait and the breeding efficiency (Ceballos et al., 2013; Njenga et al., 2014; Njoku et al, 2015). Molecular marker technologies have been reported but have not been used in marker-assisted selection (Esuma et al., 2016; Njoku et al., 2014; Rabbi et al., 2014a).

Significant and highly heritable variation has been reported for the thickness of root peel (Pérez et al., 2011), a trait that can have relevance for maximizing yields (i.e. allocation of photosynthates to peel vs root flesh), for end uses which require peeling, for tolerance of roots to physical damage while handling, and for tolerance to root-feeding insects such as the subterranean burrowing bug. Figure 3 illustrates some of the differences that can be observed in the external and internal colour of the peel as well as in their thickness. The starch and flour industries prefer non-pigmented roots and peels. Developing cassava cultivars with non-pigmented peel, very white parenchyma and a very thin peel could be breeding objectives. This type of roots could be used for the production of flour without the need of peeling the roots, which in turn could reduce considerably the cost of production and of products in the marketplace.

While roots are by far the most important product from cassava, foliage has an excellent nutritional quality for animal and human consumption and offers a great untapped potential (Babu and Chatterjee, 1999; Benesi et al., 2010). However, at least for the time being, breeding specifically for foliage exploitation is not a major effort in the ongoing cassava projects. Stems are used for commercial multiplication, but it is not always explicitly acknowledged that there are important requirements that released varieties must meet regarding the stems. As stated earlier, erect plant architecture results in longer stems that

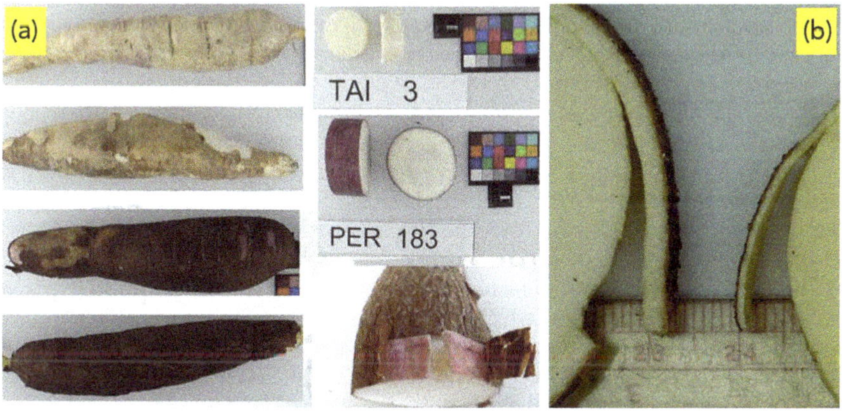

Figure 3 Illustration of different root peel traits. (a) Variation in pigmentation of external and internal peel colour. (b) Differences in thickness of the peel.

can easily be collected, transported and stored. In addition, longer stems dehydrate more slowly and, therefore, can be stored for extended periods.

3 Pests and diseases of cassava

Like most crops, cassava faces an array of pests and diseases that threaten its productivity (Bellotti et al., 1999; Bellotti, 2002; Calvert and Thresh, 2002; Hillocks and Wydra, 2002). This volume provides valuable information in different chapters regarding biotic factors affecting cassava productivity. Therefore, this chapter will not have a comprehensive list of these problems, but rather, with the use of specific examples, will illustrate strategies related to breeding.

The absence of cold winters in tropical regions and the long growing cycle of cassava offer ideal conditions for pests and diseases to cause economic damage. On the other hand, agents of biological control can more efficiently establish and control some of these biotic stresses that limit cassava productivity. The recent introduction of the mealybug *Phenacoccus manihoti* in Thailand led to a drastic reduction of yields throughout the country (Parsa et al., 2012). Thai authorities together with the private starch sector acted quickly and introduced the parasitic wasp *Anagyrus lopezi*, which had been used successfully in Africa decades earlier. In both cases, the biological control of the mealybug proved highly successful. In the case of Thailand, where production data is detailed and reliable, average yields recovered within two years after the introduction and rapid dissemination of the wasp. This illustrates the impact and power of biological control in cassava.

Hornworms (*Erinnyis ello*) can defoliate an entire cassava field within days. There is no identified source of genetic resistance to it. However, many different agents of biological control are available, ranging from baculoviruses, *Apanteles* sp. wasps, to birds (Fig. 4). This is another example where the alternative of an efficient biological control makes breeding efforts redundant.

Given the diversity of biotic problems that cassava faces, research projects have wisely given priority to the most serious pests and the most promising technologies. In the case of pests, for example, there is no breeding for tolerance or resistance to the mealybug since biological control is an efficient alternative. On the other hand, there is no reliable biological control of whiteflies currently being exploited and the main strategy has been through the use of chemical products and developing genetic resistance.

The green mite (*Mononychellus tanajoa*) offers an interesting example of synergies between breeding and biological control strategies. A predatory mite (*Typhlodromalus aripo*) is an efficient agent for its biological control. However, canopy characteristics of the host plant play a significant role in supporting the continuous survival of *T. aripo* (and other phytoseiid predatory mites) during both the rainy and dry seasons (Chalwe et al., 2015). Therefore, breeding can be directed to develop host plant characteristics that protect the populations of the beneficial mites. Canopy characteristics are also important against thrips (*Frankliniella williamsi* and *Scirtothrips manihoti*), which can cause important damage to cassava. Resistance to thrips is common and extremely efficient and durable. It is based on pubescence in the apical shoots. Selection can be done at the seedling stage, and even in the absence of the pest, since pubescence is easy to evaluate and heritability of the trait is high.

The vegetative multiplication of cassava implies that cuttings can be the source of inoculum for different diseases. Eggs and larvae can also be transported by the stems and

Figure 4 Illustration of the hornworm (a) and alternatives for its biological control. (b) *Apanteles* sp. feed on the hornworm larva. (c) Typical symptoms of a hornworm larva affected by a baculovirus. (d) Hornworm egg being parasitized by wasps.

contribute to pest dissemination. Proper management of the planting material, therefore, is a key component of pest and disease management.

There are several diseases induced by viruses, fungi, bacteria and/or phytoplasms affecting cassava. However, only a few of them are of concern because of their economic importance. The most important diseases related to viruses are CMD present in Africa and the Indian subcontinent, and cassava brown streak disease (CBSD) originally endemic to East Africa but now spreading westward (Calvert and Thresh, 2002; Monger et al., 2001). Frog skin disease (FSD) has an unknown aetiology (probably induced by a virus and/or a phytoplasm, Calvert et al., 2008; Alvarez et al., 2009) and is only found in the Americas. FSD can be controlled effectively by avoiding the use of stems from diseased plants. Finally, witches' broom disease is a new problem arising in Southeast Asia (Alvarez et al., 2013; Flôres et al., 2013).

Whiteflies are among the most widespread pests in cassava. *Bemisia tabaci* is widely distributed in tropical Africa and several Asian countries. The major impact of *B. tabaci* is through its role as a vector of the viruses that cause the devastating CMD and CBSD in Africa. Genetic resistance to whiteflies in cassava has been found, particularly for *Aleurotrachelus socialis* in several germplasm accessions from the CIAT collection (Bellotti, 2002), most notably from Ecuador (e.g. MEcu 72). The first whitefly-resistant variety of any crop was released in 2002 by ICA in Colombia – a hybrid developed at CIAT with resistance from MEcu 72 combined with high yield and good dry matter content/cooking quality.

4 Evaluating and selecting cassava for breeding

Several reviews on cassava breeding have been made over the years (Byrne, 1984; Ceballos et al., 2004, 2007b, 2010, 2012b; Gonçalves Fukuda et al., 2002; Hahn et al., 1979; Jennings and Hershey, 1985; Jennings and Iglesias, 2002; Kawano and Cock, 2005). Currently, important breeding efforts are made in Brazil, Colombia, China, Ghana, India,

Nigeria, Kenya, Mozambique, Tanzania, Thailand, Uganda and Vietnam, among tropical countries of the world. In addition, there is extensive support from advanced labs in developed countries, especially in the development and application of molecular tools, high-throughput phenotyping methods and protocols, and genetic transformation.

The cassava breeding research at CIAT initially emphasized low-input conditions in less favourable environments to provide food security and alleviate the poverty of small farmers through income generation (Kawano and Cock, 2005). From its initiation, CIAT worked in close partnership with national programmes in Latin America and Asia, and its sister institution (IITA) based in Nigeria. Throughout the years there has been a gradual shift of emphasis from the early target on achieving moderate and stable production under high stress conditions, to more intensive commercial (but still mainly small-scale) cultivation and higher inputs for cassava. Lately, more attention has been paid to reduce the environmental impact of cassava cultivation and processing towards more eco-efficient practices.

Cassava is bred through phenotypic recurrent selection, as is frequently the case for other clonally propagated crops (Burton, 1992; Grüneberg et al., 2009; Lebot, 2010). Because of the low multiplication rate of cassava, it takes several years to have enough planting material available for replicated multilocation evaluations. However, on the positive side, this slow multiplication rate somewhat obligates the breeder to include multi-year evaluations in the breeding scheme, and thus to include across-year stability as part of the selection criteria for candidate varieties. In general, most schemes rely on more or less similar approaches as described below.

An important feature of cassava breeding is that progenitors are heterozygous. Therefore large genetic variability within- as well as between-family is generated each cycle. The main unit of selection is the individual clone, contrary to what happens in recurrent selection of other non-clonal crops. In the case of maize, for example, the main target of selection is the families (e.g. half- or full-sibs, S_1, S_2, etc.) and genetic variability among individual genotypes within the family is largely ignored (Hallauer and Miranda, 1988). Only when generating the seed for a new recurrent selection cycle, individual maize genotypes within the selected families may be visually chosen. In the following sections a brief description of the different stages of the selection process is provided. The key feature of the process is that as the number of genotypes evaluated is gradually reduced through selection, the number of plants representing each genotype increases with each growing season.

4.1 Seed germination and seedling plant nurseries

Seeds may be planted directly in the field (as done at IITA) to take advantage of the availability of irrigation and high temperatures. Ideally, soil temperatures should be in the 25–35°C range for good germination. Alternatively, as done at CIAT, seeds can be germinated in greenhouse conditions for better climate control and the resulting seedlings transplanted to the field when they are about 20–25 cm tall (Jennings and Iglesias, 2002). Seedlings are relatively weak compared with plants derived from stakes. This restricts the potential use of alternative approaches to commercial production, such as those based on true seed. Root systems in plants derived from botanical seed may differ considerably compared to those from vegetative cuttings (Iglesias et al., 1994; Rajendran et al., 2000; Fig. 5). Because of this, it is difficult to correlate the root yield of clones at later stages in the evaluation/selection process, with earlier results from the plants obtained from botanical seeds. However, when seeds are germinated in containers and later transplanted, the

Figure 5 Contrast between root system in seedling (a) and cloned plants (b).

taproot often does not develop, and the seedling-derived plant may be more similar to subsequent stake-derived plants in terms of number and conformation of storage roots. Botanical seed has only been used for commercial propagation at the trial level (Iglesias et al., 1994; Rajendran et al., 2000), but it is not a generalized practice.

Propagation from true seed occurs occasionally in farmers' fields and, as such, is also the starting point for the generation of useful genetic diversity (Eke-Okoro et al., 2001; Elias et al., 2001; Pujol et al., 2002). Shamans (or efficient farmers) can be key players in this informal genetic improvement process (Salick et al., 1997; Sambatti et al., 2001). However, in commercial conditions, volunteer seedlings are normally considered to be undesirable as they would contaminate the purity of the planting material. A relatively large proportion of true seed produced in farmers´ fields may be the result of self-pollinations, particularly if a relatively large plantation (e.g. >0.5 ha) of a single clone has been planted. Inbreeding depression, together with the inherent weak nature of seedlings, works against the probabilities of survival of volunteer seedlings arising in the farmers´ fields.

The most important role of seedling plants for breeders is to provide planting material, that is, cuttings, to initiate the evaluation and selection process. Because traits expressed in single-plant plots (e.g. seedling nurseries at F_1 stage) generally have low heritability, breeders are generally reluctant to exert strong selection pressure at the seedling plant stage. There are, however, some high-heritability traits that can be selected at this stage. At IITA, for example, strong selection can be made for reaction to CMD (Egesi et al., 2007; Hahn et al., 1980a,b; Ssemakula and Dixon, 2007). Resistance to CMD is the only successful example of marker-assisted selection in cassava (Akano et al., 2002; Rabbi et al., 2014b) so far reported, which adds further options to selection for resistance at the single-plant stage. At CIAT, selection for reaction to thrips, a sucking insect that affects newly forming leaves, is effectively done at the F1 stage due to high heritability and their consistent, uniform presence at the CIAT-Palmira station. Breeders typically select seedling plants that are acceptably vigorous, with no or little pest and disease damage and of desirable plant architecture.

In response to increasing urgency to meet demands for food production and income generation, new projects have required modifications in the breeding scheme to allow for stronger selection pressure at the seedling stage, in order to accelerate breeding gains. Examples include projects related to specific breeding objectives such as high carotenoids in the roots or exploitation of the *waxy starch* (amylose-free) mutation. Both traits meet the critical requirement of high heritability. Carotenoids can be reliably measured at the seedling plant stage, allowing for significant gains through rapid cycling recurrent selection (Morante et al., 2010; Morillo Coronado, 2009; Ortiz et al., 2011). High-throughput phenotyping is now feasible through the use of near-infrared spectroscopy (NIR) (Belalcazar et al., 2016; Davrieux et al., 2016). For waxy starch, a simple and fast method (based on spraying iodine solution) can quickly and reliably detect plants with the proper genotype, which is based on a single, recessive gene (Aiemnaka et al., 2012; Morante et al., 2016).

4.2 Single row trials

In the case of CIAT the first significant selection takes place at the seedling stage, as explained earlier (except for special cases such as breeding for high carotenoids or waxy starch). The first selection based on agronomic performance and targeting traits such as fresh root yield (FRY) and DMC takes place at the single row trials (SRT), also called first clonal evaluation trials. These trials are planted in one or two sites in the respective target environment, and should be maintained in a secure research station as well, as backup and for additional evaluation. Each genotype is represented by eight plants in a single row. No replications are used at this stage. These are relatively large trials (1–2 ha) where 1000–2500 genotypes are evaluated.

Large environmental variation (particularly acute in the kinds of environments where cassava is typically grown) is to be expected. Stratified phenotypic selection (Gardner, 1961; Burton, 1992) has been successfully implemented for SRT in cassava. Each SRT is divided into a few blocks or strata. Selection is conducted within each block (Ceballos et al. 2004, 2007b, 2010); therefore, the environmental variation between the different strata does not affect the breeder's judgement. For logistic reasons only 3 or 4 strata are used. This stage requires 11–12 months for completion and at the end about 100–200 genotypes are selected based on FRY, DMC, harvest index (HIN) and plant type scores (PTS), which combine architecture and overall plant health as an indicator of integrated pest and disease resistance for the environment in question. Comprehensive data are taken from each SRT, including sprouting capacity, early vigour, pest and disease responses, branching habit and plant type, plant height, late-season leaf retention, root fresh yield, whole plant weight and root DMC. Other traits may be included depending on target growing environments and target markets, such as CNP, carotenoid content, root colour and form, and others, which allows a preliminary assessment of breeding values as described below. Variation of average performance among strata could be as high as 16% for dry matter yield (Ceballos et al., 2007b). This kind of variation, however, does not affect the selection process because it is done within each stratum.

4.3 Preliminary yield trials

Each genotype is planted in three repetitions with 10-plant plots. To reduce intra-trial variation, this stage of selection is split into three or four different experiments tracing back to the 3–4 strata in the SRT. Each experiment has 25–50 genotypes for a total of up to 100–200 genotypes. Planting of these trials follows a special design to reduce competition among genotypes with contrasting plant architecture. Row spacing is reduced from the standard 1.0 m to 0.80 m, and spacing between plants within a row is also 0.80 m. Each genotype is planted in two neighbouring rows with five plants each (for the 10-plant plot mentioned above). Spacing between plots with different genotypes is 1.6m. This design allows evaluations at a plant density close to the standard 10 000 pl ha^{-1}, while reducing competition among different genotypes and favouring competition among plants from the same genotype. The rationale is that inter-genotype competition tends to favour the more vigorous plant types in terms of yield performance; however, these same genotypes, when planted in uniform stands where only intra-genotypic competition occurs, are not necessarily the best-performing types compared to less vigorous ones. One of the critical principles of cassava breeding is to understand and manage intra- and inter-genotypic competition in the early stages of selection such that the breeder maximizes chances to select types that perform best in commercial conditions where only intra-genotypic competition comes into play.

4.4 Advanced yield trials

Plots at this stage of selection are larger than those from previous stages, with four or five rows and five plants per row. The six or nine central plants are harvested to generate the data used in the selection process. The surrounding plants in the periphery of the plot serve both as border plants that confer only intra-genotypic competition effects for the harvested plants and as a source of planting material for next season. Since only the central plants of the plot are harvested there is no need for a special planting arrangement to reduce competition between genotypes. Plant density for these trials, therefore, is the normal one (typically 10 000 plants/ha, but may vary according to the environment), with 1m between rows and 1m spacing between plants within a row. The selected clones from the different experiments at the preliminary yield trial (PYT) stage are merged to produce a single advanced yield trial (AYT) with 50–80 genotypes. Each genotype is planted in three replications. AYTs may be conducted for two consecutive years with little selection pressure between the first and second year evaluation. Second year AYTs may be planted in 2 to 3 or more locations, if goals and resources allow.

4.5 Uniform yield trials

This is the final stage in the evaluation process managed by the breeder but often involves other partners such as extension services, farmer and the processing sector. Plot size, number of repetitions and planting arrangement may be the same as those for AYTs, or may be somewhat larger. Uniform yield trials (UYTs) are planted for two consecutive years in 5–10 locations. Typically, UYTs will have 20–25 experimental clones and 5–8 local or commercial checks. Depending on the performance of the experimental clones compared with those of the checks, these trials can provide strong support for an official variety release. The testing and performance requirements for release vary among countries. Planting material of the most promising clones is shared with key farmers for semi-commercial evaluation of 0.5 to 1.0 ha. The production from these plots will be further evaluated according to the target market by processors and consumers. In general, varieties are released only after successful performance according to the criteria of multiple players in the value chain, that is, farmers, processors and consumers. Farmers and/or end users are invited to participate during the harvest of AYTs and UYTs, and their opinion is informally incorporated into the selection process.

4.6 Use of checks

At each stage of selection, beginning with the SRT, the use of check varieties with known performance characteristics is highly desirable. These checks should include popular local varieties as well as the best-bred varieties which the breeder wishes to outperform. In the case of the SRT, where there are no replications, checks may be interspersed regularly throughout the trial, for example, every ten rows. For subsequent selection stages, checks can be included among the list of entries and handled the same way the experimental materials are (e.g. same plot size and number of replications).

5 Selection index in cassava breeding

Selection of the best, among thousands of genotypes that formed the original F_1 population, is a daunting task. As is frequently the case, many different variables are relevant and need to be considered in shaping the kind of germplasm that the selection process has to achieve. In addition, selecting for high FRY is often more efficiently achieved through indirect selection for correlated traits such as HIN (Kawano, 2003; Kawano et al., 1998). To facilitate this process, CIAT has implemented a selection index (Baker, 1986) approach combining key variables:

$$SIN = (FRY * C_1) + (DMC * C_2) + (HIN * C_3) - (PTS * C_4)$$

The 'C_i' factors are subjective weights given to each variable by the breeder (in this case ranging from 1 to 10, where 1 indicates less importance and 10 for highest relevance). To avoid the differences in the magnitudes by which different variables are measured (e.g. 0–1 for HIN, 1–5 for PTS, 0–30 for FRY and 25–42 for DMC), values for each trait are standardized ($x_i - X/\sigma$, where x_i is the value for a given plot and trait; X is the average across genotypes for that particular trait; and σ is the measured standard deviation). By definition, after the standardization each trait has an average of zero and a standard deviation of 1.

The weights can be easily changed, depending on the desired outcome. The breeder should continually evaluate the SIN with regard to goals and progress. C_1 is usually assigned the maximum weight (10). C_2 was originally assigned a value of 8 but upgraded to 10 in recent years. Depending on the variation observed in each trial, the weight associated with harvest index (C_3) would vary. Over the years its importance was gradually reduced (from 5 to 3). For reasons explained below, starting in 2017 HIN will no longer be part of the selection index. Depending on the specific pressure by pests and diseases the coefficient for plant type (C_4) would also range between 3 and 5.

PTS is a combination of plant architecture and health. In most, but not all, cases cassava farmers prefer erect, non-branching types. This phenotype facilitates cultural practices as well as the harvest and handling of stems that are going to be used as planting material for the following season. In areas prone to water stress, PTS will take into consideration variation in leaf retention that results from different levels of drought tolerance. In areas where foliar diseases (e.g. bacterial blight or super elongation) or arthropod pests (such as whiteflies, thrips or mites) are prevalent, PTS will also take into consideration these plant health traits. If one or more particular diseases or pests are of prime relevance, a score for it/them may be added to the SIN formula and in that case it will not be taken into consideration for the PTS score. Of particular relevance for Africa is the incidence of CMD and CBSD.

In general, disease, pest and plant architecture scores are used based on a 1 to 5 scale using a simple approach. Before scoring, the breeder walks around the trial to have a general impression of the condition of the material to evaluate. A score of 3 is assigned to phenotypes that could represent the average of the trial. A score of 2 is given to phenotypes whose performance is probably better than the average and a score of 1 to materials that are clearly better than the average. Similarly, a score of 4 is given to materials whose performance is probably worse than the average and a score of 5 to those that are without a doubt poorer than the average. For pest and disease scoring, it is important to understand the effects of variability in pest or disease pressure, and if possible to apply techniques to make this pressure uniform across plots. This can be done by a combination of use of 'hot spots', inoculation or susceptible spreader plots, for example. Because a good PTS is 1 and an undesirable one is 5, the SIN formula above has a negative sign associated with the PTS term. In a broad sense, the PTS score answers the question that the breeder might pose: 'On a scale of 1 to 5, how much do I like the overall, integrated apparent performance of this clone based on the traits that can be observed visually, and their expected relative importance.' While clearly this is a highly subjective evaluation, it gives the breeder the opportunity to apply the *art* of breeding to the selection process, and to integrate many criteria in a simplified yet powerful way. The value of this method will depend on a breeder having considerable experience and a clear sense of the significance of the variation in observed traits.

The selection index is used to sort the genotypes from best to worse. The breeder will then determine a cut-off point and make minor adjustments of the selected and rejected genotypes by looking at other variables that are not included in the selection index.

6 Phenotypic correlations of cassava traits

Before the year 2000, selection at early stages was done visually and the only information available was if a genotype was selected or not. This was an efficient way to process large segregating populations. A decision was made in 2000 to change the strategy and start registering performance data of all genotypes evaluated, regardless of whether they were selected or not. For about 15 years, therefore, CIAT has conducted several cycles of a fully documented phenotypic recurrent selection process. One of the valuable results is the ability to relate the performance of clones reaching the multilocation stage of evaluation with complete data from all the previous stages of selection. This information (generated mainly from the sub-humid environment in Colombia´s northern coast) has been crucial for determining which variables are more efficiently selected for at different stages of selection (Ceballos et al., 2016a; Joaqui Barandica et al., 2016).

More than 20 000 genotypes were evaluated in SRTs over the indicated 14-year period. A unique feature of this dataset is that information of each genotype is available, regardless of whether they were selected or not. Only 2866 (about 14%) of these genotypes were selected and passed to the following stage of PYT. The number of genotypes reaching the AYT stage was further reduced to 615, or about 21%. Finally, 114 genotypes were evaluated in UYT during those years, that is, 19% of the AYT selected for next stage. Overall, only about 0.6% of the clones from the original SRTs passed to the UYT, a normal part of the systematic narrowing that takes place in any breeding programme. Table 2 presents the correlation coefficients for each of the four traits used in the selection index, measured in the same genotypes but at different stages of the selection process.

In general, correlations for DMC and PTS were the highest. In fact, the highest coefficient was observed for PTS between AYT and UYT ($r = 0.72$). HIN and FRY had lower coefficients. The correlation for FRY between PYT and AYT was atypically low. Correlations for HIN were generally disappointing. The repeatability of HIN across different types of trials was only slightly better than that for FRY, particularly if the atypical coefficient for FRY between PYT and AYT is ignored (Table 2). These results contrast sharply with those reported by Kawano et al. (1998).

Correlations for DMC were generally high (Table 2) in agreement with Kawano et al. (1987) who reported high-heritability values for this trait. The correlations for FRY presented in Table 2 illustrate a major challenge that cassava breeders face: SRT/PYT correlation was negligible (0.12), although it increased for SRT/AYT (0.32) and SRT/UYT (0.29). Correlations of PYT data with AYT were negligible (−0.02) and mediocre for PYT/UYT (0.29). As expected, however, correlation for FRY between AYT and UYT was relatively high (0.63). This is the reason why cassava breeders have always stated that, at early stages, selection should focus on traits with higher heritability (e.g. HIN) rather than on FRY (Kawano et al., 1998; Kawano, 2003). Table 3 presents the phenotypic correlations between HIN and FRY using data from different stages of the selection process. The correlation between HIN in SRT with FRY in UYT was disappointing (0.14). The justification for including HIN in the selection index in SRT was that it contributed to select materials with high FRY. Data presented in Table 3, however, challenges such justification.

Table 2 Linear correlation coefficients for each variable evaluated in different stages of the selection process (within parentheses are number of observations used to estimate the correlation)

Type of comparison	Plant type score 1–5	Harvest index 0.0–1.0	Dry matter content %	Fresh root yield t ha^{-1}
SRT vs PYT	0.32** (2817)	0.42** (2740)	041** (2865)	0.12** (2690)
SRT vs AYT	0.29** (590)	0.36** (591)	0.37** (614)	0.32** (579)
SRT vs UYT	0.35** (114)	0.17^{NS} (114)	0.37** (114)	0.29** (114)
PYT vs AYT	0.45** (603)	0.37** (615)	0.43** (615)	− 0.02^{NS} (615)
PYT vs UYT	0.34** (114)	0.32** (114)	0.70** (114)	0.29** (114)
AYT vs UYT	0.72** (114)	0.46** (114)	0.53** (114)	0.63** (114)

** Significant at the 5%; ^{NS} Non significant.

Table 3 Usefulness of HIN as indirect selection in early stages of evaluation for increased FRY in later stages of selection. Pearson's correlation coefficients of HIN and FRY with FRY assessed in the successive evaluation stages for cassava breeding at CIAT

Type of comparison (sample size)	Correlation coefficient (Harvest index [0.0–1.0])
HIN in SRT vs FRY in PYT (2740)	0.12**
HIN in SRT vs FRY in AYT (591)	0.11*
HIN in SRT vs FRY in UYT (114)	0.14 NS
HIN in PYT vs FRY in AYT (615)	0.27**
HIN in PYT vs FRY in UYT (114)	0.20*
HIN in AYT vs FRY in UYT (114)	− 0.11 NS

*, **Significant at 5 and 1% probability level, respectively; NS not significant.

Table 4 Pearson's correlation coefficients of different variables with FRY assessed in the same trial (within parentheses are number of observations used to estimate the correlation)

Trial type	DMC	PTS	HIN
SRT[1]	0.21** (19455)	− 0.30** (19498)	0.47** (19498)
PYT[2]	− 0.13** (2866)	− 0.23** (2866)	0.02NS (2866)
AYT[2]	− 0.14** (615)	− 0.09* (603)	0.12** (615)
UYT[2]	− 0.42** (114)	− 0.04 NS (114)	0.24* (114)

[1] Correlations on a plot basis.
[2] Correlations based on averages of genotypes across reps (and locations/years in UYT).
*, **Significant at 5 and 1% probability level, respectively; NS not significant.

One last set of comparisons using the phenotypic data from 14 years of agronomic evaluations in the sub-humid environment is provided in Table 4. In this case, phenotypic correlations are presented between different traits and FRY measured in different stages of the selection process. An interesting and striking trend can be observed for the correlations between DMC and FRY. In SRT, it shows a positive value (0.21) but gradually becomes negative in the following trials. In UYT the correlation was clearly negative (−42). This is a very interesting reflection of how cassava responds to selection. As materials are gradually selected for higher productivity, genotypes may excel in either high FRY or DMC but it is biologically very difficult to have simultaneously an outstanding performance for both traits.

7 Significance of breeding value

The same data from evaluation trials conducted from 2000 through 2013 presented above was used to answer a different question: what factor(s) at the SRT stage can be used to identify the chances of a genotype reaching the last stage of the selection process (UYT). A total of 1038 full- or half-sib families were evaluated in these trials involving more than 20 000 genotypes evaluated in the SRTs (9286 from full-sib families and 11 462 from half-sib families). A selection index was calculated for each individual genotype using the formula described above, and using the same four key traits. The analysis focused on the initial 20,748 genotypes evaluated in SRT and the 114 reaching the UTY. PYT and AYT data were not included.

Data from all the individual genotypes belonging to a given full- or half-sib family was consolidated to obtain the respective averages and other statistical parameters for the key variables: FRY, DMC, PTS, HIN and SIN. Since progenitors are used to generate more than one family, averages for each progenitor across all the families that it had generated were estimated. The phenotypic average of all the progeny (across different families) generated by a given progenitor can be considered as an estimate of the breeding value of that progenitor. Phenotypic data from the 9286 full-sib genotypes was used twice: for the estimation of breeding values of the progenitor used as female, and for the breeding value when used as a male. Falconer (1981) defined breeding value of an individual as the mean value of its progeny, a simple yet powerful concept in plant and animal breeding. The breeding value is the deviation of the progeny generated by a given progenitor from the average of a reference population. Breeding value depends on the average performance of the reference population as well as on the value of the alleles that each progenitor can transfer to its progeny (Falconer, 1981). Typically, breeding value is related to additive genetic effects, although some dominance effects (e.g. a single dominant source of resistance to a given disease or pest) can influence breeding values.

Progenitors represented by fewer than 50 genotypes among the progeny, across all families in which they had been used, were discarded from the analysis to reduce error. A sample size of less than 50 individuals was considered too small to properly represent the breeding value of the respective progenitor. The initial number of progenitors (218) was therefore reduced to 108. Table 5 presents the list of 42 (out of 108 total) progenitors represented by one or more progeny that were selected through to the UYT. Ceballos et al. (2004) reported considerable variation for breeding values at SRTs, and proposed that it could be a useful parameter to take into consideration in the selection

Table 5 Progenitors of clones that reached the UYT and were represented by more than 50 progeny in the SRT. The number of clones (#) from each of these progenitors reaching UYT is also provided

Progenitor	#	Progenitor	#	Progenitor	#	Progenitor	#
SM1411-5	20	SM805-15	5	SM1422-4	2	NGA19	1
SM1665-2	15	CM6754-8	4	SM1511-6	2	SM1201-5	1
CM8027-3	12	SM1427-1	4	SM1656-7	2	SM1210-10	1
CM9067-2	12	SM1759-29	4	SM1669-5	2	SM1565-15	1
CM7514-8	10	SM890-9	4	SM1778-45	2	SM1637-22	1
SM1433-4	8	CM523-7	3	SM1789-20	2	SM1657-12	1
CM4365-3	7	SM1219-9	3	SM2629-36	2	SM2081-34	1
TAI8	7	SM1565-17	3	SM737-38	2	SM2773-32	1
SM1521-10	6	CM7985-24	2	CG1141-1	1	SM643-17	1
SM1438-2	5	CM8475-4	2	CM4574-7	1		
SM2192-6	5	KU50	2	CM7395-5	1		

process. It is clear that there is a large asymmetry and variation in the number of progeny representing progenitors at UYT. SM1411-5 was represented by a large number of progeny (20) suggesting an outstanding performance as a progenitor. SM 1665-2, CM 8027-3, CM 9067-2 and CM 7514-8 were represented by ten or more progeny at UYT. On the other hand, 12 progenitors were represented by only one genotype and 66 by none at all. We used the available data to try to understand or explain the reasons for this asymmetry.

Average SIN for the progeny of each of the 108 progenitors in this analysis is presented in Table 6. This average is the best approximation for the estimation of the breeding value of each progenitor. The way SIN is estimated allows for a simple interpretation of its meaning. A SIN of around 0 indicates average performance of the genotype. A positive SIN indicates above average performance, whereas a negative one suggests a below-average performance. The variation in the averages of SIN presented in Table 6 is remarkable. The progenitor R90 was represented by 73 clones evaluated at SRTs,

Table 6 Average selection index (SIN) of the 10 best and 10 worst progenitors, minimum and maximum SIN as well as the number of their respective progeny. The list is ordered from higher to lower average SIN

Progenitor	# of progeny	SIN (Average)	SIN (Minimum)	SIN (Maximum)
R90	73	27.5	− 3.9	58.3
SM2545-20	78	10.5	− 31.5	52.6
SM1411-5	870	8.3	− 76.8	46.8
SM2780-17	291	8.3	− 47.5	53.0
GM462-4	103	7.6	− 55.8	43.3
C-18	235	7.3	− 45.9	52.6
SM2546-54	73	7.1	− 40.6	36.5
SM1521-10	460	6.8	− 54.7	50.0
SM2619-1	51	6.3	− 26.5	30.1
SM1656-7	159	6.2	− 58.8	52.8
CM2772-3	505	− 7.9	− 63.9	37.7
CM7514-7	124	− 8.0	− 63.0	29.4
SM2615-25	67	− 8.4	− 51.7	19.9
SM1778-45	206	− 9.7	− 69.5	36.0
CM6756-13	97	− 10.1	− 48.0	31.6
SM2623-1	71	− 10.4	− 52.7	40.1
BRA496	65	− 11.1	− 46.0	23.0
TAI1	197	− 14.9	− 68.7	35.4
SM2772-2	74	− 16.8	− 85.9	31.5
COL945	58	− 17.6	− 48.2	17.7
AVERAGE	259	− 0.4	− 55.2	41.5

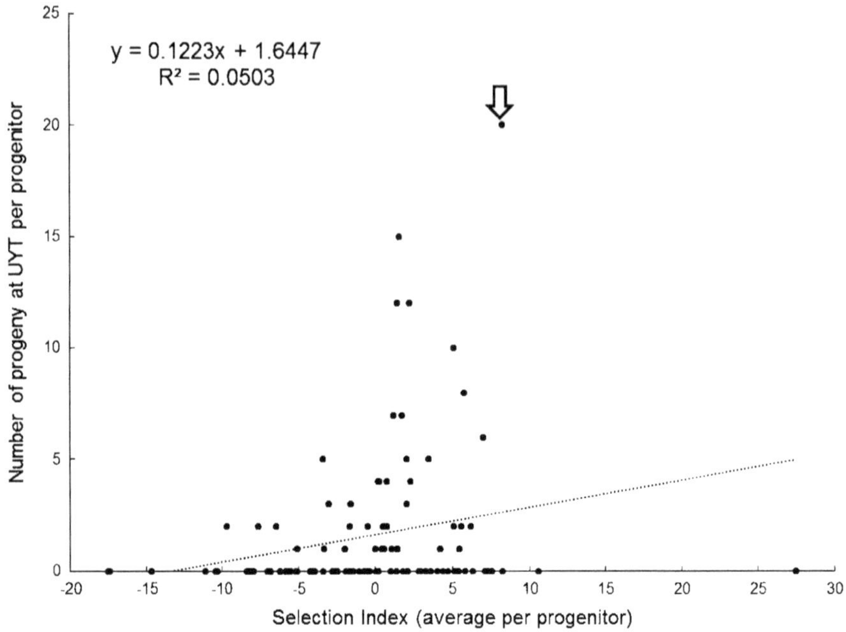

Figure 6 Relationship between average selection index (SIN) of each progenitor with the respective number of clones representing them at UYT (arrow identifies SM 1411-5).

which had an average SIN of 27.48. The range of variation for SIN among these 73 clones was from −3.91 to 58.26. Worth mentioning is the performance of SM1411-5. This is the progenitor with highest success rate at UYT, with 20 progeny (Table 5) and had the third highest average SIN (8.34, Table 6). On the bottom of Table 6 there are many progenitors with negative average SIN. For example, TAI1 had 197 clones evaluated at SRT and their average SIN was −14.88.

The next step in the analysis was to relate data from Tables 5 and 6 in a plot (Fig. 6). Results are disappointing in terms of relationship between SIN and superior progeny that a given clone produced. Although there was a promising positive association for SM 1411-5 (positive breeding value and many of its progeny reaching the UYT stage), many progenitors with positive average SIN failed to contribute with progeny in UYT. This is, for example, the case of R90 in the far right of the plot. It is clear, therefore, that breeding value estimated using data from SRT is not a good predictor of the chances of clones from a given progenitor reaching the UYT stage.

Alternative associations between different factors at SRT and number of clones from a given progenitor reaching UYT were explored. Figure 7 presents the relationship between the number of families or genotypes from a given progenitor evaluated at SRT and the number of clones from them reaching UYT. Coefficients of determination were much better (0.40 and 0.48, respectively) than that presented in Fig. 6 (0.05). Results therefore would indicate that the best predictor for the chances of the progeny from a given progenitor reaching the UYT is the number of clones screened at SRT. This is of little help to breeders,

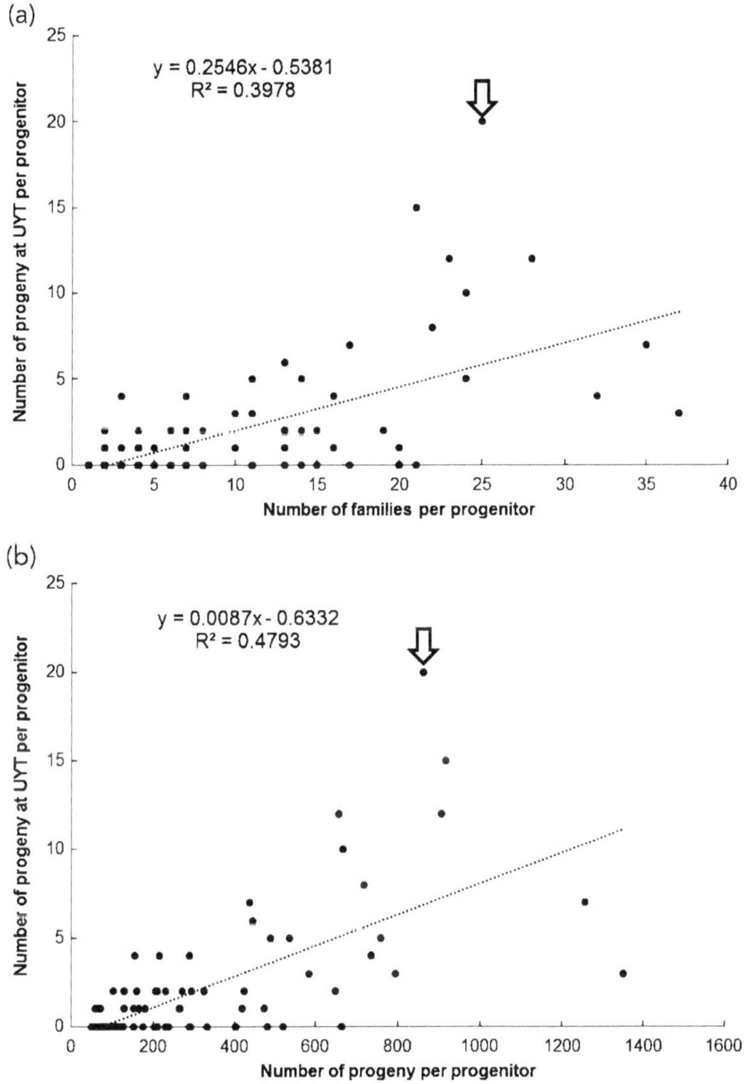

Figure 7 Number of clones derived from each progenitor reaching UYT as a function of number of families (a) and progeny (b) per progenitor (arrow identifies SM 1411-5).

other than to indicate that the best chances of success are through increasing the numbers of progeny to evaluate.

Figure 8 illustrates average selection index for the progeny of each progenitor, as well as the maximum and minimum values. This plot provides an insight into how large within-family genetic variation is in cassava relative to variation in breeding values. The consequences of this variation are apparent when looking at maximum selection indices for individual

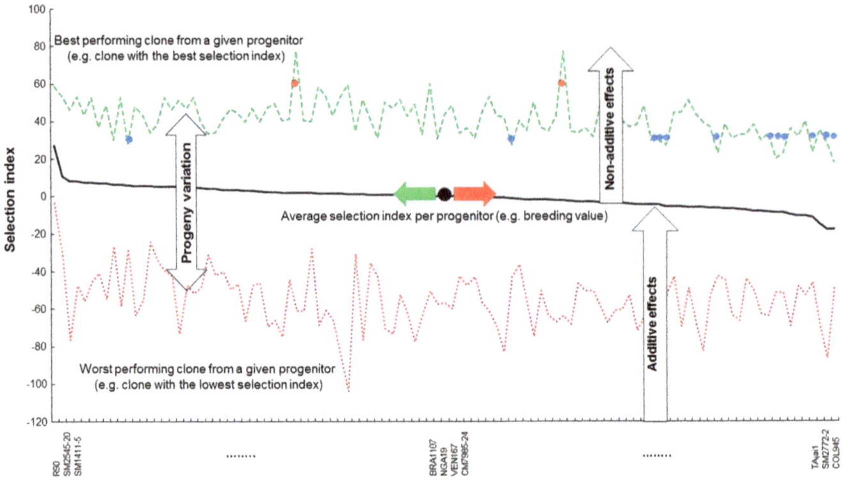

Figure 8 Average selection index (e.g. breeding value) and range of variation for progeny of different progenitors (not all progenitors are named but data from all 108 progenitors is presented). Two clones showed selection index above 60 (red dots). A group of 12 progenitors failed to generate at least one clone with selection index above 30 (blue dots).

clones. It is the best individual clones that breeders select for release, and that farmers will hopefully grow. Among the clones presented in Figure 8, only two had a selection index above 60 (red dots). One of them came from a progenitor with positive breeding value (e.g. average selection index above zero). The second clone, however, comes from a progenitor with a clearly negative breeding value. Only 12 progenitors failed to generate at least one clone with a minimum selection index of 30 (blue dots in Fig. 8). Of this group, one progenitor had a clearly positive breeding value, while the other 11 had negative ones. The relative probabilities of outstanding clones from progenitors with positive and negative breeding values indicate a positive association (in agreement with Fig. 6). However, this association is far from absolute, supporting the general idea that good performing clones can be found in most families. There were 50 progenitors with a negative breeding value; however, 39 of them produced at least one clone with a selection index above 30. This fact relates to several progenitors in Fig. 6 that, even though having a negative average selection index, managed to produce progeny that reached the UYT.

Data from Table 6 can also be used to draw similar conclusions. For example, the best clone from SM2623-1 had a SIN = 40.05 (even though the average SIN for this progenitor was −10.35). The performance of that clone was even better than those of the best clones from progenitors SM2546-54 and SM2619-1 (with average SIN values of 7.11 and 6.29, respectively) which had positive breeding values.

8 Quantitative genetics of complex traits in cassava

Several studies (mostly diallel designs) were conducted in the past two decades to dissect the genetic nature of relevant traits for cassava. Table 7 summarizes the results of three different diallel analyses conducted in three different environments in Colombia (Cach et al., 2005, 2006; Calle et al., 2005; Jaramillo et al., 2005; Pérez et al. 2005a,b). These diallel studies took advantage of the vegetative reproduction of cassava to measure the significance of epistasis. The information in Table 7 (expanded in Ceballos et al., 2015) is consistent and helps to explain the information from Table 6 and Fig. 8. A notable feature of the data is how much of the total genetic variation was allocated to the within-family component and the large non-additive effects for complex traits such as FRY.

It is stated that the large within-family genetic variation explains why good hybrids can be found even in poor-performing families. However, there are certain caveats to this generalization that need to be acknowledged. In order for this to be the case, the starting population, or set of progenitors, should have the genes for basic adaptation to the environment for which the progeny are being selected. For example, in a set of progenitors, if all are highly susceptible to CMD, it is highly unlikely that any of the progeny will be sufficiently resistant to reach the AYT stage in an environment with heavy CMD pressure. Likewise, a population of progenitors adapted to high-temperature environments is unlikely to produce progeny adapted to the highlands. In other words, good performing clones can be found in most families, so long as the progenitors of those families have the basic adaptation and resistance traits needed to succeed.

Recent experiences can illustrate this concept. The special breeding work to produce commercial varieties with amylose-free starch or high carotenoids required the use of progenitors not always adapted to the target environment. The lack of adaptation of these materials made it difficult to identify, in the first generation of segregating progeny, clones with very competitive productivity levels. In these cases, perhaps, it would be desirable to carry out a few short cycles of selection to foster genetic recombination and breaking undesirable linkages in the 'unadapted' populations.

Also relevant in data from Table 7 is the relative importance of additive vs non-additive genetic effects for key traits. Ceballos et al. (2015) reported that the relative importance

Table 7 Variance estimates (standard errors within parentheses) for FRY and DMC in three different diallel sets evaluated in the three environments for cassava production in Colombia

Genetic parameter	Fresh root yield (t ha^{-1})			Dry matter content (%)		
	Acid soil	Sub-humid	Mid-altitude	Acid soil	Sub-humid	Mid-altitude
σ^2_G (Between)	1.65	13.09	42.78	1.60	0.77	0.35
	(2.95)	(4.74)	(13.27)	(0.66)	(0.29)	(0.12)
σ^2_G (Within)	21.08	127.21	288.93	3.22	5.56	0.12
	(2.30)	(7.65)	(1918)	(0.17)	(0.31)	(0.12)
σ^2_A	− 1.49	17.82	11.88	3.38	1.45	0.99
	(6.32)	(13.75)	(24.67)	(2.40)	(0.99)	(0.47)
σ^2_D	9.03	23.87	152.11	0.87	0.77	− 0.21
	(7.93)	(11.15)	(49.08)	(0.67)	(0.50)	(0.13)
Epistasis test	15.05	100.40	168.91	0.87	4.26	− 0.32
	(6.74)	(12.74)	(39.72)	(1.29)	(0.67)	(0.92)

of GCA and SCA for FRY across many different studies was 1.6. This means that non-additive genetic effects for FRY were almost as important as the additive effects. For DMC, HIN and PTS, on the other hand, the relationship was 5.8, 3.3 and 9.1, respectively. The large influence of non-additive genetic effects for FRY can explain the disappointing results presented in Fig. 6, and the poor predictive capacity of breeding value in cassava. Breeding values, by default, are related to the average performance of progeny. However, cassava breeders, in fact, look for departures from the average performance. This was illustrated in Fig. 8. Additive effects are related to the central line representing the average SIN of the progeny from each progenitor (e.g. breeding value). However, the outstanding performance of the two best hybrids is related to non-additive genetic effects which, as stated earlier, are largely independent of the breeding value of the progenitors. It was this principle that prompted maize scientists (Crow, 2000; Hallauer and Miranda Fo 1988: 294; Mikel and Dudley, 2006) to state that what distinguishes the success of best commercial hybrids from the rest is the 'extra bit' of genetic superiority derived from epistatic effects.

Genomic selection, currently under pilot testing in cassava, brings hope of a positive impact for enhanced productivity (de Oliveira et al., 2012; Wolfe et al., 2015). Results from genomic selection already published are consistent with the information from Table 7: it can be effective at increasing DMC but it appears not to be efficient at improving FRY. The empiric validation of the predictions made based on quantitative genetics data (Ceballos et al., 2015) highlights the usefulness of these parameters when properly estimated and used.

Improving FRY in cassava is a challenging goal for conventional and molecular breeders. After the remarkable gains attained with varieties released mostly in the 1990s, further improvements in FRY in Asia and LAC became increasingly difficult (Ceballos et al., 2016b). Hopes that the use of molecular markers, including the recent implementation of genomic selection, would make selection more efficient have faced disappointing, albeit preliminary, results. Data presented in this chapter provide a consistent explanation of different experiences. Selection based on breeding value (including early attempts to select progenitors based on general combining ability and genomic selection) requires a prevalence of strong additive effects and clear differences in breeding values of progenitors. This is not the case for FRY. In addition to the genetic complexities mentioned (large non-additive genetic effects and within-family genetic variation), the changes observed for measurements of the same trait along the different stages of selection due to environmental, biotic and/or epigenetic effects complicate matters further (Joaqui Buritica et al., 2016). Molecular markers and genomic selection can be effective, on the other hand, in improving traits with higher heritability values than that of FRY (Rabbi et al., 2014a,b; Wolfe et al., 2015).

Heterosis is a key phenomenon not only for cross-pollinated crops but also, somewhat surprisingly, for an autogamous crop such as rice (Cheng et al., 2004; Spielman et al., 2013). Heterosis in crops is linked to non-additive genetic effects (dominance, over-dominance and epistasis). Cross-pollinated crops such as maize and cassava demonstrate large heterotic effects, whereas self-pollinated crops such as soya bean, rice or wheat were historically believed to show little or no heterosis. The unexpected impact of hybrid rice, however, highlights the relevance of non-additive genetic effects even for crops that were historically considered to express a relatively low level of heterosis.

Generally, crops that show strong heterotic effects also have important inbreeding depression. In fact, it has been suggested that heterosis and inbreeding depression are, to some extent, opposite sides of basically the same phenomenon (Miranda Filho, 1999). In other words, the same factors that increase performance in the heterozygous hybrid are related to inbreeding depression in the homozygous progenitors. One could perceive inbreeding depression and heterosis to be opposites for dominance and over-dominance effects but not necessarily for epistasis. It is expected, therefore, that traits showing strong heterosis (non-additive effects) should also be subjected to strong inbreeding depression. Similarly, when traits are mostly additive in nature, inbreeding depression should be negligible. Rojas-C et al. published in 2009 an analysis of inbreeding depression in eight S_1 populations derived from eight elite hybrid cassava cultivars. As expected, inbreeding depression for FRY was very high with an average of 63.9%, whereas inbreeding depression for DMC was considerably lower (average of 5.3%) because additive genetic effects probably play a major role in the latter. These results have been validated by other authors (Kawano et al., 1978; Kawuki et al., 2011; de Freitas et al., 2016).

9 Future trends

The heterozygous nature of the progenitors in cassava, along with vegetative propagation, places the crop in a unique situation for breeding, among major staple crops. Thousands of hybrids are produced and evaluated every year. Outstanding clones are then selected and released for farmers to grow. A great advantage is that farmers can, indeed, exploit all genetic effects (additive, dominance and epistasis) that determine the outstanding performance of these hybrids. However, the heterozygosity of cassava progenitors makes it impossible for the breeder to equally exploit all genetic effects when using elite clones as progenitors. Only a fraction of these effects can be exploited, particularly for complex traits such as FRY. A key objective to improve the efficiency of cassava breeding, therefore, is to partition all genetic variation in such a way that breeders can better exploit it.

Additive effects are linked to the presence or absence of alleles, whereas the non-additive heterotic effects depend on specific combinations of alleles. Gametes can only transmit alleles, not combinations of alleles (unless they come from homozygous progenitors). The exploitation of heterosis, which depends on the combination of alleles in different loci, therefore requires special breeding approaches that will be described in the following sections.

9.1 Exploitation of heterosis through reciprocal recurrent selection

Reciprocal recurrent selection (RRS) has been used to exploit heterosis in many different crops (Bernardo, 2014), including sweet potato (Grüneberg et al., 2009). The main objective of this method is to gradually build up complementary allelic frequencies in two '*reciprocal*' populations. This method has proven to be efficient at increasing heterotic responses in maize, even when based on the use of non-inbred progenitors (Hallauer and Miranda, 1988; Stuber, 1994). RRS allows gradual and consistent exploitation of non-additive genetic effects in crosses among individuals from the reciprocal populations. Additive genetic effects are also improved through changes in the allelic frequencies within each population (Hallaurer and Miranda, 1988).

The first step in RRS implementation is the identification or development of reciprocal populations or heterotic groups. Unfortunately, genetic distances in cassava have proven not to be an efficient way to identify strong specific combining ability effects that are related to heterosis (Ceballos et al., 2016c). Strategic efforts need to be implemented to develop a population structure that would facilitate the creation or identification of heterotic groups in cassava. Diverse gene pools that have evolved isolated from each other over a long period could be a reasonable starting point (Melchinger, 1999; Melchinger and Gumber, 1998). However, it is unknown whether this approach would yield positive results for cassava. Certainly, the research to answer some of the fundamental questions about heterosis in cassava is more costly and time-consuming compared to seed-propagated crops due to the difficulties of producing large numbers of seeds from specific crosses, and the long growth cycle of the crop. In part, because of the minimal interest showed by private sector breeding companies in cassava, along with the shorter-term applied focus of nearly all current breeding programmes, the funding for answering some of the basic questions about cassava genetics has been inadequate.

It is expected that RRS will improve the average performance of the crosses between genotypes from the two reciprocal populations. However, in the case of cassava this would not be the end of the breeding process because each cross would be a full-sib family with the large within-family variation already described. The most challenging problem in cassava breeding is probably how to identify the best genotype out of the billions that two (heterozygous) progenitors could hypothetically produce. Certainly, there is a limit to the number of genotypes that can be used to represent a given full-sib family, and this is a major limitation that both genomic selection and conventional cassava breeding face. It is the selection of a single outstanding genotype out of tens or even hundreds representing the same full-sib family, however, that makes cassava breeding so challenging.

Implementing an RRS scheme, however, would clearly improve the cassava breeders' ability to exploit heterosis more consistently. RRS, rather than exploiting all genetic effects simultaneously, separates (to a certain extent) the improvement of additive and non-additive genetic effects, an advantage that counterbalances its added complexity. The separation of a breeding project into two or more heterotic groups is the only way to guarantee a gradual improvement of non-additive genetic effects. Development of a population structure can be achieved through various strategies. In vegetable breeding (S. Knapp, pers. comm.), heterotic populations can be used as pools for different traits. One population, for example, can be the source for defensive traits while the other could provide desirable quality traits to the resulting hybrids.

9.2 Advantages of homozygous parents

The use of heterozygous progenitors in RRS must still deal with the fact that the optimum combination of alleles in different loci achieved in one cycle will be lost in the following cycle of selection. If there was a possibility of producing fully inbred lines, the exploitation of heterosis could be further maximized and the combination of different traits facilitated. Breeding many cross-pollinated crops, therefore, relies on the use of inbred progenitors, which offers key advantages. Crosses among inbred lines do not generate within-family variation. More importantly, inter-loci allelic combinations can be fixed and, if necessary, gradually improved. In a way, the use of inbred progenitors in the production of commercial hybrids is an extreme case of RRS where the reciprocal populations have been reduced to two outstanding, highly heterotic, inbred lines. A venerable example is the maize inbred lines Mo17 and B73, parents of a successful and predominant cross in the commercial production of maize in the 1980s.

The use of inbred progenitors offers several advantages. Several seeds for the same genotype can be produced. This is convenient because the evaluation of hybrids could begin at the preliminary or AYT stage, thus reducing almost in half the lengthy current evaluation system (Ceballos et al. 2012b, 2015). Heterotic responses are also more clearly identified when homozygous progenitors are used. The performance of inbred lines *per se* represents twice the σ^2_A present in the original F_1 cross and does not expose any dominance genetic effect. Selection among inbred lines within each reciprocal population can therefore only act on additive effects ($\sigma^2_A + \sigma^2_{AA} + \sigma^2_{AAA} + \ldots$). The performance of the inbred lines *per se* would allow reducing the frequency of undesirable alleles (genetic load), and resulting in the improvement of their breeding value.

The development of inbred progenitors in cassava, however, is time consuming and difficult. Because of the reproductive biology of cassava, it would take almost a decade to produce inbred lines through successive self-pollinations. CIAT began this process in the mid-2000s. A result of this activity was the development of AM 560-2 (an S_3 genotype derived from MCOL 1505), which was used to sequence cassava's genome. However, successive self-pollinations unavoidably encourage the use of early flowering genotypes, which are undesirable. There is ongoing research to achieve homozygosity through the production of doubled haploids by andro- or gynogenesis. During the past few years, considerable progress has been achieved in inducing cell division in gametic tissue (Perera et al., 2012, 2013) and research is now focused on regenerating plants. Simultaneously, there are also ongoing efforts to induce flowering in cassava. These efforts, if successful, would allow for the production of S_5 or S_6 lines in as little as 3–4 years through successive self-pollinations, and would generally facilitate cassava breeding by allowing seed production at a much lower cost, and among clones that are currently difficult to cross because of shy flowering.

9.3 Use of partially inbred progenitors

An alternative to the use of homozygous progenitors is partial inbreeding (e.g. S_1 or S_2 lines); this approach was initiated at CIAT, and will be consolidated in the next few years, until a protocol for fully inbred genotypes becomes available. Partial inbreeding allows identifying more clearly the breeding value of progenitors by reducing the within-family genetic variation. It would also facilitate the development or identification of heterotic groups. Moreover, partial inbreeding could also be the way to improve breeding value. For example, resistance to CMD has been linked to a single dominant factor (Rabbi et al., 2014b). If an S_1 genotype, homozygous for resistance to CMD, was used (CC in Fig. 9) instead of the (putatively) heterozygous S_0 progenitor from which it was derived, its breeding value would double (e.g. 100% of the progeny rather than 50% of the progeny would be resistant to CMD). This process is described on the left side of Figure 9. In addition to homozygous resistance to CMD, segregating S_1 genotypes could be selected for agronomic performance as well. Similarly, a 'complementary' breeding population may be developed for increased levels of DMC (right side of Fig. 9).

Let us assume that a new recessive source of tolerance to a given herbicide has been identified. The source of tolerance is already partially inbred and homozygous (hh) for tolerance to the herbicide. The initial products (e.g. S_1 genotypes) from the first step of selection in the two complementary populations presented in Fig. 9 are susceptible (HH) to the herbicide. The S_1 genotypes selected for resistance to CMD or high DMC are then crossed with the source of tolerance to herbicide. The resulting crosses will be heterozygous for the monogenic traits and intermediate for dry matter content. Self-pollination of the resulting crosses will allow the recovery of S_1 genotypes that are homozygous for CMD and for tolerance to the herbicide (left side of Fig. 9) or have improved levels of DMC combined with tolerance to the herbicide (right side of Fig. 9). Crossing the second-step products generates progenies that are 100% resistant to CMD (Cc), tolerant to the herbicide (hh) and have excellent levels of DMC.

The key principle behind the idea presented here is that the gametes produced by the selected S_1 genotypes should carry a higher frequency of desirable alleles. This is clearly the case for the traits these genotypes had been selected for. For other traits, the frequency of desirable alleles should be, on average, similar in the S_1 genotypes as that of the S_0 progenitors from which they were derived (in fact, it should be somewhat better, because inbreeding reduces genetic load). Crosses among the selected partially inbred lines, because of their enhanced breeding value, will generate (by default) better-performing hybrids. A second and fundamental advantage of the proposed scheme is that it allows for the gradual, consistent, stepwise fixation of simply inherited traits in the partially inbred selected genotypes. Eventually, partially inbred lines from different heterotic groups (when identified or developed) would allow the implementation of conventional RRS schemes.

Several root quality traits have relatively simple inheritance (carotenoids and dry matter contents; amylose-free or *waxy* starch, small starch granules, perhaps low CNP and 'cooking quality' for some of the many different ethnic end uses of cassava). High heritability is also found in the resistance to pests and diseases (thrips and whiteflies, bacterial blight, super elongation disease, CMD and probably cassava brown streak virus), and plant architecture traits (such as reduced branching for enhanced storage capacity of the stems and facilitated mechanization). Introgression and gradual fixation of these traits can be more easily achieved through the use of partial inbreeding. Certainly, another group of traits that would benefit from this approach are those from genetic transformation. Future advances in our knowledge of plant biology arising from *Arabidopsis* research will foster the need and intensity of trait introgression as they are identified in cassava. The reduced within-family variation in progeny derived from partially inbred parents could also contribute to improvement in more complex traits such as FRY.

10 Conclusion

Information provided in this chapter highlights some key features of cassava breeding. There is a need to shift the current system based on crossing elite germplasm, with the hope of identifying even better progeny, into a system based on the improvement of progenitors (through partial or full inbreeding) with enhanced breeding values and aiming at the development of population structures that can facilitate the conventional breeding work as well as molecular approaches. This would be a rational way to assure sustained and predictable genetic gains and improve the efficiency of cassava breeding.

As a less desirable alternative, cassava breeders could opt to focus on breeding easier-to-improve traits such as carotenoids or starch mutants that are characterized by higher heritabilities. Ultimately, however, breeders owe their clients – the farmers and consumers who rely on cassava for their livelihoods and nutrition – the full range of options, including better yields through optimized breeding approaches.

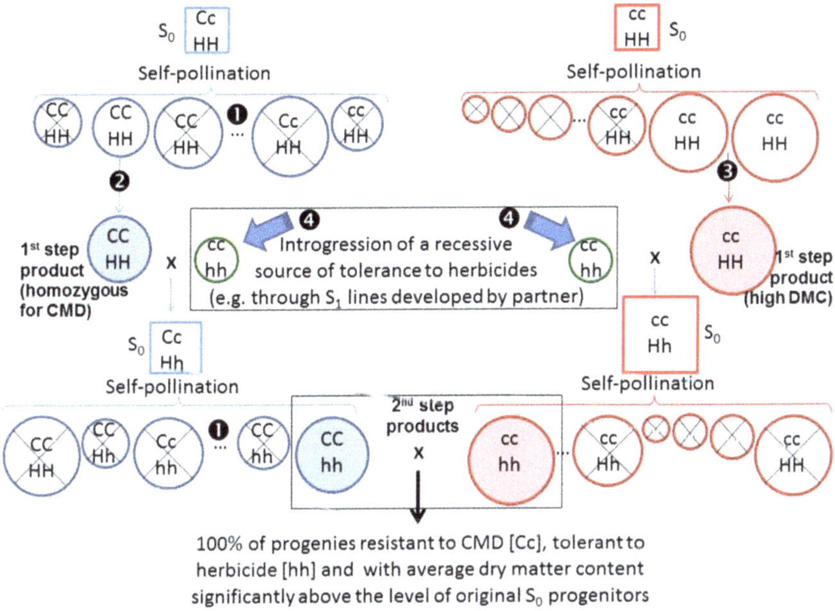

Figure 9 Illustration of the way breeding values could be consistently improved in a stepwise fashion in two *'complementary'* breeding populations. Squares indicate S_0 genotypes, whereas circles indicate partial inbreds. On the left, selections are made for resistance to CMD. Molecular markers can be used to distinguish homozygous [CC] from heterozygous [Cc] genotypes (1). In addition to homozygous resistance to CMD, segregating S_1 genotypes are selected for agronomic performance (2). Diameters of the circles (or size of squares for S_0) in both left and right diagrams represent levels of DMC (larger circles or squares, higher DMC). On the right, selections in the *'complementary'* population are made for increased dry matter content (3). This population does not carry resistance to CMD so the genotype for this trait [cc] has not been included in every genotype. The selected products (S_1 genotypes) from these first steps of selection are shaded. Both products, however, are susceptible to a target herbicide. In a parallel process (perhaps from a partner), S_1 genotypes have been generated that are homozygous for a recessive source for tolerance to an herbicide (4). The S_1 genotypes selected for resistance to CMD or high DMC are then crossed with the source of tolerance to herbicides. The resulting crosses will be heterozygous for monogenic traits and intermediate for dry matter content. Self-pollination of the resulting crosses will allow the recovery of S_1 genotypes that are homozygous for CMD and for tolerance to the herbicide (left side), or have improved levels of DMC combined with tolerance to the herbicide (right side). The second-step products are also shaded. Crossing the second-step products generate progeny that are 100% resistant to CMD [Cc], tolerant to the herbicide [hh] and have excellent levels of DMC (Ceballos et al., 2016a).

11 Where to look for further information

Progress in cassava research and breeding is continuously updated in CIAT´s (https://ciat.cgiar.org) and IITA`s (www.iita.org) webpages. The NextGen project (www.nextgencassava.org) includes a large phenotypic database (CassavaBase).

12 References

Abraham, A. (1957). Breeding of tuber crops in India. *Indian J. Gen. Plant Breed.* 17:212–17.

Aiemnaka, P., A. Wongkaew, J. Chanthaworn, S. K. Nagashima, S. Boonma, J. Authapun, S. Jenweerawat, P. Kongsila, P. Kittipadakul, S. Nakasathien, T. Sreewongchai, V. Vichukit, L. A. Becerra López-Lavalle, H. Ceballos, C. Rojanaridpiched and C. Phumichai (2012). Molecular Characterization of a Spontaneous Waxy Starch Mutation in Cassava (*Manihot esculenta* Crantz). *Crop Sci.* 52:2121–30.

Akano A. O., A. G. O. Dixon, C. Mba, E. Barrera and M. Fregene (2002). Genetic mapping of a dominant gene conferring resistance to cassava mosaic disease. *Theor Appl Genet* 105:521–5.

Akinbo, O., M. Labuschagne and M. Fregene (2012). Increased storage protein from interspecific F1 hybrids between cassava (*Manihot esculenta* Crantz) and its wild progenitor (*M. esculenta* ssp. *flabellifolia*). *Euphytica* 185:303–11.

Allem, A. C. (2002). The origins and taxonomy of cassava. In: R. J. Hillocks, J. M.Tresh, and A. C. Bellotti (eds), *Cassava: Biology, Production and Utilization.* CABI Publication. Wallingford, UK and New York, USA, pp. 1–16.

Alvarez, E., J. F. Mejía, G. A. Llano, J. B. Loke, A. Calari, B. Duduk and A. Bertaccini.(2009). Characterization of a Phytoplasma associated with frogskin disease in dassava. *Plant Dis.* 93:1139–45.

Alvarez, E., J. M. Pardo, J. F. Mejia, A. Bertaccini, N. D. Thanh and T. X. Hoat (2013). Detection and identification of 'Candidatus Phytoplasma asteris'-related phytoplasmas associated with a witches' broom disease of cassava in Vietnam. *Phytopathogenic Mollicutes.* 3(2):77–81.

Anyanwu, C. N., C. N. Ibeto, S. L. Ezeoha and N. J. Ogbuagu (2015).Sustainability of cassava (*Manihot esculenta* Crantz) as industrial feedstock, energy and food crop in Nigeria. *Renewable Energy* 81:745–52.

Andersen, M. D., P. K. Busk, I. Svendsen and B. L. Møller (2000). Cytochromes P-450 from cassava (*Manihot esculenta* Crantz) catalyzing the first steps in the biosynthesis of the cyanogenic glucosides linamarin and lotaustralin. *J. Biol. Chem.* 275:1966–75.

Ashby, J. A., C. A. Quirós and Y. M. Rivera (1987). Farmer participation in on-farm variety trials. Agricultural Administration (Research and Extension) Network Discussion Paper No. 22. London. Overseas Development Institute.

Babu, L., and S. R. Chatterjee (1999). Protein content and amino acid composition of cassava tubers and leaves. *J. Root Crops.* 25(20):163–8.

Bakayoko, S., A. Tschannen, C. Nindjin, D. Dao, O. Girardin and A. Assa (2009). Impact of water stress on fresh tuber yield and dry matter content of cassava (*Manihot esculenta* Crantz) in Côte d'Ivoire. *Afr. J. Agric. Res.* 4(1):21–7.

Baker, R. J. (1986). *Selection Indices in Plant Breeding.* CRC Press, Boca Raton, FL.

Beeching J. R., P. Marmey, M. C. Gavalda, M. Noirot, H. R. Haysom, M. A. Hughes and A. Charrier (1993). An assessment of genetic diversity within a collection of cassava (*Manihot esculenta* Cranz) germplasm using molecular markers. *Ann. Bot.* 72:515–20.

Belalcazar, J., D. Dufour, M. S. Andersson, M. Pizarro, J. Luna, L. Londoño, N. Morante, A. M. Jaramillo, L. Pino, L. A. Becerra López-Lavalle, F. Davrieux, E. F. Talsma and H. Ceballos (2016). High-throughput phenotyping and improvements in breeding cassava for increased carotenoids in the roots. *Crop Sci.* 56:2916–25.

Bellotti, A. C., L. Smith and S. L. Lapointe (1999). Recent advances in cassava pest management. *Annu. Rev. Entomol.* 44:343–70.

Bellotti, A. C. (2002). Arthopod pests. In: R. J. Hillocks, J. M. Tresh and A. C.Bellotti (eds), *Cassava: Biology, Production and Utilization.* CABI Publication. Wallingford, UK and New York, USA, pp. 209–36.

Benesi, I. R. M., M. T. Labuschagne, A. G. O. Dixon and N. M. Mahungu (2004). Stability of native starch quality parameters, starch extraction and root dry matter of cassava genotypes in different environments. *J. Sci. Food Agric.* 84:1381–8.

Benesi, I. R. M., M. T. Labuschagne, L. Herselman, N. M. Mahungu and J. K. Saka (2008). The effect of genotype, location and season on cassava starch extraction. *Euphytica* 160:59–74.
Benesi, I. R. M., M. T. Labuschagne, L. Herselman and N. Mahungu (2010). Ethnobotany, morphology and genotyping of cassava germplasm from Malawi. *J. Biol. Sci.* 10(7):616–23.
Bernardo, R. (2014). *Essentials of Plant Breeding*. Stemma Press Woodbury, Minnesota, USA. 252p.
Bolhuis, G. G. (1953). A survey of some attempts to breed cassava varieties with a high content of proteins in the roots. *Euphytica* 2:107–12.
Bueno, A. (1991) Estimación de los parámetros genéticos en la yuca. In C. H. Hershey (ed.), *Mejoramiento genético de la yuca en América Latina*. CIAT Publication, Colombia, CA, pp. 197–220.
Burns, A., R. Gleadow, J. Cliff, A. Zacarias and T. Cavagnaro (2010). Cassava: The drought, war and famine crop in a changing world. *Sustainability* 2: 3572–3607.
Burton, G. W. (1992). Recurrent restricted phenotypic selection. *Plant Breed. Rev.* 9:101–13.
Byrne, D. (1984). Breeding cassava. *Plant Breed. Rev.* 2:73–134.
Caccamisi, D. S. (2010). Cassava: Global production and market trends. *Chronica Hort.* 50(2):15–18.
Cach, N. T., J. C. Perez, J. I. Lenis, F. Calle, N. Morante and H. Ceballos (2005). Epistasis in the expression of relevant traits in cassava (*Manihot esculenta* Crantz) for subhumid conditions. *J. Hered.* 96(5):586–92.
Cach, T. N., J. I. Lenis, J. C. Perez, N. Morante, F. Calle and H. Ceballos (2006). Inheritance of relevant traits in cassava (*Manihot esculenta* Crantz) for sub-humid conditions. *Plant Breed.* 124:1–6.
Calle, F., J. C. Perez, W. Gaitán, N. Morante, H. Ceballos, G.Llano and E. Alvarez (2005). Diallel inheritance of relevant traits in cassava (*Manihot esculenta* Crantz) adapted to acid-soil savannas. *Euphytica* 144(1–2):177–86.
Calvert, L. A., and J. M. Thresh (2002). The viruses and virus diseases of cassava. In R. J. Hillocks, J. M. Tresh and A. C. Bellotti (eds), *Cassava: Biology, Production and Utilization*. CABI Publication, Wallingford, UK and New York, USA, pp. 237–60.
Calvert, L. A., M. Cuervo, I. Lozano, N. Villareal and J. Arroyave (2008). Identification of three strains of a virus associated with cassava plants affected by frogskin disease. *J. Phytopathol.* 156:647–53.
Ceballos H., C. A. Iglesias, J. C. Pérez and A. G. O. Dixon (2004). Cassava breeding: opportunities and challenges. *Plant Mol. Biol.* 56:503–15.
Ceballos H., T. Sánchez, A. L. Chávez, C. Iglesias, D. Debouck, G. Mafla and J. Tohme (2006). Variation in crude protein content in cassava (*Manihot esculenta* Crantz) roots. *J. Food Compos. Anal.* 19:589–93.
Ceballos, H., T. Sánchez, N. Morante, M. Fregene, D. Dufour, A. M. Smith, K. Denyer, J. C. Pérez, F. Calle and C. Mestres (2007a). Discovery of an amylose-free starch mutant in cassava (*Manihot esculenta* Crantz). *J. Agric. Food Chem.* 55(18):7469–76.
Ceballos, H., M. Fregene, J. C. Pérez, N. Morante and F. Calle (2007b). Cassava Genetic Improvement. In M. S. Kang and P. M. Priyadarshan (eds), *Breeding Major Food Staples*. Blackwell Publication, Ames, IA, pp. 365–91.
Ceballos, H., T. Sánchez, K. Denyer, A. P. Tofiño, E. A. Rosero, D. Dufour, A. Smith, N. Morante, J. C. Pérez and B. Fahy (2008). Induction and identification of a small-granule, high-amylose mutant in cassava (*Manihot esculenta* Crantz). *J. Agric. Food Chem.* 56:7215–22.
Ceballos, H., and F. Calle. (2010). Cassava. In J. Fajardo, L. NeBambi, M. I. Larinde, C. Rosell, I. Barker, W. Roca and E. Chujoy (eds), *Quality Declared Planting Material*. Protocols and Standards for Vegetatively Propagated Crops (FAO Plant Production and Protection Paper 195), pp. 33–40.
Ceballos, H., E. Okogbenin, J. C. Pérez, L. A. Becerra and D. Debouck (2010). Cassava. In J. Bradshaw (ed.), *Root and Tuber Crops*. Springer Publishers, New York, pp. 53–96.
Ceballos, H., J. Ramirez, A. C. Bellotti, A. Jarvis and F. Alvarez (2011). Adaptation of cassava to changing climates. In S. S. Yadav, B. Redden, J. S. Hatfield, G. Lotze-Campen and A. Hall (eds), *Crop Adaptation to Climate Change*. Wiley–Blackwell. Hoboken, NJ, pp. 411–25.
Ceballos, H., J. Luna, A. F. Escobar, J. C. Pérez, D. Ortiz, T. Sánchez, H. Pachón and D. Dufour (2012a). Spatial distribution of dry matter in yellow fleshed cassava roots and its influence on carotenoids retention upon boiling. *Food Res. Int.* 45:52–9.

Ceballos, H., C. Hershey and L. A. Becerra-López-Lavalle (2012b). New approaches to cassava breeding. *Plant Breed. Rev.* 36:427–504.
Ceballos, H., N. Morante, T. Sánchez, D. Ortiz, I. Aragón, A. L. Chávez, M. Pizarro, F. Calle and D. Dufour (2013). Rapid cycling recurrent selection for increased carotenoids content in cassava roots. *Crop Sci.* 53:2342–51.
Ceballos, H., R. S. Kawuki, V. E. Gracen, G. C. Yencho and C. H. Hershey (2015). Conventional breeding, marker assisted selection, genomic selection and inbreeding in clonally propagated crops: a case study for cassava. *Theor. Appl. Genet.* 9:1647–67.
Ceballos, H., J. C. Pérez, J. B. Orlando, J. I. Lenis, N. Morante, F. Calle and C. Hershey (2016a). Cassava breeding I: the value of breeding value. *Front. Plant Sci.* doi: 10.3389/fpls.2016.01227.
Ceballos, H., and C. Hershey (2016b). Road map for cassava genetic improvement. Proceedings of the World Congress on Root and Tuber Crops and Third Scientific Conference of the Global Cassava Partnership for the 21st Century. Nanning, China.
Ceballos, H., L. A. Becerra, F. Calle, N. Morante, T. Ovalle, J. C. Pérez and C. Hershey (2016c). Genetic distance and heterosis in cassava. *Euphytica* 210: 79–92. DOI 10.1007/s10681–016-1701-7.
Chalwe, A., R. Melis, P. Shanahan and M. Chiona (2015). Inheritance of resistance to cassava green mite and other useful agronomic traits in cassava grown in Zambia. *Euphytica* 205:103–19.
Chávez, A. L., T. Sánchez, G. Jaramillo, J. M. Bedoya, J. Echeverry, E. A. Bolaños, H. Ceballos and C. A. Iglesias (2005). Variation of quality traits in cassava roots evaluated in landraces and improved clones. *Euphytica* 143:125–33.
Chen, H., M.-l. Xu, Q. Guo, L. Yang and Y. Ma (2016). A review on present situation and development of biofuels in China. *J. Energy Inst.* 89:248–55.
Cheng S.-H., L.-Y. Cao, S.-H. Yang and H.-Q. Zhai (2004). Forty years' development of hybrid rice: China's experience. *Rice Sci.* 11(5–6):225–30.
Chipeta, M. M., J. M. Bokosi, V. W. Saka and I. R. M. Benesi (2013) Combining ability and mode of gene action in cassava for resistance to cassava green mite and cassava mealy bug in Malawi. *J. Plant Breed. Crop Sci.* 5(9):195–202.
CIAT (2001). Centro Internacional de Agricultura Tropical, Project IP3, Improved Cassava for the Developing World, Annual Report 2001. Apdo Aéreo 6713, Colombia, CA.
Cock, J. (1985). *Cassava: New Potential for a Neglected Crop*. Westview Press. Boulder, CO.
Cours, G . (1951). Le manioc a Madagascar. Memoires de L'Institut Scientifique de Madagascar . Serie B 3: 203–400.
Crow, J. F. (2000). The rise and fall of overdominance. *Plant Breed. Rev.* 17:225–57.
Davrieux, F., D. Dufour, P. Dardenne, J. Belalcazar, M. Pizarro, J. Luna, L. Londoño, A. Jaramillo, T. Sanchez, N. Morante, F. Calle, L. A. Becerra and H. Ceballos (2016). LOCAL regression algorithm improves NIRS predictions when the target constituent evolves in breeding populations. *JNIRS* 24:109–17.
De Carvalho, R. D., and M. Guerra (2002). Cytogenetics of *Manihot esculenta* Crantz (cassava) and eight related species. *Hereditas* 136:159–68.
de Freitas, J. P. X., V. da Silva Santos and E. J. de Oliveira (2016). Inbreeding depression in cassava for productive traits. *Euphytica* 209:137–45.
de Oliveira E. J., M. D. Vilela de Resende, V. da Silva Santos, C. Fortes Ferreira, G. Alvarenga Fachardo Oliveira, M. Suzarte da Silva, L. Alves de Oliveira and C. I. Aguilar-Vildoso (2012). Genome-wide selection in cassava. *Euphytica* 187:263–76.
Dixon, A. G. O., R. Asiedu and M. Bokanga (1994). Breeding of cassava for low cyanogenic potential: problems, progress and perspectives. *Acta Hort.* 375:153–61.
Du, L., M. Bokanga, B. L. Møller and B. A. Halkier (1995). The biosynthesis of cyanogenic glucosides in roots of cassava. *Phytochemistry* 39:323–6.
Dufour, D., J. Ricci, T. Sanchez, N. Morante, G. Morel, M. Reynes, C. Hershey and H. Ceballos (2012). A reappraisal of nitrogen-to-protein conversion factors in cassava roots. 2nd Conference of the Global Cassava Partnership for the 21st Century. Kampala, Uganda.
Duputié, A., J. Salick and D. McKey (2011). Evolutionary biogeography of Manihot (Euphorbiaceae), a rapidly radiating Neotropical genus restricted to dry environments. *J. Biogeogr.* 38(6):1033–43.

Easwari Amma, C. S., N. Sheela and P. K. Thankamma Pillai (1995). Combining ability analysis in cassava. *J. Root Crops* 21(2):65–71.

Egesi, C. N., F. O. Ogbe, M. Akoroda, P. Ilona and A. Dixon. 2007. Resistance profile of improved cassava germplasm to cassava mosaic disease in Nigeria. *Euphytica* 155:215–24.

Eke-Okoro, O. N., O. U. Okereke and J. E. Okeke. (2001). Effect of stake sizes on some growth indices and yield of three cassava cultivars (*Manihot esculenta*). *J. Agric. Sci.* 137:419–26.

Elias, M., L. Penet, P. Vindry, D. McKey O. Panaud and T. Robert (2001). Unmanaged sexual reproduction and the dynamics of genetic diversity of a vegetatively propagated crop plant, cassava (*Manihot esculenta* Crantz) in a traditional farming system. *Mol. Ecol.* 10:1895–1907.

El-Sharkawy, M. A. (2016). Prospects of photosynthetic research for increasing agricultural productivity, with emphasis on the tropical C_4 Amaranthus and the cassava C_3-C_4 crops. *Photosynthetica* 54(2):161–84.

Esuma, W., L. Herselman, M. T. Labuschagne, P. Ramu and F. Lu (2016). Genome-wide association mapping of provitamin A carotenoid content in cassava. *Euphytica* DOI 10.1007/s10681-016-1772-5.

Falconer, D. S. (1981). *Introduction to Quantitative Genetics* (2nd Edition). Longman Inc. New York.

Flôres, D.H., I. C. Canale, M. C. Bedendo, I, Paulo (2013). Molecular identification of a 16SrIII-B phytoplasma associated with cassava witches' broom disease. *European Journal of Plant Pathology* 137: 237–42.

Gardner, C. O. (1961). An evaluation of effects of mass selection and seed irradiation with thermal neutrons on yields of corn. *Crop Sci.* 1:241–45.

Gonçalvez Fukuda, W. M. and N. Saad (2001). Participatory research in cassava breeding with farmers in Northeastern Brazil. Document CNPMF 99. EMBRAPA, Cruz das Almas. Bahia, Brazil.

Gonçalves Fukuda, W. M., S. de Oliveira and Silva and C. Iglesias (2002). Cassava breeding. *Crop Breed. Appl. Biote*, Ch. 2(4):617–38.

Graner, E. A. (1935). Contribuição para o estudo citológico da mandioca . Publication of the Escola Superior de Agricultura Luiz de Queiroz, Piracicaba, S. Paulo, Brazil. pp. 1–28.

Grüneberg, W., R. Mwanga, M. Andrade and J. Espinoza (2009). Breeding clonally propagated crops. In S. Ceccarelli, E. P Guimarães and E. Weltzien (eds), *Plant Breeding and Farmer Participation*. FAO, Rome, pp. 175–322.

Hahn, S. K., K. V. Bai and R. Asiedu (1990). Tetraploids, triploids, and 2n pollen from diploid interspecific crosses with cassava. *Theor. Appl. Gen.* 79:433–9.

Hahn, S. K., A. K. Howland and E. R. Terry (1980a). Correlated resistance to cassava to mosaic and bacterial blight diseases. *Euphytica* 29: 305–11.

Hahn, S. K., E. R.Terry, and K. Leuschner (1980b). Breeding cassava for resistance to cassava mosaic disease. *Euphytica* 29:673–83.

Hahn, S. K., E. R. Terry, K. Leuschner, I. O. Akobundu, C. Okali and R. Lal (1979). Cassava improvement in Africa. *Field Crops Res.* 2:193–226.

Hallauer, A. R. and J. B. Miranda Fo. (1988). *Quantitative Genetics in Maize Breeding*. Second Edition. Iowa State University Press, Ames.

Hershey, C., G. Henry, R. Best, K. Kawano, R. Howeler and C. Iglesias (2001). Cassava in Asia–Expanding the competitive edge in diversified markets. In *A Review of Cassava n Asia with Country Case Studies on Thailand and Vietnam*. FAIO/IFAD, Rome, Italy, pp. 1–62.

Hillocks, R. J., and K. Wydra (2002). Bacterial, fungal and nematode diseases. In R. J. Hillocks, J. M. Tresh and A. C. Bellotti (eds), *Cassava: Biology, Production and Utilization*. CABI Publication. Wallingford, UK and New York, USA, pp. 261–80.

Iglesias, C. A., C. Hershey, F. Calle and A. Bolaños (1994). Propagating cassava (*Manihot esculenta* Crantz) by sexual seed. *Exp. Agric.* 30:283–90.

Iglesias, C., J. Mayer, L. Chávez and F. Calle (1997). Genetic potential and stability of carotene content in cassava roots. *Euphytica* 94:367–73.

Jaramillo, G., N. Morante, J. C. Pérez, F. Calle, H. Ceballos, B. Arias and A. C. Bellotti (2005). Diallel analysis in cassava adapted to the midaltitude valleys environment. *Crop Sci.* 45:1058–63.

Jennings, D. L. (1957). Further studies in breeding cassava for virus resistance. *E. Afr. Agric. J.* 22: 213–19.

Jennings, D. L. (1963). Variation in pollen and ovule fertility in varieties of cassava, and the effect of interspecific crossing on fertility. *Euphytica*:12:69–76.

Jennings D. L., and C. Hershey (1985). Cassava breeding: a decade of progress from international programmes. In G. E. Russel (ed.), *Progress in Plant Breeding*. Butterworths Press. London and Boston, pp. 89–116.

Jennings, D. L., and C. A. Iglesias (2002). Breeding for crop improvement. In R. J. Hillocks, J. M. Thresh and A. C. Bellotti. (eds), *Cassava: Biology, Production and Utilization*. CABI Publication. Wallingford, UK and New York, USA, pp. 149–66.

Joaqui Barandica, O., J. I. Lenis, F. Calle, N. Morante, J. C. Pérez, C. Hershey and H. Ceballos (2016). Cassava breeding II: phenotypic correlations through the different stages of selection. *Front. Plant Sci.* (in press).

Karlström, A., F. Calle, S. Salazar, N. Morante, D. Dufour and H. Ceballos (2016). Biological implications in cassava for the production of amylose-free starch: impact on root yield and related traits. *Front. Plant Sci.* 7 doi: 10.3389/fpls.2016.00604.

Kawano, K. (1980). Cassava. In W. R. Fehr and H. H. Hadley (eds), *Hybridization of Crop Plants*. ASA, CSSA. Madison, WI, pp. 225–33.

Kawano, K. (2003). Thirty years of cassava breeding for productivity – biological and social factors for success. *Crop Sci.* 43:1325–35.

Kawano, K., A. Amaya, P. Daza and M. Ríos (1978). Factors affecting efficiency of hybridization and selection in cassava (*Manihot esculenta* Crantz). *Crop Sci.* 18:373–6.

Kawano K., F. W. M. Gonçalves Fukuda and U. Cednpukdee. (1987), Genetics and environmental effects on dry matter content of cassava root. *Crop Sci.* 27:69–74.

Kawano, K., K. Narintaraporn, P. Narintaraporn, S. Sarakarn, A. Limsila, J. Limsila, D. Suparhan, V. Sarawat and W. Watananonta (1998). Yield improvement in a multistage breeding program for cassava. *Crop Sci.* 38:325–32.

Kawano K., and J. H. Cock (2005). Breeding cassava for underprivileged: institutional, socio-economic and biological factors for success. *J. Crop Improv.*14:197–219.

Kawuki, R. S., E. Nuwamanya, M. T. Labuschagne, L. Herselman and M. E. Ferguson (2011). Segregation of selected agronomic traits in six S1 cassava families. *J. Plant Breed. Crop Sci.* 3(8):154–60.

Koehorst-van Putten, H. J. J., E. Sudarmonowati, M. Herman, I. J. Pereira-Bertram, A. M. A. Wolters, H. Meima, N. de Vetten, C. J. J. M. Raemakers and R. G. F. Visser (2012). Field testing and exploitation of genetically modified cassava with low-amylose or amylose-free starch in Indonesia. *Transgenic Res.* 21:39–50.

Kulembeka, H. P., M. Ferguson, L. Herselman, E. Kanju, G. Mkamilo, E. Masumba, M. Fregene and M. T. Labuschagn (2012) Diallel analysis of field resistance to brown streak disease in cassava (*Manihot esculenta* Crantz) landraces from Tanzania. *Euphytica* 187:277–88.

Kvitschal, M. V., P. Soares Vidigal Fo., C. A. Scapim, M. C. Gonçalves-Vidigal, E. Sagrilo, M. G. Pequeno and F. Rimoldi (2009). Comparison of methods for phenotypic stability analysis of cassava (*Manihot esculenta* Crantz) genotypes for yield and storage root dry matter content. COMPLETAR REFERENCIA.

Lebot, V. (2010). Sweet potato. In J. Bradshaw (ed.), *Root and Tuber Crops*. Springer Publication, New York. pp. 97–125

Lokko Y., A. G. O. Dixon, S. K. Offei and E. Y. Danquah (2006). Combinning ability analysis of resistance to mosaic virus disease in cassava. *Afr. Crop Sci.* 14(3):221–30.

Lozano, J. C., J. H. Cock and J. Castaño (1978). New developments in cassava storage. In T. Brekelbaum, A. Bellotti and J. C. Lozano (eds), Proceedings Cassava Protection Workshop. CIAT, Cali, Colombia, pp. 135–41.

Magoon, M. L., R. Krishnan and K. V. Bai (1969). Morphology of the pachytene chromosomes and meiosis in *Manihot esculenta* Crantz. *Cytologia* 34:612–26.

Manu-Aduening, J. A., R. I. Lamboll, G. Ampong Mensah, J. N. Lamptey, E. Moses, A. A. Dankyi and R. W. Gibson (2006). Development of superior cassava cultivars in Ghana by farmers and scientists: the process adopted, outcomes and contributions and changed roles of different stakeholders. *Euphytica* 150:47–61.

Maziya-Dixon, B., A. G. O. Dixon and A. R. A. Adebowale (2007). Targeting different end uses of cassava: genotypic variations for cyanogenic potentials and pasting properties. International *J. Food Sci. Technol.* 42:969–76.

McMahon, J. M., W. L. B. White and R. T. Sayre (1995). Cyanogenesis in cassava (*Manihot esculenta* Crantz). *J. Expt. Bot.* 46:731–41.

Melchinger, A. E. (1999) Genetic diversity and heterosis. In J. G. Coors and S. Pandey (ed.), *Genetic and Exploitation of Heterosis in Crops,*. American Society of Agronomy, Inc. Madison, WI, USA, pp. 99–118.

Melchinger, A. E., and R. K. Gumber (1998). Overview of heterosis and heterotic groups in agronomic crops. In K. R. Lamkey and J. E. Staub (eds), *Concepts and Breeding of Heterosis in Crop Plants.* Crop Science Society of America, Madison, WI, USA, pp. 29–56.

Mikel, M. A., and J. W. Dudley (2006). Evolution of North American dent corn from public to proprietary germplasm. *Crop Sci.* 46:1193–1205.

Miranda Filho, J. B. (1999). Inbreeding depression and heterosis. In J. G. Coors and S. Pandey (eds), *The Genetic Exploitation of Heterosis in Crops.* American Society of Agronomy, Madison, WI, USA, pp. 69–80.

Mkumbira J., L. Chiwona-Karltun, U. Lagercrantz, N. Meso Mahungu, J. Saka, A. Mhone, M. Bokanga, L. Brimer, U. Gullberg and H. Rosling (2003). Classification of cassava into 'bitter' and 'cool' in Malawi: from farmers' perception to characterisation by molecular markers. *Euphytica* 132:7–22.

Monger W. A., S. Seal, A. M. Isaac and G. D. Foster (2001). Molecular characterization of the cassava grown streak virus coat protein. *Plant Pathol.* 50:527–34.

Morante, N., T. Sánchez, H. Ceballos, F. Calle, J. C. Pérez, C. Egesi, C. E. Cuambe, A. F. Escobar, D. Ortiz and A. L. Chávez (2010). Tolerance to Post-harvest Physiological Deterioration in Cassava Roots. *Crop Sci.* 50:1333–8.

Morante, N., H. Ceballos, T. Sánchez, A. Rolland-Sabaté, F. Calle, C.Hershey, O. Gibert and D. Dufour (2016). Discovery of new spontaneous sources of amylose-free cassava starch and analysis of their structure and techno-functional properties. Accepted for publication in *Food Hydrocolloids* 56: 383–95.

Morillo Coronado, Y. (2009). Herencia del contenido de carotenos en raíces de yuca (*Manihot esculenta* crantz). Ph D. Diss. National University. Colombia. Palmira Campus.

Nassar, N. M. A., and R. Ortiz (2008). Cassava genetic resources: Manipulation for crop Improvement. *Plant Breed. Rev.* 31:247–75.

Nguyen T. L. T., S. H. Gheewala and S. Garivait (2007). Full chain energy analysis of fuel ethanol from cassava in Thailand. *Environ. Sci. Technol.* 41:4135–42.

Nichols, R. F. W. (1947). Breeding cassava for virus resistance. *E. Afr. Agric. J.* 12:184–94.

Njenga, P., R. Edema and J. Kamau (2014). Combining ability for beta-carotene and important quantitative traits in a cassava F1 population. *J. Plant Breed. Crop Sci.* 6(2):24–30.

Njenga, P., E. Richard and J. Kamau (2014). Combining ability for beta-carotene and important quantitative traits in a cassava F1 population. *J. Plant Breed. Crop Sci.* 6(2):24–30.

Njoku, D. N., V. E. Gracen, S. K. Offei, I. K. Asante, E. Y. Danquah, C. N. Egesi and E. Okogbenin (2014). Molecular marker analysis of F1 progeny and their partent for carotenoids inheritance in African cassava (*Manihot esculenta* Crantz). *Afr. J. Biotechnol.* 13: 3999–4007.

Njoku, D. N., V. E. Gracen, S. K. Offei, I. K. Asante, C. N. Egesi, P. Kulakow and H. Ceballos (2015). Parent-offspring regression analysis for total carotenoids and some agronomic traits in cassava. *Euphytica* DOI 10.1007/s10681-015-1482-4.

Normanha, E. S. (1970). Cassava breeding work at the São Paulo State. Agronomic Institute, Campinas, Brazil. In Trabalhos do Encontro de Engenheiros Agrônomos do Estado de São Paulo, Brazil, pp. 40–7.

Olsen, K. M., and B. A. Schaal (2001). Microsatellite variation in cassava (Manihot esculenta, Euphorbiaceae) and its wild relatives: further evidence for a southern Amazonian origin of domestication. *Am. J. Bot.* 88(1):131–42.

Ortiz, D., T. Sánchez, N. Morante, H. Ceballos, H. Pachón, M. C. Duque, A. L. Chávez and A. F. Escobar (2011). Sampling strategies for proper quantification of carotenoids content in cassava breeding. *J. Plant Breed. Crop Sci.* 3(1):14–23.

Owolade, O. F., A. G. O. Dixon and A. Y. A. Adeoti (2006). Diallel analysis of cassava genotypes to anthracnose disease. *World J. Agric. Sci.* 2(1):98–104.

Parkes, E. Y., M. Fregene, A. Dixon, B. B. Peprah and M. T. Labuschagne (2013). Combining ability of cassava genotypes for cassava mosaic disease and cassava bacterial blight, yield and its related components in two ecological zones in Ghana. *Euphytica* 194:13–24.

Parsa, S., T. Kondo and A. Winotaii (2012). The cassava mealybug (*Phenacoccus manihoti*) in Asia: first records, potential distribution, and an identification key. *PLoS ONE* 7(10):e47675.

Perera, P. I. P., B. Dedicova, C. Ordoñez, J. D. J. S., Kularatne, M. Quintero and H. Ceballos (2012). Recent advances in androgenesis induction of cassava (*Manihot esculenta* Crantz). *Acta Hortic.* 961:319–25.

Perera, P. I. P., C. A. Ordoñez, L. A. Becerra López-Lavalle and B. Dedicova (2013). A milestone in the doubled haploid pathway of cassava. *Protoplasma* 251:233–46.

Pérez J. C., H. Ceballos, G. Jaramillo, N. Morante, F. Calle, B. Arias and A. C. Bellotti (2005a). Epistasis in cassava adapted to mid-altitude valley environments. *Crop Sci.* 45:1491–6.

Pérez, J. C., H. Ceballos, F. Calle, N. Morante, W. Gaitán, G. Llano and E. Alvarez (2005b). Within-family genetic variation and epistasis in cassava (*Manihot esculenta* Crantz) adapted to the acid-soils environment. *Euphytica* 145(1–2):77–85.

Pérez, J. C., J. I. Lenis, F. Calle, N. Morante, T. Sánchez, D. Debouck and H. Ceballos (2011). Genetic variability of root peel thickness and its influence in extractable starch from cassava (Manihot esculenta Crantz) roots. *Plant Breed.* 130:688–93.

Pujol, B., G. Gigot, G. Laurent, M. Pinheiro-Kluppel, M. Elias, M. Hossaert-McKey and D. McKey (2002). Germination ecology of cassava (*Manihot esculenta* Crantz, Euphorbiaceae) in traditional agroecosystems: seed and seedling biology of a vegetatively propagated domesticated plant. *Econ. Bot.* 56:366–79.

Rabbi, I., M. Hamblin, M. Gedil, P. Kulakow, M. Ferguson, A. S. Ikpan, D. Ly and J.-L. Jannink (2014a). Genetic Mapping Using Genotyping-by-Sequencing in the Clonally Propagated Cassava. *Crop Sci.* 54:1384–96.

Rabbi, I. Y., M. T. Hamblin, P. Lava Kumar, M. A. Gedil, A. S. Ikpan, J.-L. Jannink, and P. A. Kulakow (2014b) High-resolution mapping of resistance to cassava mosaic geminiviruses in cassava using genotyping by sequencing and its implications for breeding. *Virus Res.* 186:87–96.

Rajendran, P. G., C. S. Ravindran, S. G. Nair and T. V. R. Nayar (2000). True cassava seeds (TCS) for rapid spread of the crop in non-traditional areas. Central Tuber Crops Research Institute, Indian Council of Agricultural Research Thiruvananthapuram, Kerala, India.

Reilly, K., D. Bernal, D. F. Cortes, R. Gomez-Vasquez, J. Tohme and J. R. Beeching (2007). Towards identifying the full set of genes expressed during cassava post-harvest physiological deterioration. *Plant Mol. Biol.* 64:187–203.

Rogers, D. J., and S. G. Appan (1973). Manihot and manihotoides (Euphorbiaceae). *A Computer-Assisted Study.* Flora Neotropica. Monograph No. 13. Hafner Press, New York, 272p.

Rojas-C. M., J. C. Pérez, H. Ceballos, D. Baena, N. Morante and F. Calle (2009). Introduction of inbreeding and analysis of inbreeding depression in eight S1 cassava families. *Crop Sci.* 49:543–8.

Rolland-Sabaté, A., T. Sánchez, A. Buléon, P. Colonna, B. Jaillais, H. Ceballosand D. Dufour (2012). Structural characterization of novel cassava starches with low and high-amylose contents in comparison with other commercial sources. *Food Hydrocolloids* 27:161–74.

Rolland-Sabaté, A., T. Sanchez, A. Buléon, P. Colonna, H. Ceballos, S.-S. Zhao, P. Zhang and D. Dufour (2013). Molecular and supra-molecular structure of waxy starches developed from cassava (*Manihot esculenta* Crantz). *Carbohydrate Polym.* 92:1451–62.

Sagrilo, E., P. Soares Vidigal Fo., M. G. Pequeno, M. C. Gonçalves-Vidigal, and M. V. Kvitschal (2008). Dry matter production and distribution in three cassava (*Manihot esculenta* Crantz) cultivars during the second vegetative plant cycle. *Braz Arch Biol. Technol.* 51(6):1079–87.

Salick, J., N. Cellinese and S. Knapp (1997). Indigenous diversity of cassava: generation, maintenance, use and loss among Amuesha, Peruvian upper Amazon. *Econ. Bot.* 51:6–19.

Sambatti, J. B. M., P. S. Martins and A. Ando (2001). Folk taxonomy and evolutionary dynamics of cassava: a case study in Ubatuba, Brazil. *Econ. Bot.* 55:93–105.

Sánchez, T., D. Dufour, I. X. Moreno and H. Ceballos (2010). Commparison of pasting and gel stability of waxy and normal starches from cassava, potato, maize, and rice under thermal, chemical and mechanical stress. *J. Agric. Food Chem.* 58:5093–9.

Sánchez, T., D. Dufour, J. L. Moreno, M. Pizarro, I. Arango, M. Domínquez and H. Ceballos (2013). Changes in extended shelf life of cassava roots during storage in ambient conditions. *Postharvest Biol. Technol.* 86:520–8.

Sánchez, T., H. Ceballos, D. Dufour, D. Ortiz, N. Morante, F. Calle, T. Zum Felde and F. Davrieux (2014). Carotenoids and dry matter Prediction by NIRS and hunter color in fresh cassava roots. *Food Chem.* 151:444–51.

Sanni, L. O., A. A. Adebowale, B. Mayiza-Dixon and A. G. O. Dixon. (2008). Chemical composition and pasting properties of CMD resistant cassava clones planted at different locations. *J. Food Agric. Environ* 6(2):97:104.

Spielman, D. J., D. E. Kolady and P. S. Ward (2013). The prospects for hybrid rice in India. *Food Secur.* 5:651–65.

Sriroth, K., V. Santisopasri, C. Petchalanuwat, K. Kurotjanawong, K. Piyachomkwan and C. G. Oates (1999). Cassava starch granule structure-function properties: influence of time and conditions at harvest on four cultivars of cassava starch. *Carbohydr. Polym.* 38:161–70.

Sriroth, K., K. Piyachomkwan, S. Wanlapatit and S. Nivitchanyong (2010). The promise of a technology revolution in cassava bioethanol: From Thai practice to the world practice. *Fuel* 89:1333–8.

Ssemakula, G., and A. Dixon (2007). Genotype x environment interaction, stability and agronomic performance of carotenoid-rich cassava clones. *Sci. Res. Essay* 2(9):390–9.

Stapleton, G. (2012). Global starch market outlook and competing starch raw materials for 732 starches by product segment and region. Cassava Starch World 2010. Centre for Management Technology (CMT). Phnom Penh, Cambodia 22–24 February.

Stuber, C. W. (1994). Heterosis in plant breeding. *Plant Breed. Rev.* 12:227–51.

Uarrota, V. G., E. Nunes, L. Peruch, E. Neubert, B. Coelho, R. Moresco, M. Domínguez, T. Sánchez, J. Meléndez, D. Dufour, H. Ceballos, L. Becerra Lopez-Lavalle, C. Hershey, M. Rocha and M. Maraschin (2015). Toward better understanding of postharvest deterioration: Biochemical changes in stored cassava (*Manihot esculenta* Crantz) roots. *Food Sci. Nutr.* DOI: 10.1002/fsn3.303.

van Oirschot, Q. E. A., G. M. O'Brien, D. Dufour, M. A. El-Sharkawy and E. Mesa (2000). The effect of pre-harvest pruning of cassava upon root deterioration and quality characteristics. *J. Sci. Food Agric.* 80:1866–73.

Vandegeer, R., R. E. Miller, M. Bain, R. M. Gleadow and T. R. Cavagnaro (2013). Drought adversely affects tuber development and nutritional quality of the staple crop cassava (*Manihot esculenta* Crantz). *Funct. Plant Biol.* 40:195–200.

Wang, C., Z. Lentini, E. Tabares, M. Quintero, H. Ceballos, B. Dedicova, C. Sautter, C. Olaya and Z. Peng (2011). Microsporogenesis and pollen formation in cassava (*Manihot esculenta* Crantz). *Biologia Plantarum* 55(3):469–78.

Were, W. V., P. Shanahan, R. Melis, O. O. Omari (2012) Gene action controlling farmer preferred traits in cassava varieties adapted to mid-altitude tropical climatic conditions of western Kenya. *Field Crop Res.* 133:113–18.

Westwood, N. N. (1990). Maintenance and storage: clonal germplasm. *Plant Breed. Rev.* 7:111–28.

Wheatley, C. C., and G. Chuzel (1995). Cassava: the nature of the tuber and use as a raw material. In R. Macrae, R. K. Robinson and M. J. Sadler (eds), *Encyclopaedia of Food Science, Food Technology and Nutrition*. Academic Press, San Diego, CA, pp. 734–43.

Wolfe M. D., P. Kulakow, I. Y. Rabbi and J.-L. Jannink (2015). Marker-based estimates reveal significant non-1 additive effects in clonally propagated cassava (*Manihot esculenta*): implications for the prediction of total genetic value and the selection of varieties. *bioRxiv* doi: http://dx.doi.org/10.1101/031864.

Zacarias, A. M. and M. T. Labuschagne (2010). Diallel analysis of cassava brown streak disease, yield and yield related characteristics in Mozambique. *Euphytica* 176:309–20

Zehntner, L. (1919). Estudo sobre algumas variedades de mandiocas Brasileiras. Sociedad Nacionale de Agricultura, Impresa Ingleza-Camerino 61, Rio de Janeiro, Brazil.

Zhao, S., H. Ceballos, D. Dufour, T. Sánchez and P. Zhang (2011). Development of waxy cassava with different biological and physico-chemical characteristics of starches for industrial applications. *Biotechnol. Bioeng.* 108(8):1925–35.

Zidenga, T., E. Leyva-Guerrero, H. Moon, D. Siritunga and R. Sayre (2012). Extending cassava root shelf life via reduction of reactive oxygen species production. *Plant Physiol.* 159:1396–1407.

Chapter 4

Molecular approaches in cassava breeding

Luis Augusto Becerra Lopez-Lavalle, International Center for Tropical Agriculture (CIAT), Colombia

1 Introduction

In cassava, as for other major crops, molecular markers can reveal the genetic basis for both qualitative and quantitative phenotypic changes. These molecular markers are chromosome landmarks that tag genes or gene regions that to a large extent contribute to a desired phenotype. They indicate differences in nucleotide sequences that can be uncovered by interrogating the crop's DNA using PCR and non-PCR approaches. DNA markers are abundant and many of them show high polymorphism levels. They are extremely useful to breeders because they are not influenced by the environment, thus allowing the indirect selection for a trait in the absence of ideal conditions, that is, where the trait is not expressed reliably due to absence of an impact factor such as pests or pathogens, or because of genotype-by-environment (G x E) interactions.

Since the advent of DNA-based molecular marker technologies, cassava scientists have used them to describe the crop's genetic diversity and how these markers link genetically to define chromosome maps. Thus gene sequence variation such as amplified fragment length polymorphism (AFLPs), and short sequence repeats (SSRs) have been extensively used to elucidate and describe the crop's diversity, domestication and ancestry (Carvalho and Schaal 2001; Colombo et al. 2000; Duputié et al. 2011; Fregene et al. 1994, 2003; Olsen and Schaal 2001; Roa et al. 1997, 2000). Molecular markers such as SSRs and AFLPs have been used in the construction of several genetic linkage maps (Fregene et al. 1997; Mba et al. 2012; Okogbenin et al. 2006). However, these technologies in cassava have been recently superseded by the identification of a large number of single-nucleotide polymorphism (SNP) variants via next-generation sequencing (Rabbi et al. 2014; Soto et al. 2015).

Both genetic mapping and gene tagging in cassava have been undertaken with the purpose of providing tools that can increase the cost-effectiveness and efficiency of cassava

http://dx.doi.org/10.19103/AS.2016.0014.16

breeding. Target traits include resistance to pests and diseases, traits expressed only at the end of the crop's growing cycle (e.g. root quality factors), and those for which phenotype is difficult to measure (e.g. where environments are unpredictable, as for drought). Molecular markers have been used to tag three different sources of cassava mosaic disease (CMD) resistance originally found in *Manihot glaziovii*, TME3 and TMS97/2205 (Akano et al. 2002; Fregene et al. 2000). Two SSR markers have been found associated with CGM resistance (NS1009 and NS346). About six markers were found to be associated with cassava bacterial blight (CBB), explaining 9–27% of the phenotypic variance of response to five Xam strains (Jorge et al. 2001). Early root bulking, a trait of increasing priority for many programmes, is a quantitative trait in cassava where molecular markers could help breeders accelerate selection. Towards this goal, three QTLs explaining 25–33% of phenotypic variance were found for foliage dry weight production, while five other QTLs associated with harvest index, explaining 18–27% of phenotypic variance, were identified (Okogbenin and Fregene 2002). Deployment will require further investment. Several other gene tagging projects have since been conducted or are on-going in cassava at CIAT for other traits such as PPD, whiteflies, and β-carotene. Except for resistance to CMD and cassava green mites (CGM), markers identified for other traits are yet to be further tested and validated. These markers need to be fine mapped to the genomic regions enhance their application in MAS (Rabbi et al. 2014).

The rapid change in DNA-screening technology and the reduction in the costs of DNA sequencing – by several orders of magnitude – is greatly increasing the scale at which cassava geneticists can pursue SNP discovery, at levels previously accessible only to major commodity crops. Recently, two initiatives towards completing the saturation of the cassava genetic map have resulted in the incorporation of expressed sequence tags (ESTs) and next-generation sequencing SNPs in a consensus linkage map using the K family of Fregene et al. 1997 (Soto et al. 2015) and the development of a consensus 2412-cM map using 22 403 SNP markers that integrate 10 biparental maps (comprising 3480 meiosis). Furthermore, this multiple biparental consensus map was used to anchor 71.9% of the draft genome assembly and 90.7% of the predicted protein-coding genes. Chromosome-anchored genome sequences will be useful for cassava breeding by assisting in the rapid identification of markers linked to important traits (ICGMC 2015).

2 Genetic diversity

The first DNA phylogenetic relationships within *Manihot* species was obtained using RFLPs (Haysom et al. 1994). This study revealed little variation within species and a close relationship between *M. chlorosticta* and *M. esculenta*. Later, Olsen and Schaal (2001) used microsatellites and hypothesized *M.e. flabellifolia* as a possible ancestor of cultivated *Manihot*. Similar work conducted at CIAT by Roa and co-workers (1997), using AFLP markers, assessed the crop's diversity against six wild taxa, and also concluded that not only *M.e. flabellifolia*, but also *M. trisitis* are more closely related to cassava than its Mexican relative *M. aesculifolia,* indicating that cassava's origin might be linked to these two species. Independently, through AFLP evaluation of a large collection of cassava from the South American centre of diversity found at CIAT, Second and collaborators (1997) proposed *M. peruviana* and *M.e. flabellifolia* as the closest wild ancestors. Their work also showed evidence of introgression into cassava from *Ceará* or *India rubber* (*M. glaziovii*). While the history of systematic introgression of *M. glaziovii* into the African cassava germplasm by breeders is well documented, the genomic consequences have only recently been reported (Bredeson et al. 2016).

Molecular markers have also been used to obtain a quantitative assessment of genetic similarity among cassava clones (Beeching et al. 1993; Elias et al. 2000). Similarly, markers have been used to study genetic structure and the basis of genetic differentiation of cassava landraces in Africa (Mkumbira et al. 2003; Rabbi et al. 2015), and the genetic structure of germplasm resistant to diseases (Fregene et al. 2000; Sanchez et al. 1999).

Markers have also revealed intra-varietal polymorphism, indicating that a variety can be made up of more than one genotype (Elias et al. 2000; Rabbi et al. 2015). These multi-clonal varieties probably usually develop through farmers incorporating new genotypes that arise from true seed, or are introduced from other farms or village, into existing varieties having similar appearance and behaviour. Markers have been used to study the effect of disease on genetic diversity in cassava. Kizito and co-workers (2005) reported the loss of rare alleles in areas with high CMD incidence in Uganda, and finding no evidence for a differentiation of the cassava gene pool into a western and an eastern genetic lineage. Similarly, CIAT compared the genetic diversity of New World and African cultigens, using an SSR diversity kit of 36 SSR markers previously developed by Fregene and collaborators (2004), and presently superseded by a Fluidigm Dynamic Array™ 180 SNP-array (Cassava SNPY-array) (Becerra Lopez-Lavalle, unpublished). The Cassava SNPY-array is currently being used by CIAT's cassava programme to identify landraces and improved cultigens in farmers' fields and in breeding programmes. Other applications include the use of Cassava SNPY-array for the identification of duplicates in CIAT's germplasm collection (previously done using SSRs), and the diversity analysis of germplasm from the littoral and Amazonian regions of Colombia (Peña-Venegas et al. 2014).

3 Marker-assisted selection (MAS)

The use of MAS in cassava breeding has the sole purpose to facilitate the selection and introgression of traits more effectively and more economically compared to phenotypic selection. There are especially comparative advantages of MAS for traits that are difficult (e.g. postharvest physiological deterioration) or expensive (e.g. resistance to whiteflies or green mites) to evaluate in field conditions. Thus molecular markers are particularly useful where pathogen or pest pressure is low, variable, erratic or just absent. To be useful for MAS, markers must adequately account for a large proportion of the genetic variance and must increase the efficiency of the breeding schemes by shortening the time for identification of plants that carry genes for the desired trait. These conditions enable breeders to make decisions using much smaller populations in the field, which in turn reduces the cost of breeding, but more importantly, enables larger populations to be produced and analysed to achieve a higher rate of genetic gain.

3.1 Case study: cassava mosaic disease (CMD)

CMD and its variants are prevalent both in Africa and South Asia (Swanson and Harrison 1994). For cassava in Africa, CMD is the most important disease, given the high evolutionary capacity of the virus, the widespread prevalence of the whitefly vector, Yield losses from CMD have reached over a billion US dollars in Africa alone. In the early 2000s, at CIAT, the discovery of a molecular marker linked to a dominant gene (CMD2) allowed for the first time the selection of CMD-resistant genotypes in the absence of the pathogen (Akano et al. 2002). Five molecular markers were initially identified for this gene, with the closest being RME1 (a SCAR marker) and NS 158 (an SSR marker) at distances of four and seven cM respectively. MAS has been carried out using multiple flanking markers involving these two markers with 68% efficiency (Okogbenin et al. 2007). Currently at CIAT, we are in the process of fine mapping of the CMD2 region to identify closely linked markers and attempt positional cloning of the gene. This approach is possible thanks to the construction of BAC libraries using the African landrace TME3 (highly resistant to CMD), which is available at the Clemson University Genomic Institute (CUGI). The BAC clones containing the region with the flanking markers have been screened and characterized at CIAT. Preliminary results allowed the identification of SSCP-SNP markers which have been located between RME1 and the CMD2. Current efforts are being directed towards screening for new sources of CMD resistance in African germplasm (Rabbi et al. 2014). Where different sources of CMD resistance exist, this is indistinguishable phenotypically, and markers are being used to screen identified genotypes in the breeding populations in Africa. Under the Genotyping Support Services activities of the Generation Challenge Program (a discontinued initiative of the CGIAR system), crosses were carried out using six parents, resulting in the development of nine F_1 segregating populations. Four of the populations were selected and phenotypically evaluated and analysed with 530 SSR markers to identify new sources of CMD resistance different from CMD2.

3.2 Case study: cassava green mite (CGM)

Two markers were found to be associated with the resistant response against CGM attack (NS1009 and NS346) and have been used in MAS as part of various validation studies (Macea Choperena et al. 2012). At CIAT, four families were derived from a cross between cassava and an accession of *Manihot esculenta* sub spp. *flabellifolia* (CW68, CW65, CW67, and CW66). Preliminary evaluations showed that progenies selected with these markers showed good resistance to the pest in East Africa but not in Nigeria where only moderate tolerance was observed for CGM based on the markers. The phenotypic differences between these African sub-regions might be due to variation in the CGM pressure, which is higher in Umudike, Nigeria, where a long dry season of 4–5 months in the humid transition ecology often results in high CGM incidence. By comparison pressure is less severe in Tanzania and Uganda. In Tanzania, local cassava varieties selected by farmers from the three major agro-ecologies, namely Eastern, Lake and Southern Zones have been crossed with genotypes from CIAT with resistance to the CMD, CBB and CGM. Molecular markers for CMD, CBB and CGM will be employed to identify the resistant F_1 progenies.

3.3 Case study: introgression of useful traits from wild relatives

Wild *Manihot* germplasm theoretically offers a wide array of useful genes frequently related to tolerance to biotic and abiotic stresses (Nassar 1980, 2000). However, the use of wild relatives is cumbersome and will require reducing or completely eliminating a large proportion of the wild donor genome, apart from the few genes of interest. This unavoidably will create *linkage drag* – the difficulty of breaking linkages between favourable and undesired genes after crossing between wild and cultivated species. This phenomenon often creates more problems for the breeder than the gain from transfer of new traits.

In cassava, a desirable trait may sometimes be seen as early as in the F_1 progeny, indicating that dominant or additive genes are controlling the traits. Introgressing desired traits back into the cultivated cassava will require intensive backcrossing to the cultivated type. The situation in cassava is further complicated by the fact that breeders cannot use a single variety as a recurrent parent, as it would result in rapid approach to homozygosity and the concomitant inbreeding depression (see Chapter 15). A modified backcrossing is required, where different adapted genotypes are used as the recurrent parent base.

Simulation by Stam and Zeven (1981) indicated that the use of molecular markers could reduce linkage drag and the number of generations required in a backcross scheme. Hospital and collaborators (1992) experimentally corroborated this in achieving a reduction of two backcross generations with the use of molecular marker selection. Similarly, Frisch and co-workers (Frisch et al. 1999), through a simulation study, found that use of molecular markers for the introgression of a single target allele saved two to four backcross generations. Both authors agreed that marker-assisted selection had the potential to reach the same level of recurrent parent genome representation in generation BC_3 as is reached in BC_7 without molecular markers. However, the above studies are based on schemes where the recurrent progenitor is homozygous, a condition that is not yet possible for cassava, as described above.

4 Genome sequencing of cassava

The first draft of the cassava genome sequence was completed in 2009 with the support of 454 Life Sciences using the Genome Sequencer FLX platform with long read GS FLX Titanum chemistry to rapidly generate the DNA sequence data (Prochnik et al. 2012). The cassava sequencing project included several participating institutions such as the International Laboratory for Tropical Agricultural Biotechnology at the Danforth Plant Science Center in St Louis, USDA laboratory in Fargo, ND; Washington University St Louis; University of Chicago; The Institute for Genomic Research (TIGR); Missouri Botanical Garden; the Broad Institute; CIAT, IITA and Ohio State University.

More than 61 million reads were generated and assembled in a draft genome, which covers 69% of the predicted genome and contains an estimated 96% of cassava's protein-coding gene space. The predicted 30 666 genes and 3485 alternate splice forms are supported by 1.4 M ESTs available from the wide cassava research community. The annotated draft genome sequence is available at DOE JGI's phytozome website (https://phytozome.jgi.doe.gov/pz/portal.html#!info?alias=Org_Mesculenta). It is one of the first large genome projects to primarily use 454 Life Sciences long read sequencing platform which enabled both improved quality of the sequence and its rapid generation.

The availability of a cassava reference genome allows researchers to take advantage of comparative genomic approaches. Moreover, molecular breeding tool development will be greatly accelerated by building upon the genome sequence. It will open a unique opportunity to address an enhanced use of the genetic resources across both cultivated varieties and wild relatives that, once re-sequenced, can be used to unravel and exploit novel traits. This invaluable resource not only will help implement more classical marker-assisted selection, but will also allow better exploitation of genomic selection. Genomic selection approaches for cassava are under development through an initiative funded by the Bill and Melinda Gates Foundation to Cornell University, IITA, NRCRI (Nigeria) NaCRRI (Uganda) and others (Ly et al. 2013). The reference map is also a key tool for genome-wide association studies (Wolfe et al. 2015). These resources provide better understanding of key traits and processes such as starch and protein biosynthesis, photosynthate partitioning for storage roots, and stress control mechanisms. The tools will enable cassava improvement, and will also shed light on such mechanisms shared by other important related species, including the rubber tree, castor bean and Jatropha (oil crop).

The first attempt to elucidate the genome organization of cassava, its evolution and the mechanisms involved was undertaken by Wang and co-workers (2014). They identified 1584 and 1678 gene models specific to the wild relatives and domesticated varieties, respectively, and discovered high heterozygosity and millions of single-nucleotide variations. Also, they identified genes involved in photosynthesis, starch formation and abiotic stresses being positively selected, while those involved in cell wall biosynthesis and secondary metabolism, including cyanogenic glucoside formation, have been negatively selected in the cultivated varieties, reflecting the result of natural selection and domestication.

Since the first release of the cassava genome, CIAT – mainly through a project supported by the Root, Tuber and Banana programme of CGIAR – has undertaken a large resequencing approach to understand the overall genetic diversity of the landraces found in Latin America. This research provides the basis for developing molecular tools for accelerating breeding for this crop. The array of tens of thousands of megabases sequenced currently is expected to increase our knowledge of the molecular interactions mediating growth and development processes in cassava. Functional genomics research will be further accelerated by the wealth of information to be made available. The first application will be in the development of SNPs for mapping with huge potential for the genetic analysis of complex traits. Current efforts are underway to identify a large collection of high-quality SNPs which will be integrated with genome sequence and annotation in order to make a comprehensive genomic resource available to the cassava community.

These resources will be made available via the cassava genome hub (http://www.cassavagenome.org/) providing a number of tools and resources that allows harnessing the cassava genome sequencing efforts. The cassava genome hub provides essential bioinformatics tool to improve Community Annotation System, phylogenomics resources and metabolic pathways.

5 Genetic engineering in cassava

The prerequisite for successful genetic engineering of cassava is having at hand an efficient transformation and tissue culture system that allows the selection and regeneration of transgenic plants. In 1996, Li and collaborators at ETH Zurich, Zentrum successfully developed the first protocol for cassava transformation after co-cultivation with *Agrobaterium tumefaciens*. Regeneration of transgenic shoots was achieved via organogenesis from somatic embryogenesis explants grown on hygromycin or geneticin expressing the *uidA*. Schöpke and co-workers (1996) published a successful protocol for cassava transformation after plasmid bombardment with paramomycin-containing genes encoding for *nptII* and *uidA*. In the same year Remarker and co-workers (1996) successfully transferred the luciferin marker. Four distinct transformation systems have been developed for cassava. The pros and cons for each have been discussed in some depth (Fregene and Puounti-Kaerlas 2002; Raemakers et al. 1997; Taylor et al. 2002).

Regardless of the method used for achieving transgenic events, a critical step towards successful adoption of transgenic cassava by smallholder farmers is the adequate selection of traits that farmers themselves consider to be beneficial, and that are transferred to the appropriate germplasm. In cassava there is a pressing need to develop capacity to transform the most important landraces and elite materials for each of the most important cassava growing regions (Taylor et al. 2002). One of the main remaining challenges to overcome is the development of an efficient genotype-independent transformation system. This goal remains elusive for cassava. Today, transgenic plants have been obtained from nine cultigens (COL22, COL1505, COL2215, PER183, ICA-Negrita, TMS60444, Bonoua Rouge, L12 and Adira-4).

This technology has been used to develop genetically transformed cassava expressing traits with agronomic and economic value for cassava growers or consumers. Progress has been made with traits such as reduced cyanogenic content (CYP79DI and CYP79D2), insect resistance (cry genes), CMD resistance, Cassava brown streak disease resistance, Herbicide resistance, manipulation of the starch content, high protein and b-carotenoid content.

6 References

Akano, A. O., Dixon, A. G. O., Mba, C., Barrera, E. and Fregene, M. 2002. Genetic mapping of a dominant gene conferring resistance to cassava mosaic disease. *Theor. Appl. Genet.* 105(4): 521–5.

Beeching, J. R., Marmey, P., Gavalda, M.-C., Noirot, M., Haysom, H. R., Hughes, M. A. and Charrier, A. 1993. An assessment of genetic diversity within a collection of cassava (*Manihot esculenta* Crantz) germplasm using molecular markers. *Ann. Bot.* 72(6): 515–20. doi:10.1006/anbo.1993.1139.

Bredeson, J. V., Lyons, J. B., Prochnik, S. E., Wu, G. A., Ha, C. M., Edsinger-Gonzales, E., Grimwood, J., Schmutz, J., Rabbi, I. Y., Egesi, C., Nauluvula, P., Lebot, V., Ndunguru, J., Mkamilo, G., Bart, R. S., Setter, T. L., Gleadow, R. M., Kulakow, P., Ferguson, M. E., Rounsley, S. and Rokhsar, D. S. 2016. Sequencing wild and cultivated cassava and related species reveals extensive interspecific hybridization and genetic diversity. *Nat. Biotech.* 34(5): 562–70. doi:10.1038/nbt.3535. http://www.nature.com/nbt/journal/v34/n5/abs/nbt.3535.html#supplementary-information.

Carvalho, L. J. C. B. and Schaal, B. A. 2001. Assessing genetic diversity in the cassava (*Manihot esculenta* Crantz) germplasm collection in Brazil using PCR-based markers. *Euphytica* 120(1): 133–42

Colombo, C., Second, G. and Charrier, A. 2000. Genetic relatedness between cassava (*Manihot esculenta* Crantz) and *M. flabellifolia* and *M. peruviana* based on both RAPD and AFLP markers. *Genet. Mol. Biol.* 23(2): 417–23.

Duputié, A., Salick, J. and McKey, D. 2011. Evolutionary biogeography of *Manihot* (Euphorbiaceae), a rapidly radiating Neotropical genus restricted to dry environments. *J. Biogeogr.* 38(6): 1033–43. doi:10.1111/j.1365-2699.2011.02474.x.

Elias, M., Panaud, O. and Robert, T. 2000. Assessment of genetic variability in a traditional cassava (*Manihot esculenta* Crantz) farming system, using AFLP markers. *Heredity* 85(3): 219–30.

Fregene, M., Angel, F., Gomez, R., Rodriguez, F., Chavarriaga, P., Roca, W., Tohme, J. and Bonierbale, M. 1997. A molecular genetic map of cassava (*Manihot esculenta* Crantz). *Theor. Appl. Genet.* 95(3): 431–41. doi:10.1007/s001220050580.

Fregene, M., Bernal, A., Duque, M., Dixon, A. and Tohme, J. 2000. AFLP analysis of African cassava (*Manihot esculenta*; Crantz) germplasm resistant to the cassava mosaic disease (CMD). *Theor. Appl. Genet.* 100(5): 678–85. doi:10.1007/s001220051339.

Fregene, M., Matsumura, H., Akano, A., Dixon, A. and Terauchi, R. 2004. Serial analysis of gene expression (SAGE) of host-plant resistance to the cassava mosaic disease (CMD). *Plant Mol. Biol.* 56(4): 563–71.

Fregene, M. and Puounti-Kaerlas, J. 2002. Cassava Biotechnology. In Hillocks, R. J., Thresh, J. M. and Bellotti, A. C. (Eds), *Cassava: Biology, Production and Utilization*. CABI Publishing: Wallingford, pp. 179–207.
Fregene, M. A., Suarez, M., Mkumbira, J., Kulembeka, H., Ndedya, E., Kulaya, A., Mitchel, S., Gullberg, U., Rosling, H., Dixon, A. G. O., Dean, R. and Kresovich, S. 2003. Simple sequence repeat marker diversity in cassava landraces: Genetic diversity and differentiation in an asexually propagated crop. *Theor. Appl. Genet.* 107(6): 1083–93. doi:10.1007/s00122-003-1348-3.
Fregene, M. A., Vargas, J., Ikea, J., Angel, F., Tohme, J., Asiedu, R. A., Akoroda, M. O. and Roca, W. M. 1994. Variability of chloroplast DNA and nuclear ribosomal DNA in cassava (*Manihot esculenta* Crantz) and its wild relatives. *Theor. Appl. Genet.* 89(6): 719–27. doi:10.1007/bf00223711.
Frisch, M., Bohn, M. and Melchinger, A. E. 1999. Comparison of Selection Strategies for Marker-Assisted Backcrossing of a Gene. *Crop Sci.* 39(5): 1295–301.
Haysom, H. R., Chan, C. T. L. and Hughes, M. A. 1994. Phylogenetic relationships of *Manihot* species revealed by restriction fragment length polymorphism. *Euphytica* 76(3): 227–34. doi:10.1007/bf00022167.
Hospital, F., Chevalet, C. and Mulsant, P. 1992. Using markers in gene introgression breeding programs. *Genetics* 132(4): 1199–210.
ICGMC, I. C. G. M. C. 2015. High-Resolution Linkage Map and Chromosome-Scale Genome Assembly for Cassava (Manihot esculenta Crantz) from 10 Populations. *Genes Genom. Genet.* 5(1): 133–44. doi:10.1534/g3.114.015008.
Jorge, V., Fregene, M., Velez, C. M., Duque, M. C., Tohme, J. and Verdier, V. 2001. QTL analysis of field resistance to *Xanthomonas axonopodis* pv. *manihotis* in cassava. *Theor. Appl. Genet.* 102(4): 564–71.
Kizito, E., Bua, A., Fregene, M., Egwang, T., Gullberg, U. and Westerbergh, A. 2005. The effect of cassava mosaic disease on the genetic diversity of cassava in Uganda. *Euphytica* 146(1–2): 45–54. doi:10.1007/s10681-005-2959-3.
Ly, D., Hamblin, M., Rabbi, I., Melaku, G., Bakare, M., Gauch, H. G., Okechukwu, R., Dixon, A. G. O., Kulakow, P. and Jannink, J.-L. 2013. Relatedness and Genotype × Environment Interaction Affect Prediction Accuracies in Genomic Selection: A Study in Cassava. *Crop Sci.* 53(4): 1312–25. doi:10.2135/cropsci2012.11.0653.
Macea Choperena, E. d. P., Ospina, C., Fregene, M., Montoya-Lerma, J. and Bellotti, A. C. 2012. Identificación de microsatélites en yuca asociados con la resistencia al ácaro Mononychellus tanajoa (Acari: Tetranychidae). *Rev. Colomb. Entomol.* 38: 70–5.
Mba, C., Guimaraes, E. P. and Ghosh, K. 2012. Re-orienting crop improvement for the changing climatic conditions of the 21st century. *Agr. Food Sec.* 1(7).
Mkumbira, J., Chiwona-Karltun, L., Lagercrantz, U., Meso Mahungu, N., Saka, J., Mhone, A., Bokanga, M., Brimer, L., Gullberg, U. and Rosling, H. 2003. Classification of cassava into 'bitter' and 'cool' in Malawi: From farmers' perception to characterisation by molecular markers. *Euphytica* 132(1): 7–22. doi:10.1023/a:1024619327587.
Nassar, N. 1980. Attempts to hybridize wild *Manihot*; species with cassava. *Econ. Bot.* 34(1): 13–15. doi:10.1007/bf02859550.
Nassar, N. 2000. Wild cassava, Manihot spp.: Biology and potentialities for genetic improvement. *Genet. Mol. Biol.* 23(1): 201–12.
Okogbenin, E. and Fregene, M. 2002. Genetic analysis and QTL mapping of early root bulking in an F_1 population of non-inbred parents in cassava (*Manihot esculenta* Crantz). *Theor. Appl. Genet.* 106(1): 58–66 https://doi.org/10.1007/s00122-002-1068-0.
Okogbenin, E., Marin, J. and Fregene, M. 2006. An SSR-based molecular genetic map of cassava. *Euphytica* 147(3): 433–40.
Okogbenin, E., Porto, M. C. M., Egesi, C., Mba, C., Espinosa, E., Santos, L. G., Ospina, C., Marín, J., Barrera, E., Gutiérrez, J., Ekanayake, I., Iglesias, C. and Fregene, M. A. 2007. Marker-Assisted Introgression of Resistance to Cassava Mosaic Disease into Latin American Germplasm for the Genetic Improvement of Cassava in Africa. *Crop Sci.* 47: 1–10.

Olsen, K. M. and Schaal, B. A. 2001. Microsatellite variation in cassava (*Manihot esculenta*, Euphorbiaceae) and its wild relatives: Further evidence for a southern Amazonian origin of domestication. *Am. J. Bot.* 88(1): 131–42.

Peña-Venegas, C., Stomph, T., Verschoor, G., Becerra Lopez-Lavalle, L. A. and Struik, P. 2014. Differences in Manioc Diversity Among Five Ethnic Groups of the Colombian Amazon. *Diversity* 6(4): 792–826.

Prochnik, S., Marri, P., Desany, B., Rabinowicz, P., Kodira, C., Mohiuddin, M., Rodriguez, F., Fauquet, C., Tohme, J., Harkins, T., Rokhsar, D. and Rounsley, S. 2012. The Cassava Genome: Current Progress, Future Directions. *Trop. Plant Biol.* 5(1): 88–94. doi:10.1007/s12042-011-9088-z.

Rabbi, I. Y., Hamblin, M. T., Kumar, P. L., Gedil, M. A., Ikpan, A. S., Jannink, J.-L. and Kulakow, P. A. 2014. High-resolution mapping of resistance to cassava mosaic geminiviruses in cassava using genotyping-by-sequencing and its implications for breeding. *Virus Res.* 186: 87–96. doi:http://dx.doi.org/10.1016/j.virusres.2013.12.028.

Rabbi, I. Y., Kulakow, P. A., Manu-Aduening, J. A., Dankyi, A. A., Asibuo, J. Y., Parkes, E. Y., Abdoulaye, T., Girma, G., Gedil, M. A., Ramu, P., Reyes, B. and Maredia, M. K. 2015. Tracking crop varieties using genotyping-by-sequencing markers: A case study using cassava (Manihot esculenta Crantz). *BMC Genetics* 16(1): 1–11. doi:10.1186/s12863-015-0273-1.

Raemakers, C. J. J. M., Sofiari, E., Jacobsen, E. and Visser, R. G. F. 1997. Regeneration and transformation of cassava. *Euphytica* 96(1): 153–61. doi:10.1023/a:1002970925897.

Raemakers, C. J. J. M., Sofiari, E., Taylor, N., Henshaw, G., Jacobsen, E. and Visser, R. G. F. 1996. Production of transgenic cassava (Manihot esculenta Crantz) plants by particle bombardment using luciferase activity as selection marker. *Mol. Breed.* 2(4): 339–49. doi:10.1007/bf00437912.

Roa, A. C., Chavarriaga-Aguirre, P., Duque, M. C., Maya, M. M., Bonierbale, M. W., Iglesias, C. and Tohme, J. 2000. Cross-species amplification of cassava (*Manihot esculenta*) (Euphorbiaceae) microsatellites: Allelic polymorphism and degree of relationship. *Am. J. Bot.* 87(11): 1647–55.

Roa, A. C., Maya, M. M., Duque, M. C., Tohme, J., Allem, A. C. and Bonierbale, M. W. 1997. AFLP analysis of relationships among cassava and other *Manihot* species. *Theor. Appl. Genet.* 95(5/6): 741–50. doi:10.1007/s001220050620.

Sanchez, G., Restrepo, S., Duque, M.-C., Fregene, M., Bonierbale, M. and Verdier, V. 1999. AFLP assessment of genetic variability in cassava accessions (Manihot esculenta) resistant and susceptible to the cassava bacterial blight (CBB). *Genome* 42(2): 163–72.

Schöpke, C., Taylor, N., Carcamo, R., Konan, N. D. K., Marmey, P., Henshaw, G. G., Beachy, R. N. and Fauquet, C. 1996. Regeneration of transgenic cassava plants (Manihot esculenta Crantz) from microbombarded embryogenic suspension cultures. *Nat. Biotech.* 14(6): 731–5. doi:10.1038/nbt0696-731.

Second, G., Allem, A. C., Mendes, R., Carvalho, L., Emperaire, L., Ingram, C. and Colombo, C. 1997. Molecular marker (AFLP)-based Manihot and cassava numerical taxonomy and genetic structure analysis in progress: Implications for their dynamic conservation and genetic mapping. *Afr. J. Root Tuber Crops* 2(1–2): 140–7.

Soto, J. C., Ortiz, J. F., Perlaza-Jiménez, L., Vásquez, A. X., Becerra Lopez-Lavalle, L. A., Mathew, B., Léon, J., Bernal, A. J., Ballvora, A. and López, C. E. 2015. A genetic map of cassava (*Manihot esculenta* Crantz) with integrated physical mapping of immunity-related genes. *BMC Genomics* 16(1): 190.

Stam, P. and Zeven, A. C. 1981. The theoretical proportion of the donor genome in near-isogenic lines of self-fertilizers bred by backcrossing. *Euphytica* 30(2): 227–38. doi:10.1007/bf00033982.

Swanson, M. and Harrison, B. 1994. Properties, relationships and distribution of cassava mosaic geminiviruses. *Trop. Sci.* 34(1): 15–25.

Taylor, N. J., Masona, M., Schopke, C. and Fauquet, C. 2002. *Transgenic Cassava for Food Security and Economic Development*. Transgenic Plants and Crops: Marcel-Dekker, New York, pp. 523–46.

Wang, W., Feng, B., Xiao, J., Xia, Z., Zhou, X., Li, P., Zhang, W., Wang, Y., Møller, B. L., Zhang, P., Luo, M.-C., Xiao, G., Liu, J., Yang, J., Chen, S., Rabinowicz, P. D., Chen, X., Zhang, H.-B., Ceballos, H., Lou, Q., Zou, M., Carvalho, L. J. C. B., Zeng, C., Xia, J., Sun, S., Fu, Y., Wang, H., Lu, C.,

Ruan, M., Zhou, S., Wu, Z., Liu, H., Kannangara, R. M., Jørgensen, K., Neale, R. L., Bonde, M., Heinz, N., Zhu, W., Wang, S., Zhang, Y., Pan, K., Wen, M., Ma, P.-A., Li, Z., Hu, M., Liao, W., Hu, W., Zhang, S., Pei, J., Guo, A., Guo, J., Zhang, J., Zhang, Z., Ye, J., Ou, W., Ma, Y., Liu, X., Tallon, L. J., Galens, K., Ott, S., Huang, J., Xue, J., An, F., Yao, Q., Lu, X., Fregene, M., Becerra Lopez-Lavalle, L. A., Wu, J., You, F. M., Chen, M., Hu, S., Wu, G., Zhong, S., Ling, P., Chen, Y., Wang, Q., Liu, G., Liu, B., Li, K. and Peng, M. 2014. Cassava genome from a wild ancestor to cultivated varieties. *Nat. Comm.* 5(5110): 1–5. doi:10.1038/ncomms6110.

Wolfe, M., Rabbi, I. Y., Egesi, C., Hamblin, M., Kawuki, R., Kulakow, P., Lozano, R., Pino Del Carpio, D., Ramu, P. and Jannink, J.-L. 2015. Genome-wide association and prediction reveals the genetic architecture of cassava mosaic disease resistance and prospects for rapid genetic improvement. Plant Genome 9(2). https://doi.org/10.3835/plantgenome2015.11.0118.

Chapter 5

Marker-assisted selection in cassava breeding

Ismail Y. Rabbi, International Institute of Tropical Agriculture (IITA), Nigeria

1 Introduction

Cassava breeding through phenotypic recurrent selection is a lengthy process, owing to its 12-month cropping cycle and the low clonal multiplication rate of about 5–10 clones/generation for every parent clone propagated. A breeding cycle takes five to six years to obtain enough planting material to conduct replicated trials across multiple locations (Kawano 2003). Although there are several tools and methods to increase multiplication rate, these are typically not practical for application to the large number of genotypes managed in a breeding programme. Moreover, cassava field plots require relatively large spacing between plants, which limits breeders' ability to adequately phenotype large numbers of entries during early stages of the selection cycle. By reducing the number of seedlings using marker-assisted selection (MAS), field breeders can allocate their limited phenotyping resources to a smaller number of selection candidates for further phenotypic evaluation of complex traits such as yield and drought tolerance.

In MAS, individuals are selected based on the presence of favourable alleles at trait-linked markers rather than the actual traits such as disease resistance, nutritional value and starch content. To realize the potential of molecular markers in cassava genetics and breeding, researchers have developed substantial genomic resources, including dense linkage maps, an annotated reference genome, catalogues of common genetic variants and identified numerous genomic regions linked to key traits. Application of MAS in cassava is still at its infancy compared to other crops and so far only very few varieties have been effectively developed through a MAS breeding scheme (Ceballos et al. 2012).

http://dx.doi.org/10.19103/AS.2016.0014.33

Despite its low use in cassava breeding, MAS remains relevant particularly since the onset of the next-generation sequencing era and the related explosion of genomic data.

Effective deployment of MAS in the cassava breeding pipeline requires careful consideration of target traits and the appropriate stage of selection during the breeding cycle. It is well-known that MAS is suitable for traits that are controlled by few major genes that explain a sufficient proportion of total genetic variation. For highly polygenic traits with low heritability, genomic selection (GS), which uses genome-wide molecular markers, is a more suitable approach (Meuwissen et al. 2001). The GS approach considers the entire complement of quantitative trait effect loci to produce genomic estimated breeding values (Poland et al. 2012). GS has become practical thanks to the increasing abundance of molecular markers that can capture many genetic loci with small effects. MAS and GS are complementary and can be used in tandem.

This chapter describes the strengths and limitations of MAS, the resources that are required and outlines the practical considerations for its implementation. It also summarizes the published quantitative trait loci (QTL) studies in cassava and gives a preview of new efforts underway.

1.1 Advantages of MAS and appropriate target traits for cassava

Since MAS can be performed at the seedling stage, it is an appropriate tool for traits that are expressed late in plant development stages, as well as those that are laborious, time-consuming and expensive to measure through conventional phenotyping. Examples include dry matter content, starch content, micronutrient density (such as carotenoid (provitamin A) content), root diseases such as cassava brown streak disease (CBSD) and post-harvest physiological deterioration.

MAS is also suitable for traits whose selection depends on specific environmental conditions for expression of the target phenotype. The conditions for good expression of such traits often occur irregularly or non-uniformly. Typical cases in cassava are pests and diseases that require presence of transmitting vectors, which in turn depend on external factors such as favourable weather conditions. For example, cassava green mite (CGM) resistance can only be detected during the dry season, when infestation levels are high. Likewise, pre-emptive breeding for resistance to virus diseases, particularly cassava mosaic disease (CMD), CBSD and cassava frog-skin disease in the absence of causal pathogens should benefit from MAS.

Traits that are determined by recessive alleles are also suitable targets for MAS. For instance, such genes include those at the downstream of the carotenoid biosynthesis pathway. Less efficient forms of these genes, when present in homozygous state, is expected to increase the amount of provitamin A in storage roots.

Finally, MAS is also suitable for stacking multiple traits in the same background or pyramiding multiple genes. Ability to genotype multiple markers linked to one or more traits on the same DNA makes MAS more cost effective. For example, multiple single genes for resistance to CMD could be combined together, or could be combined with QTLs through MAS, with the likelihood of increasing the durability and stability of resistance.

1.2 Potential shortcomings of MAS application in cassava breeding

The possibility of having false positives due to potential recombination between the marker and the gene of interest is a major challenge during implementation of MAS. However, this can be mitigated by using flanking markers on either side of the locus of interest. Another possible shortcoming includes the lack of precision in locating of QTLs, which may result in slower progress than expected, especially when sparse genetic maps and small population sizes are used. Finally, one should keep in mind that markers discovered in one population may not be transferrable to other populations, either due to lack of appropriate polymorphism or the absence of a marker-trait association.

Marker-assisted backcrossing (MAB), a form of MAS, is useful for introgressing major genes into pure lines such as inbred lines. Due to the highly heterozygous nature of cassava, MAB is not directly applicable to current cassava breeding systems (i.e. inbred lines do not yet exist). Since all cassava clones are highly heterozygous, it is impossible to reproduce the original clone due to segregation at heterozygous loci (Ceballos et al. 2012). This notwithstanding, MAS can be integrated into the routine recurrent selection breeding pipeline and not necessarily in the sense of MAB. For cassava, MAS can be used to introgress major genes into populations but not into specific genotypes as in self-pollinated crops or crops where inbred lines are developed.

2 Molecular markers as genomic resources of cassava

At the core of MAS are molecular markers. Before the advent of these markers, early breeders relied on morphological characteristics of plants and used the observable features as a marker to track the inheritance pattern of any given trait. As technologies evolved, many different ways to identify markers came into play. Variations in DNA sequences in different individuals could be detected and identified using numerous techniques. Ferguson et al. (2012a) provided a thorough review of the molecular markers in cassava. For efficient MAS, molecular markers should be inexpensive and easy to use, require small amounts of DNA, give reproducible results, occur throughout the genome and be co-dominant. The following section introduces commonly used molecular markers in cassava and describes emerging genomic resources that are contributing towards realizing the promise of marker application in the crop's improvement.

2.1 Restriction fragment length polymorphisms

Restriction fragment length polymorphisms (RFLP) are one of the earliest types of markers, which detect differences among individuals based on restriction enzyme sites. In RFLP analysis, the DNA sample is digested into fragments of varying sizes depending on the presence/absence of restriction sites. The restriction fragments are separated according to their sizes by gel electrophoresis. Though RFLP markers are co-dominant, this approach is not only low throughput but also laborious, and has subsequently become largely obsolete.

2.2 Amplified fragment length polymorphisms

Amplified fragment length polymorphisms (AFLP) represent an advancement of the RFLP approach and rely on PCR to amplify the digested fragments for subsequent detection. Typically, DNA adapters are ligated to sticky ends of restriction fragments and a subset of these fragments are then selected for PCR amplification. Amplified fragments are separated and visualized on denaturing polyacrylamide. AFLPs are dominant markers, and alleles are scored as present or absent.

2.3 Random amplification of polymorphic DNA

Random amplification of polymorphic DNA (RAPD) is an anonymous type of marker that is derived, as the name suggests, from random PCR amplification. Short arbitrary primers of 8–12 nucleotides are used in a PCR to amplify matching sequences. These are then separated in gels (either agarose or polyacrylamide) and the resulting pattern is scored as dominant markers. A major drawback of this method is its low reproducibility.

2.4 Microsatellite or simple sequence repeats

Microsatellite or simple sequence repeats (SSR) are tandem repeats of DNA motifs of two to seven nucleotides that typically recur 7–50 times. SSRs are co-dominant markers that occur at thousands of genomic locations. Due to their high mutation rates, SSRs are generally multi-allelic and have been widely deployed in kinship testing, forensic identification and population genetics. Microsatellites are usually obtained from screening sequences of genomic DNA and/or expressed sequence tags (ESTs) in databases. Alternatively, *de novo* discovery of SSRs is done through identification and sequencing of a small-insert clone library. Unique primers flanking single-copy SSRs are used to amplify SSRs and the resulting fragments are separated in agarose or polyacrylamide gels. Capillary electrophoresis can also be used to obtain accurate size of the amplified fragments and therefore resolve alleles that differ by as few as single repeats. Fluorescently labelled primers allow multiplexing of up to four different SSR markers per sample, thereby reducing the cost of a single-marker assay. A few thousand SSR markers have been developed in cassava from ESTs and enriched genomic DNA libraries (Chavarriaga-Aguirre et al. 1998; Mba et al. 2001; Raji et al. 2009; Sraphet et al. 2011). Ferguson et al. (2012b) compiled and curated these SSRs into a non-redundant set of 3367 markers anchored in the cassava reference genome.

2.5 Single-nucleotide polymorphisms

Single-nucleotide polymorphisms (SNPs) represent the most common form of naturally occurring genetic variation within the genome, occurring on average about every 100–300 bases. SNPs have rapidly gained the centre stage of molecular genetics in recent years due to their amenability to high-throughput detection chemistries and platforms. First-generation SNP markers in cassava were derived from sequencing of candidate genes, ESTs and bacterial artificial chromosome libraries (Lopez et al. 2005; Kawuki et al. 2009; Rabbi et al. 2012; Ferguson et al. 2012b). A total of 1740 SNP markers have been converted to competitive allele-specific PCR (KASP™) assays and Illumina GoldenGate™ fixed array (Ferguson et al. 2012a). A subset of these markers that segregated in a bi-parental population was subsequently placed on a genetic linkage map (Rabbi et al. 2012). An advantage of the allele-specific SNP genotyping methods is that they can accommodate varying dimensions of genotyping (i.e. any number of markers and samples) and therefore are highly amenable for MAS projects compared to fixed array methods.

2.6 Next-generation sequencing SNP markers

In recent years, a large number of next-generation SNP markers have been obtained from genotyping-by-sequencing (GBS). First described by Elshire et al. (2011), GBS is a method that enables simultaneous discovery and genotyping of a large number of SNPs on a genome-wide scale more affordably. The dense marker coverage from GBS has enabled novel genetic analyses such as population genomics and genome-wide association studies (GWAS). Typically, GBS takes advantage of the high capacity of next-generation sequencing platforms by employing restriction enzymes to reduce genome complexity and sample multiplexing. After digestion, the fragments are ligated to a unique barcode of few nucleotides that identify a specific DNA sample, pooled in equimolar amounts. This GBS library is then amplified using PCR and sequenced using next-generation sequencing technologies, usually resulting in about 100 bp single-end reads. As a result, the approach is relatively inexpensive and has become increasingly popular, even for species that do not have sufficient genomic resources such as reference genomes. Moreover, the degree of complexity reduction can be varied depending on project objectives and species by swapping the restriction enzyme and/or the level of sample multiplexing per given unit of sequencing throughput. For example, a frequent cutting enzyme like *Apek*I that recognizes 4.5 nucleotides (GCWGC, where W can be A or T) will result in more fragments being amplified for sequencing compared to a rare cutting enzyme such as *Pst*I (CTGCAG) and is therefore more suitable for organisms with relatively smaller genomes. Similarly, multiplexing fewer samples, while holding other factors constant, will result in higher sequencing depth for each amplified fragment, thereby increasing the accuracy of SNP calling. Another method that is similar to GBS is the restriction-site-associated DNA sequencing (RADseq), but which differs from GBS in that it includes a step of random shearing digested DNA fragments, followed by adapter ligation, PCR amplification and sequencing. The first GBS SNPs in cassava were generated from a *Pst*I-digested sequencing library of a bi-parental population and yielded 770 genetically mapped markers (Rabbi et al. 2014a). Subsequent GBS using an *Apek*I-based library, also in a bi-parental population, produced 6756 mapped SNPs, almost ten-fold more markers that those from *Pst*I (Rabbi et al. 2014c).

3 Other cassava genomic resources

3.1 Development of high-density genetic linkage maps

Until recently, cassava was one of the major staple crops of the world that lacked sufficient genomic resources, including genome-wide SNP marker coverage, genetic linkage maps and a reference genome – all of which are prerequisites for dissecting the number and locations of genes/QTLs that underlie traits of interest in breeding. To address this challenge, several genetic maps with between 100 and 22 403 markers have been generated (Fig. 1). The early maps with up to a few hundred markers were constructed using AFLP, RFLP, RAPD and SSRs. The rapid advances in next-generation sequencing in recent years has enabled construction of dense genetic maps with thousands of SNPs obtained from GBS. For example, a consensus map from 10 bi-parental populations with 22 403 SNPs anchored on 18 linkage groups was developed by the International Cassava Genetic Map Consortium (ICGMC), a collaboration of the University of California – Berkeley, CGIAR and National Agricultural Research Systems in Africa (International Cassava Genetic Map Consortium (ICGMC 2014).

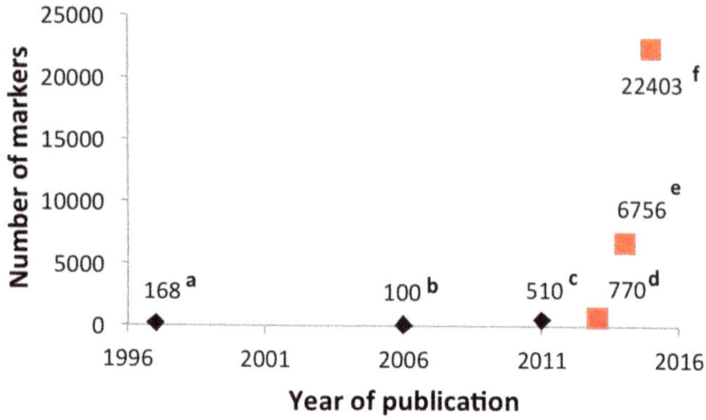

Figure 1 Graph showing the progress in development of cassava genetic linkage maps over the last two decades. The superscripts **a** to **e** denote the respective publications associated with these linkage maps in the following order: a: Fregene et al. 1997; b: Okogbenin et al. 2006; c: Sraphet et al. 2011; d: Rabbi et al. 2014a; e: Rabbi et al. 2014c; f: International Cassava Genetic Map Consortium (ICGMC) 2014. Note the increase in number of markers since 2012. This graph illustrates the progress made in the recent past and is not a comprehensive list of all cassava maps.

3.2 Reference genome

The first cassava draft reference genome of a partially inbred clone from Latin America, AM560-2, was published in 2012 (Prochnik et al. 2012). This draft genome assembly, consisting of 12 977 scaffolds along with gene annotations from mRNA sequences, was made available through the Phytozome genome browser (https://phytozome.jgi.doe.gov/pz/). This comparative genomics portal provided access for browsing, searching and downloading all available cassava sequence and annotation data and for comparative plant genomic analysis. Since the initial draft genome was highly fragmented, further efforts were expended to improve it and organize the scaffolds into the 18 cassava chromosomes. This was partially achieved through anchoring of 71.9% of the draft genome and 90.7% of predicted protein-coding genes on the consensus genetic linkage map (International Cassava Genetic Map Consortium (ICGMC) 2015). A *higher quality genome assembly* (version 6.1) with improved contiguity, linkage and completeness has been developed (Bredeson et al. 2016) and also released on Phytozome. More than 96% of 33 033 predicted protein-coding genes, based on homology and transcriptome data, are anchored to a chromosomal position.

3.3 Cassava haplotype map

Besides the reference genome, dozens of diverse global accessions of cultivated and wild *Manihot* species have been sequenced at the whole-genome level to generate a catalogue of common genetic variants in cassava germplasm (Bredeson et al. 2016). Each clone was sequenced at >30x coverage resulting in >30 million variants of which 28.38 million are SNPs and 2.14 million are indels (Ramu et al. 2016). This cassava haplotype map (HapMap) is a key community resource for GWAS, MAS and GS in cassava. To enhance its utility and impact, this database has been made publicly available through the Phytozome Plant Comparative Genomics portal, and a new database has been dedicated

(a)

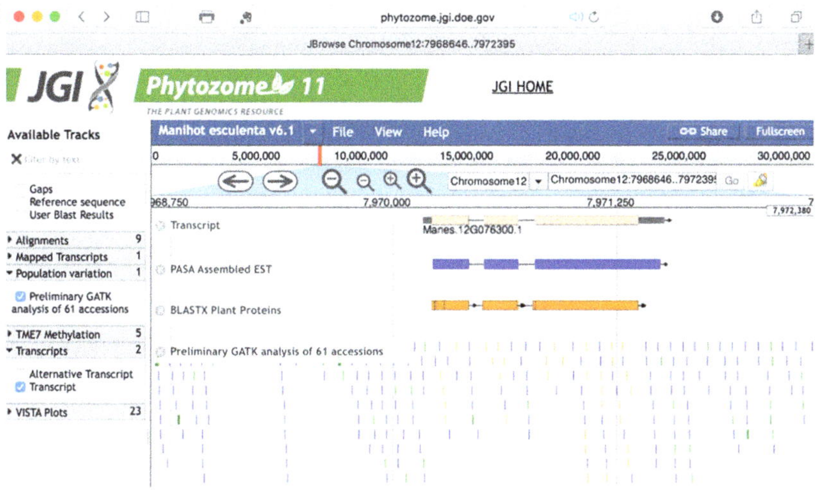

(b) **Variant Call ss.88403934 at position 7971418: SNV T -> C (score: 44832)**

Genotyped Alleles:

Sequence	Called Fraction
T (ref)	0.643
C	0.357

Called Fraction is the fraction of each allele in the total called dataset, post-filter

Genotypes (61):

T/T	31.1%
T/C	55.7%
C/C	4.9%
NA	8.2%

Sample	GT	GQ	PL	AD	DP
AM560-2	T/T	99	0, 120, 1800	T = 108	108
AR37-80	C/C	66	1446, 66, 0	T = 1; C = 36	37
AR40-6	T/C	99	740, 0, 1126	T = 33; C = 23	56
AVOCA	T/C	99	768, 0, 482	T = 15; C = 21	36
Akena	T/C	99	239, 0, 300	T = 8; C = 7	15
Albert	T/C	99	521, 0, 674	T = 29; C = 21	50
Aulizaye_Mjinga	T/T	69	0, 69, 945	T = 23	23
BRA856-5	T/C	99	3395, 0, 157	T = 14; C = 98	112
CM3306-4-5	T/T	99	0, 120, 1800	T = 49	49

Figure 2 A screen capture showing the cassava Phytozome jbrowse: (a) tracks showing chromosome and zoomed-in region, identified transcript, plant peptide annotation blast and HapMap SNP variants; (b) a pop-up screen showing the genotype alleles at a specific variant, its overall allele frequencies and genotype calls in the sequenced clones.

to cassava breeding and genomics (ftp://ftp.cassavabase.org/HapMapII/). One can, for example, browse for SNP variation in a chromosomal region or gene of interest among the sequenced clones (Fig. 2).

4 QTL mapping strategies

4.1 Mapping populations

Genetic mapping in cassava is generally carried out using segregating populations derived from crossing of heterozygous parents that differ for one or more traits of interest. The number of progenies in a bi-parental population usually ranges from 100 to 300 individuals, though this number can be larger when high-resolution mapping is desired. Alternatively, a mapping population can be derived from self-pollination of the same clone to produce an S1 segregating population. An advantage of the vegetative propagation in cassava is that the segregating progenies can be genetically fixed and multiplied at the F1 or S1 stage through cloning. In seed propagated species, both inbred or outbred, full homozygosity is required to produce homozygous or 'true-breeding' recombinant inbred lines that can be multiplied and produced without genetic changes occurring. In recent years, the process of inbreeding has been shortened in some species through use of doubled-haploids. Although there has been intensive research to produce cassava doubled-haploid plants, this landmark event has not yet been achieved.

4.2 QTL mapping

Genetic mapping of a trait of interest is based on the principle of detecting significant association between the phenotype and marker genotypes. Basically, the population is partitioned into segregating genotypic classes (2 or 3 depending on marker system and population type) at a particular marker locus. A significant difference between phenotypic means of the groups indicates that the marker locus is linked to a QTL controlling the trait. The shorter the distance between a marker and the QTL, the lower the chance of recombination (i.e. tight linkage). On the other hand, random segregation occurs when a marker is loosely linked to a QTL or even located on a different chromosome. In this case, there are significant differences between means of the genotype groups.

Four methods widely used for QTL mapping are bulk segregant analysis (BSA), single-marker analysis, interval mapping and composite interval mapping. In BSA, two bulks of DNA are prepared by pooling equal numbers of individuals from a segregating population which differ in a trait of interest (Michelmore et al. 1991). The bulks are then screened by between 300 and 500 markers. It is expected that all loci are randomized, except for the region containing the gene of interest where polymorphism between the bulks can be detected. Subsequent genotyping of individual plants in the population with polymorphic markers results in a localized linkage map (Collard et al. 2005). Most of the early QTL mapping efforts in cassava relied on this method (Akano et al. 2002; Lokko et al. 2005; Akinbo et al. 2012; Okogbenin et al. 2012).

The single-marker analysis relies on a simple method for detecting QTLs that uses several statistical tests such as *t*-test, analysis of variance (ANOVA) and linear regression, where coefficient of determination (R^2) is obtained. Like BSA, this method does not require genetic linkage maps. Its main disadvantages include its inability to identify the location of a QTL and underestimation of true QTL effects. When sparse markers are used, the power to detect QTLs is dramatically reduced.

The simple interval mapping (SIM) method relies on linkage maps and analyses intervals between adjacent markers (Lander and Botstein 1989). In this method, each locus is considered one at a time and the logarithm of the odds ratio (LOD score) is calculated for

Table 1 A summary of QTL studies carried out in cassava

Study/Traits	Parents	Markers used	Comments/Key findings
Scaphet et al. 2016; Root yield and starch content	HuayBong60× Hanatee F1 (*n* = 100)	199 SSRs and 168 EST-SSRs	Eight QTLs were associated with root yield, underlying 12.9–40.0% of the phenotypic variation (PVE). Nine QTLs were associated with starch content, underlying 11.3–27.3% of the PVE
Boonchanawiwat et al. 2016; Resistance to cassava anthracnose disease (CAD)	HuayBong 60× Hanatee F1 (*n* = 200)	512 SSRs	3 QTLs underlying the trait were identified. 23 candidate resistance genes to CAD infection were identified
Olasanmi et al. 2014; Early bulking	TMS 30572, TMS 97/2205, TMS 98/0505, TMS 30555, NR 8212 and NR 8083 F1	542 SSRs	7 markers with 10% or more coefficient of determination (R^2) were linked to major quantitative trait loci associated with early bulking in cassava
Rabbi et al. 2014c; Resistance to CMD	TMS011412× TMS4(2)1425 F1 (*n* = 180)	6756 SNPs	There was a single source of monogenic resistance in the crop's gene pool, tracing back to a common ancestral clone
Okogbenin et al. 2012; Resistance to CMD	TMS972205× NR8083 F1	530 SSRs	A marker was associated with a QTL for CMD resistance with 11% of the phenotypic variance. A new QTL (CMD3) for CMD resistance was discovered
Whankaew et al. 2011; Cyanogenic potential (CNP)	Hanatee× HuayBong60 F1 (*n* = 100)	303 SSRs	Five QTLs underlying CNP were detected. Among all the identified QTLs, CN09R1 has the greatest association with the CNP trait also with 26% phenotypic variation explained.
Boonchanawiwat et al. 2011; Plant and first branch height	Huaybong60× Hanatee F1(*n* = 100)	640 SSRs	Seven QTLs found for plant height. Five QTLs located on five linkage groups were identified for first branch height
Okogbenin et al. 2008, 43; Yield components	TMs30572× CM2177-2 F2 (*n* = 268)	122 SSRs	Single QTL model approach identified three QTLs each for dry root yield (DRY), fresh foliage yield (FFY) and harvest index. The two QTL model approach identified groups of QTLs that together explained 33% for FFY.5% for DRY and 36% for harvest index
Lopez et al. 2007; Cassava bacterial blight	TMS30572× CM2177-2 F1 (*n* = 150)	10 defence-related genes, 2 BACs from resistance candidate genes and 340 regular markers.	Uncovered two new QTLs. The largest QTL in the BAC containing resistance candidate genes explained 61.6% of phenotypic variance

(*Continued*)

Table 1 (*Continued*)

Study/Traits	Parents	Markers used	Comments/Key findings
Kizito et al. 2007; Cyanogenic glucoside potential (CNP) and high dry matter content.	MCOL1684× THAI1 S1 (*n* = 199)	110 SSRs	Two QTLs on two different linkage groups controlling CNP and six QTLs on four different linkage groups controlling DMC were identified
Akinbo et al. 2007; Cassava anthracnose disease (CAD)	TME117× TMS92/0326 F1 (*n* = 60)	53 RAPDs	Chi-square test revealed a monogenic dominant gene action. Using bulk segregant analysis, found two RAPD markers (OPAF2 and OPF06) explained 33.7 and 27.4% (P<0.05) of the CAD resistance
Wydra et al. 2004; Cassava bacterial blight	TMS30572× CM2177-2 BCI (*n* = 111)	63 SSRs	Eleven markers were identified, explaining between 16% and 33.3% of phenotypic variance of area under disease progress curve
Okogbenin and Fregene 2003; Plant architecture and productivity traits	TMS30572× CM2177-2 F1 (*n* = 144)	Fregene et al. 1997 (250 markers)	A total of 30 primary QTLs. Several genomic segments QTLs for multiple traits co-localize
Akano et al. 2002; CMD	TME3× TMS30555 F1 (*n* = 80)	186 SSRs	Used bulk segregant analysis (BSA) approach, with 40 resistant. One SSR (SSR Y28) was tightly linked with QTL. Trait was scored qualitatively
Jorge et al. 1999; Cassava bacterial blight	TMS30572× CM2177-2 F1 (*n* = 150)	68 RFLPs (female-derived map) and 15 (male-derived map)	Identified 8 QTLs in female-derived map (explaining 9–20%) and 4 QTLS on male-derived map (explaining 10.7–27.1% of variance)

the model that the given locus is a true QTL. The composite interval mapping (CIM) has become popular for mapping QTLs. This method combines interval mapping with linear regression and includes additional genetic markers in the statistical model in addition to an adjacent pair of linked markers for interval mapping (Collard et al. 2005). CIM is more precise and effective at mapping QTLs compared to single-point analysis and interval mapping, particularly when linked QTLs are involved. Commonly used software for SIM and CIM are *MapQTL* (Kyazma®) and *R/qtl* (Broman et al. 2003). Table 1 presents a summary of QTL studies carried out in cassava, population types and sizes, target traits and major findings.

In cassava, most QTL mapping has been done using the F1 segregating populations derived from bi-parental crosses. Though useful, classical QTL methods provide low resolution for the position of a QTL. In addition, only the allelic diversity sampled in the two parents of the cross can be analysed, thus the QTLs mapped in a single two-parent population often have little relevance to QTLs segregating in other populations, limiting the scope of inference of QTL studies and the application of marker-based selection in crops (Holland 2007).

5 Genome-wide association mapping

With the advent of high-density genotyping, GWAS have emerged as a powerful alternative approach for identifying QTLs. This mapping method relies on existing linkage disequilibrium between genes coding for a trait and closely linked markers to map these genes at relatively high resolution. Association mapping is usually implemented in existing germplasm collections, unlike traditional QTL mapping that requires development of segregating bi-parental populations. It is therefore possible to exploit existing datasets that are normally obtained in the course of routine germplasm characterization. Moreover, the use of broader genetic variations in GWAS enables a number of alleles and traits to be evaluated simultaneously. For these reasons, association mapping has been hailed as a more efficient way of determining the genetic basis of complex traits.

Already, GWAS has been used in cassava to study the genetic architecture of disease resistance, micronutrient density and dry matter content (Esuma et al. 2016; Wolfe et al. 2016). In an effort to determine the number and location of major loci underlying resistance to CMD, a large GWAS study using more than 6000 African breeding lines, each genotyped at more than 42 000 genome-wide SNP markers was carried out (Wolfe et al. 2016). This study revealed that resistance to this disease is controlled by gene(s) from a single genomic region that accounted for up to 66% of genetic variation. Other loci with small effects were also detected. In another study, Esuma et al. (2016) used GWAS to identify QTLs associated with provitamin A accumulation in storage roots using 591 clones and 179 310 SNP markers. They found four SNPs that were linked to a single major locus on chromosome 1 that was responsible for this trait. Similarly, a region on chromosome 11 was found to be associated with resistance to cassava virus brown streak disease (Kawuki et al. 2016). The trait-linked markers from these studies are currently being converted to singleplex SNP assays for MAS.

It is anticipated that more traits will be analyzed in the next few years, including dry matter and starch content, resistance to CGM, plant type, stay-green during the dry season and a number of morphological traits. These efforts will generate a comprehensive catalogue of major QTLs that can be used for forward breeding in MAS.

6 Conclusion and future trends

Molecular markers have been widely used for mapping of QTLs underlying various agronomically important traits in cassava, including cyanogenic potential and dry matter content (Balyejusa Kizito et al. 2007), plant architecture and productivity (Okogbenin and Fregene 2003; Okogbenin et al. 2008; Sraphet et al. 2011), bacterial blight (Jorge et al. 2000; Wydra et al. 2004; López et al. 2007) and CMD (Akano et al. 2002; Rabbi et al. 2014b; Wolfe et al. 2016). Table 1 provides a summary of published QTL studies in cassava, including target traits, parents and population type, and markers used. Despite the large efforts in identification of these QTLs, their use in breeding is very limited (Ferguson et al. 2012a), except for a few cases where MAS was applied to develop CMD-resistant varieties (Okogbenin et al. 2012; Parkes et al. 2014). Some of the technical challenges that are holding back the use of trait-linked markers in breeding pipelines and how can they be overcome are presented below.

1 Most studies used small sample sizes and a limited number of markers, which has led to poor mapping resolution in the QTL regions and an equally low proportion of variation explained by linked markers. Newer QTL mapping efforts are addressing this challenge by using larger bi-parental populations and high-density genotyping of thousands of SNP markers. Moreover, the resolution of QTLs from ongoing GWAS is expected to be higher.
2 The molecular markers used in most of the historical QTL studies, such as RFLPs, RAPDs and SSRs are either not reproducible or not amenable to high-throughput genotyping. This is set to change, thanks to the increased access to SNP markers from sequencing-based genotyping and robust singleplex genotyping platforms. Indeed genotyping needs for MAS can now be outsourced to commercial service providers, thereby reducing or eliminating the need for capital outlay for establishing in-house genotyping capacity. For example, the KASP™ genotyping assay has become the standard genotyping technology for MAS because it allows greater numbers of assays to be rapidly developed and genotyped in a highly flexible manner. In some cases, all that a breeder is required to do is to ship leaf samples to a service lab which in turn extracts DNA, genotypes the samples using selected diagnostic markers and sends the results to the breeders for field selection. A comprehensive list of genotyping service providers that supports projects related to marker development and validation, genetic mapping and MAS is available on the Integrated Breeding Platform website (https://www.integratedbreeding.net).
3 Large-scale MAS application requires DNA extraction capacity to process hundreds to thousands of selection candidates. This can be facilitated by protocols that handle 96 samples in parallel. A short turnaround time from DNA extraction to data is necessary. Several commercial genotyping labs today provide complete service, from receiving leaf tissue, DNA extraction, genotyping and marker scoring. All that is required from a breeder is to collect leaf tissue in 96-well formats and ship to service providers.

7 Where to look for further information

Further reading on markers, QTL mapping and marker-assisted selection:

1 Collard, B. C. Y., Jahufer, M. Z. Z., Brouwer, J. B. and Pang, E. C. K. (2005). An introduction to markers, quantitative trait loci (QTL) mapping and marker-assisted selection for crop improvement: the basic concepts. *Euphytica*, 142(1-2): 169–96.
2 Collard, B. C., & Mackill, D. J. (2008). Marker-assisted selection: an approach for precision plant breeding in the twenty-first century. *Philosophical Transactions of the Royal Society B: Biological Sciences*, 363(1491): 557–72.
3 Broman, K. W., & Sen, S. (2009). *A Guide to QTL Mapping with R/qtl (Vol. 46)*. New York: Springer.

8 References

Akano, O., Dixon, O., Mba, C. and et al. (2002). Genetic mapping of a dominant gene conferring resistance to cassava mosaic disease. *Theor. Appl. Genet.* 105:521–5. doi:10.1007/s00122-002-0891-7.

Akinbo, O., Labuschagne, M. T., Marín, J. and et al. (2012). QTL analysis for root protein in a backcross family of cassava derived from manihot esculenta ssp flabellifolia. *Trop. Plant. Biol.* 5:161–72. doi:10.1007/s12042-012-9095-8.

Balyejusa Kizito, E., Rönnberg-Wästljung, A.-C., Egwang, T. and et al. (2007). Quantitative trait loci controlling cyanogenic glucoside and dry matter content in cassava (Manihot esculenta Crantz) roots. *Hereditas* 144:129–36. doi:10.1111/j.2007.0018-0661.01975.x.

Bredeson, J. V., Lyons, J. B., Prochnik, S. E. and et al. (2016). Sequencing wild and cultivated cassava and related species reveals extensive interspecific hybridization and genetic diversity. *Nat. Biotechnol.* 34:562–70. doi:10.1038/nbt.3535.

Broman, K. W., Wu, H., Sen, Ś. and Churchill, G. A. (2003). R/qtl: QTL mapping in experimental crosses. *Bioinformatics* 19:889–90. doi:10.1093/bioinformatics/btg112.

Ceballos, H., Kulakow, P. and Hershey, C. (2012). Cassava breeding: Current status, bottlenecks and the potential of biotechnology tools. *Trop. Plant Biol.* 5:73–87. doi:10.1007/s12042-012-9094-9.

Chavarriaga-Aguirre, P., Maya, M. M., Bonierbale, M. W. and et al. (1998) Microsatellites in cassava (Manihot esculenta Crantz): Discovery, inheritance and variability. *Theor. Appl. Genet.* 97:493–501. doi:10.1007/s001220050922.

Collard, B. C.Y., Jahufer, M. Z. Z., Brouwer, J. B. and Pang, E. C. K. (2005). An introduction to markers, quantitative trait loci (QTL) mapping and marker-assisted selection for crop improvement : The basic concepts. *Euphytica* 142:169–96. doi:10.1007/s10681-005-1681-5

Elshire, R. J., Glaubitz, J. C., Sun, Q. and et al. (2011). A robust, simple genotyping-by-sequencing (GBS) approach for high diversity species. *PLoS One* 6:1–10. doi:10.1371/journal.pone.0019379.

Esuma, W., Herselman, L., Labuschagne, M. T. and et al. (2016). Genome-wide association mapping of provitamin A carotenoid content in cassava. *Euphytica* 212:97–110. doi:10.1007/s10681-016-1772-5

Ferguson, M., Rabbi, I., Kim, D. J. and et al. (2012a). Molecular markers and their application to cassava breeding: Past, present and future. *Trop. Plant Biol.* 5:95–109. doi:10.1007/s12042-011-9087-0.

Ferguson, M. E., Hearne, S. J., Close, T. J. and et al. (2012b) Identification, validation and high-throughput genotyping of transcribed gene SNPs in cassava. *Theor. Appl. Genet.* 124:685–95. doi:10.1007/s00122-011-1739-9.

Fregene, M., Angel, F., Gomez, R. and et al. (1997). A molecular genetic map of cassava (Manihot esculenta crantz). *Theor. Appl. Genet.* 95:431–41. doi:10.1007/s001220050580.

Holland, J. B. (2007). Genetic architecture of complex traits in plants. *Curr. Opin. Plant Biol.* 10:156–61. doi:10.1016/j.pbi.2007.01.003.

International Cassava Genetic Map Consortium (ICGMC). (2015). High-resolution linkage map and chromosome-scale genome assembly for cassava (Manihot esculenta Crantz) from 10 populations. *G3 Genes Genom. Genet.* 5:133–44. doi:10.1534/g3.114.015008.

Jorge, V., Fregene, M. A., Duque, M. C. and et al. (2000). Genetic mapping of resistance to bacterial blight disease in cassava (Manihot esculenta Crantz). *TAG Theor. Appl. Genet.* 101:865–72. doi:10.1007/s001220051554.

Kawano, K. (2003). Thirty years of cassava breeding for productivity–biological and social factors for success. *Crop Sci.* 43:1325–35.

Kawuki, R. S., Ferguson, M., Labuschagne, M. and et al. (2009). Identification, characterisation and application of single nucleotide polymorphisms for diversity assessment in cassava (Manihot esculenta Crantz). *Mol. Breed.* 23:669–84. doi:10.1007/s11032-009-9264-0.

Kawuki, R. S., Kaweesi, T., Esuma, W. and et al. (2016) Eleven years of breeding efforts to combat cassava brown streak disease. *Breed. Sci.* 66:560–71. doi:10.1270/jsbbs.16005.

Lander, E. S. and Botstein, D. (1989). Mapping mendelian factors underlying quantitative traits using RFLP linkage maps. *Genetics* 121:185–99. doi:10.1038/hdy.2014.4.

Lokko, Y., Danquah, E. Y., Offei, S. K. and et al. (2005). Molecular markers associated with a new source of resistance to the cassava mosaic disease. *Afr. J. Biotechnol.* 4:873–81.

López, C. E., Piégu, B., Cooke, R. and et al. (2005). Using cDNA and genomic sequences as tools to develop SNP strategies in cassava (Manihot esculenta Crantz). *Theor. Appl. Genet.* 110:425–31. doi:10.1007/s00122-004-1833-3.

López, C. E., Quesada-Ocampo, L. M., Bohórquez, A. and et al. (2007). Mapping EST-derived SSRs and ESTs involved in resistance to bacterial blight in Manihot esculenta. *Genome* 50:1078–88. doi:10.1139/G07-087.

Mba, R. E. C., Stephenson, P., Edwards, K. and et al. (2001). Simple sequence repeat (SSR) markers survey of the cassava (Manihot esculenta Crantz) genome: Towards an SSR-based molecular genetic map of cassava. *Theor. Appl. Genet.* 102:21–31. doi:10.1007/s001220051614.

Meuwissen, T. H. E., Hayes, B. J. and Goddard, M. E. (2001). Prediction of total genetic value using genome-wide dense marker maps. *Genetics* 157:1819–29.

Michelmore, R. W., Paran, I. and Kesseli, R. V. (1991). Identification of markers linked to disease-resistance genes by bulked segregant analysis - a rapid method to detect markers in specific genomic regions by using segregating populations. *Proc. Natl. Acad. Sci. U. S. A.* 88:9828–32. doi:10.1073/pnas.88.21.9828.

Okogbenin, E., Egesi, C. N., Olasanmi, B. and et al. (2012). Molecular marker analysis and validation of resistance to cassava mosaic disease in elite cassava genotypes in Nigeria. *Crop Sci.* 52:2576. doi:10.2135/cropsci2011.11.0586.

Okogbenin, E. and Fregene, M. (2003). Genetic mapping of QTLs affecting productivity and plant architecture in a full-sib cross from non-inbred parents in Cassava (Manihot esculenta Crantz). *Theor. Appl. Genet.* 107:1452–62. doi:10.1007/s00122-003-1383-0.

Okogbenin, E., Marin, J. and Fregene, M. (2006). An SSR-based molecular genetic map of cassava. *Euphytica* 147:433–40. doi:10.1007/s10681-005-9042-y.

Okogbenin, E., Marin, J. and Fregene, M. (2008). QTL analysis for early yield in a pseudo F2 population of cassava. *Afr. J. Biotechnol.* 7:131–8.

Parkes, E., Fregene, M., Dixon, A. and et al. (2014). Developing Cassava Mosaic Disease resistant cassava varieties in Ghana using a marker assisted selection approach. *Euphytica* 203:549–56. doi:10.1007/s10681-014-1262-6.

Poland, J., Endelman, J., Dawson J. and et al. (2012). Genomic selection in wheat breeding using Genotyping-by-Sequencing. *Plant Genome* 5:103–13. doi:10.3835/Plantgenome2012.06.0006.

Prochnik, S., Marri, P. R., Desany, B. and et al. (2012). The cassava genome: Current progress, future directions. *Trop. Plant. Biol.* 5:88–94. doi:10.1007/s12042-011-9088-z.

Rabbi, I., Hamblin, M., Gedil, M. and et al. (2014a). Genetic mapping using genotyping-by-sequencing in the clonally propagated cassava. *Crop Sci.* 54:1384–96. doi:10.2135/cropsci2013.07.0482.

Rabbi, I. Y., Hamblin, M. T., Kumar, P. L. and et al. (2014b) High-resolution mapping of resistance to cassava mosaic geminiviruses in cassava using genotyping-by-sequencing and its implications for breeding. *Virus Res.* 186: 87–96. doi:10.1016/j.virusres.2013.12.028.

Rabbi, I. Y., Hamblin, M. T., Kumar, P. L. and et al. (2014c). High-resolution mapping of resistance to cassava mosaic geminiviruses in cassava using genotyping-by-sequencing and its implications for breeding. *Virus Res.* 186:87–96. doi:10.1016/j.virusres.2013.12.028.

Rabbi, I. Y., Kulembeka, H. P., Masumba, E. and et al. (2012). An EST-derived SNP and SSR genetic linkage map of cassava (Manihot esculenta Crantz). *Theor. Appl. Genet.* 125:329–42. doi:10.1007/s00122-012-1836-4.

Raji, A. A., Anderson, J. V., Kolade, O. A. and et al. (2009). Gene-based microsatellites for cassava (Manihot esculenta Crantz): Prevalence, polymorphisms, and cross-taxa utility. *BMC Plant Biol.* 9:118. doi:10.1186/1471-2229-9-118.

Ramu, P., Esuma, W., Kawuki, R. and et al. (2016). Cassava haplotype map highlights fixation of deleterious mutations during clonal propagation. *Nat. Genet.* Advance online publication. doi:10.1038/ng.3845.

Sraphet, S., Boonchanawiwat, A., Thanyasiriwat, T. and et al. (2011). SSR and EST-SSR-based genetic linkage map of cassava (Manihot esculenta Crantz). *Theor. Appl. Genet.* 122:1161–70. doi:10.1007/s00122-010-1520-5.

Wolfe, M. D., Rabbi, I. Y., Egesi, C. and et al. (2016). Genome-wide association and prediction reveals genetic architecture of cassava mosaic disease resistance and prospects for rapid genetic improvement. *Plant Genome*. 2(2) doi:10.3835/plantgenome2015.11.0118.
Wydra, K., Zinsou, V., Jorge, V. and Verdier, V. (2004). Identification of pathotypes of Xanthomonas axonopodis pv. manihotis in Africa and detection of quantitative trait loci and markers for resistance to bacterial blight of cassava. *Phytopathology* 94:1084–93. doi:10.1094/PHYTO.2004.94.10.1084.

Chapter 6

Advances in genetic modification of cassava

P. Zhang, Q. Ma, M. Naconsie, X. Wu, W. Zhou, National Key Laboratory of Plant Molecular Genetics, CAS Center for Excellence in Molecular Plant Sciences, Chinese Academy of Sciences, China; and J. Yang, Shanghai Chenshan Plant Science Research Center, Shanghai Chenshan Botanical Garden, China

1 Introduction

Cassava (*Manihot esculenta* Crantz) is an important food crop in the tropics and subtropics. Traditionally, hybridization has been used to breed new cassava varieties (Nassar and Oritz, 2010). Changes in global climate, food security and industrialization have accelerated the breeding of new cassava varieties with increased nutrition, high stress resistance and starch content (Ceballos et al., 2012). Genetic engineering shows great potential in germplasm innovation by improving specific traits without changes in other important traits, especially with the development of genome editing technology. The '-omics' tools have led to intensive cassava study focused especially on starchy storage root development, starch accumulation, health-promoting components (e.g. beta-carotene), and stress response and regulation (Sayre et al., 2011; Yang et al., 2011). In this chapter, we update the recent progress related to transgenic modification of cassava (Liu et al., 2011). As a vital component of an integrated breeding system, genetic engineering, together with functional genomics, proteomics, marker-assisted selection and traditional hybridization, has greatly promoted

http://dx.doi.org/10.19103/AS.2016.0014.17

the efficiency of cassava genetic improvement over the last 15 years. Hence, the role of cassava in food security, commercialization and bioenergy development can be addressed by strengthening fundamental research and applied technology.

2 Transition from model cultivars to farmer-preferred cultivars

Genetic transformation systems have been well-developed for cassava since the first reports in 1996 and much progress has been made in the development of *Agrobacterium*-mediated transformation protocols for TMS60444 friable embryogenic callus (FEC) (reviewed by Liu et al., 2011). Extending the protocol from the model cultivar to farmer-preferred cultivars, especially disease-resistant cultivars, is under way. Recently, several genetic transformation systems of farmer-preferred cassava cultivars have been successfully established using African, Asian and South American cultivars (Table 1, Fig. 1), although they still need to be optimized due to cultivar dependence.

As the key to cassava transformation, regular and efficient production of high-quality FEC for use in *Agrobacterium*-mediated transformation experiments is essential for sustained production of transgenic plants. Several laboratories had successfully induced FEC from various genotypes including TME1, TME3, TME14, T200, Ebwanantereka, TMS91/02327, Rosinha, Asira and so on (Table 1) (Rossin et al., 2011; Zainuddin et al., 2012; Chetty et al., 2013; Nyaboga et al., 2013, 2015). Rossin et al. (2011) found that FEC was successfully derived from cultured initial explants of six cultivars including T200, MCOL2261, SM707-17, BRA1183, CM523-7 and CR25-4, indicating that the FEC induction might not be necessarily produced from somatic embryos transferred to GD medium supplemented with 50 μM picloram. Under improved FEC culture conditions, Zainuddin et al. (2012) used robust transformation protocols for the production of transgenic farmer-preferred cassava landraces (TME3, TME7 and TME14). Nyaboga et al. (2013) optimized the FEC induction conditions for three East African farmer-preferred cultivars (Ebwanatereka, Kibandameno and Serere). They suggested that supplementation of L-tyrosine at 250 μM into FEC induction medium significantly enhanced the conversion of organized embryogenic structures (OES) into FEC in three cultivars (Ebwanatereka, Serere and 60444), whereas FEC production and proliferation were inhibited on L-tyrosine-based media for Kibandameno, suggesting that the L-tyrosine effect was also genotype-dependent. These cultivars also showed variable time to generate FEC from OES, ranging from 9 to 22 weeks. Other conditions, such as culture under constant darkness or fragmentation, also enabled FEC induction in cultivar T200 (Chetty et al., 2013). Nyaboga et al. (2015) described an improved, high-efficiency transformation protocol for the recalcitrant cassava cultivar TME14. They found that the FEC efficiency was enhanced by DKW medium, crushing of OES through 1–2-mm metal wire mesh, washing of crushed OES tissues and short exposure of tyrosine to somatic embryos.

Improvement of *Agrobacterium*-mediated transformation systems is also essential for high transformation efficiency. The conversion of FEC into somatic embryos has been a challenge. Zainuddin et al. (2012) reported that co-cultivation in the dark facilitated somatic embryogenesis in MSN medium. Modification or optimization of the shoot organogenesis protocol, such as frequent transfers to shoot elongation medium (CEM), incorporation of a higher concentration of the shoot-stimulating hormone, or silver nitrate (Zhang et al., 2001) improved efficiencies in T200 (Chetty et al., 2013). Gradual increase of 6-benzylamino

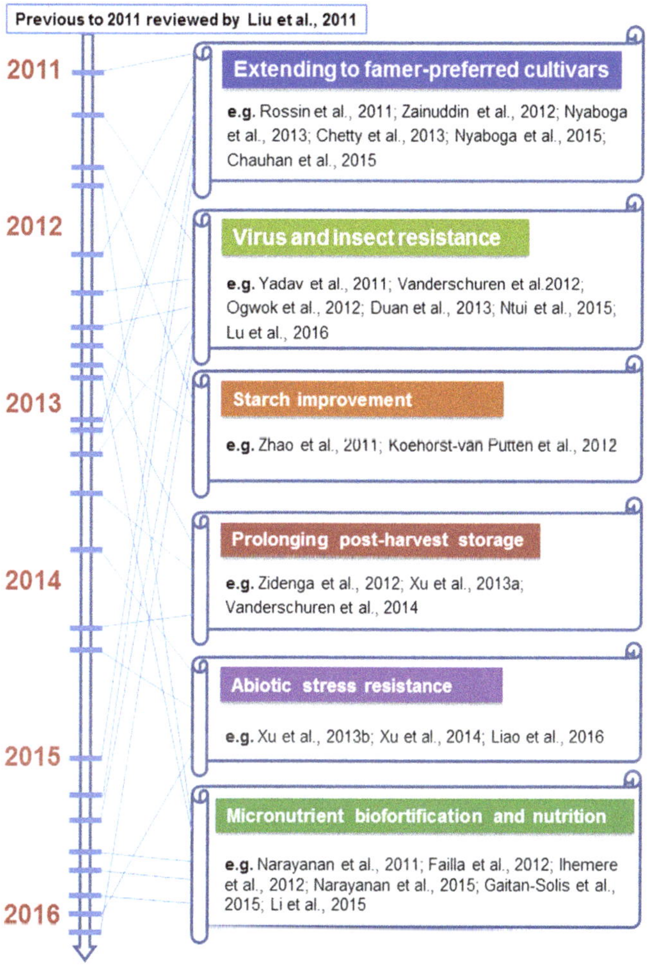

Figure 1 The development of cassava genetic engineering for technology adaption to farmer-preferred cassava cultivars and value-added traits in the years 2011–2016.

purine (BAP) concentration in the shoot elongation medium for germinating cotyledon-stage embryos enhanced the plant regeneration efficiency (Nyaboga et al., 2015). During *Agrobacterium* inoculation, in addition to immersing FEC in *Agrobacterium* suspension for 30–45 min at room temperature with gentle shaking, dropping *Agrobacterium* liquid suspension directly to FEC clusters on co-cultivation medium has been used (Bull et al., 2009). Transformation efficiency was enhanced with low *Agrobacterium* density during co-cultivation and co-centrifugation of FEC with *Agrobacterium*, possibly facilitating plant regeneration (Nyaboga et al., 2015; Chauhan et al., 2015). Pre-treatment with antibiotic moxalactam in FEC induction medium prior to inoculation with *Agrobacterium* greatly

Table 1 Progress and current status of *Agrobacterium*-mediated cassava genetic transformation and trait improvement since 2011

Cultivar	Explant	Regeneration mode	Plasmid (marker genes)	Selection	Target traits	Integration/ expression	Reference
TMS60444	FEC	SE	*P35S/P54::AtPDX1-P35S/P54::AtPDX2* (*hpt*)	Hygromycin	Increased bioavailable vitamin B6	RAP, SAP	Li et al., 2015
TME204	FEC	SE	p8012 (*npt*II)	Paromomycin	Iron accumulation	RAP, qRAP	Narayanan et al., 2015
TME204/TME7/ TMS60444	FEC	SE	p5001, p8001, pCAMBIA2300-EGFP (*npt*II)	Paromomycin	CBSD resistance, carotene enhancement	VMP, SAP	Chauhan et al., 2015
TMS60444	FEC	SE	p908 (*hpt*), p8116 (*npt*II)	Hygromycin/ paromomycin	–	SAP, VMP	Okwuonu et al., 2015
TMS60444	FEC	SE	FMV:AtZIP1, FMV:AtMTP1, PAT:AtZIP1, PAT:AtMTP1 (*npt*II)	Paromomycin	Zinc accumulation	SAP, NAP	Gaitán-Solís et al., 2015
TME14	FEC	SE	pCAMBIA2301 (*uid*Aint, *npt*II)	Paromomycin	–	RAP, SAP, VMP	Nyaboga et al., 2015
KU50	Somatic cotyledon	SO	pEKH2IN2 (*hpt*, *npt*II)	Kanamycin	SLCMV resistance	RAP, SAP, NAP	Ntui et al., 2015
TMS60444	FEC	SE	PAT-GPX	Hygromycin	Delayed PPD	RAP, PAP	Vanderschuren et al., 2014
TMS60444	FEC/ suspension	SE	P54::MeCu/ ZnSOD-35S::MeAPX2	Hygromycin	Tolerance to oxidative and chilling stresses	SAP, qRAP	Xu et al., 2014
Serere/ Ebwanateraka/ Kibandameno	FEC	SE	pCAMBIA1301 (*hpt*, *uid*A)	Hygromycin	–	SAP, RAP, VMP	Nyaboga et al., 2013
TMS60444	FEC/ suspension	SE	pC35S::Cry1Aa (*hpt*, *uid*A)	Hygromycin	Insect resistance	SAP, RAP, WAP	Duan et al., 2013

TMS60444	FEC/ suspension	SE	P54::MeCu/ZnSOD-35S::MeCAT1 (*hpt*)	Hygromycin	Delayed PPD/ cold and drought stress tolerance	RAP, SAP, PAP	Xu et al., 2013a, b
TMS60444/T200	FEC	SE	pCAMBIA1305.1 (*hpt*)	Hygromycin	–	RAP, VMP	Chetty et al., 2013
TMS60444/TME3/ TME7/TME14	FEC	SE	pCAMBIA1301 (*hpt, gusA*)	Hygromycin	–	SAP, VMP	Zainuddin et al., 2012
TMS60444	FEC	SE	Patatin-DXS, Patatin-crtB (*npt*II)	Paromomycin	Biofortified β-carotene	–	Failla et al., 2012
TMS60444	Somatic cotyledon	SO	3DF*-patatin-FEA1 (*npt*II)	Paromomycin	Iron accumulation	RAP	Ihemere et al., 2012
Adira4	FEC/ suspension	SE	p5IRTCGBa (*bar*)	Phosphinothricin (PPT)	Low amylose	SAP	Koehorst-van Putten et al., 2012
TMS60444	Somatic cotyledon	SO	3D-AtAox1A CO (*npt*II)	Paromomycin	ROS reduction	RAP	Zidenga et al., 2012
TMS60444/TME7	FEC	SE	pRNAi-CBSV-CP (*hpt*)	Hygromycin	CBSD and CMD resistance	qRAP, SAP	Vanderschuren et al., 2012
TMS60444	FEC	SE	pILTAB718, pILTAB719	Paromomycin	UCBSV resistance	NAP, RAP	Ogwok et al., 2012
TMS60444	FEC	SE	pCAMBIA2300 (*npt*II)	Paromomycin	–	SAP, DBAP, VMP	Taylor et al., 2012
TMS60444	FEC	SE	RNAi FL-CP (*npt*II)	Paromomycin	CBSUV resistance	RAP, DBAP, NAP	Yadav et al., 2011
TMS60444	FEC	SE	pCAMBIA2300-HNL (*npt*II)	Paromomycin	Protein and free amino acids, low cyanogen	RAP, WAP	Narayanan et al., 2011
TMS60444	FEC	SE	pCAMBIA2301 (*npt*II)	Paromomycin	–	SAP, VMP	Niklaus et al., 2011
TMS60444	FEC/ suspension	SE	35S::GBSSI-RNAi, p54/1.0::GBSSI-RNAi	Hygromycin	Low amylose (waxy)	RAP, SAP, PAP	Zhao et al., 2011

SE = somatic embryogenesis, SO = shoot organogenesis, SAP = Southern analysis positive, DBAP = dot blot analysis positive, NAP = Northern analysis positive, PAP = protein analysis positive, RAP = RT-PCR analysis positive, qRAP = real-time PCR analysis positive, VMP = visual marker positive, WAP = Western analysis positive. For gene abbreviations see text.

increased the recovery of transgenic plants in TME204, but not in TMS60444 (Chauhan et al., 2015).

FEC-based transformation platforms of farmer-preferred cultivars (e.g. TME7, TME14, TME204, T200, Ebwanatereka, Kibandameno and Serere) have been successfully established for routine operation in several laboratories and for genetic improvement (Table 1, Fig. 1) (Zainuddin et al., 2012; Nyaboga et al., 2013, 2015; Chauhan et al., 2015). Because of differences in genotypes, tissue culture system, transformation methods and even differences in regeneration capacity among FEC lines derived from the same genotype, optimization of the corresponding protocols is required to meet various demands. Further, understanding the regulatory mechanisms leading to the initiation of FEC from somatic embryos, and understanding somaclonal variation in FEC, are essential steps for improving the system. Our group reported the dynamic changes in cellular structure, physiological status and gene expression profiles during transition from somatic embryo to FEC in cassava (Ma et al., 2015), which enable improved FEC induction from recalcitrant cultivars as well as ensuring the quality of regenerated plants through improved culture conditions. Based on our analysis, changing the basal medium components, especially the micronutrients, might facilitate FEC induction. The increase in sucrose content also delayed the occurrence of somaclonal variation during FEC subculture. We also found that reduced DNA methylation during FEC subculture resulted in the vulnerability of somaclonal variation (Ma et al., 2015).

3 Tools of gene expression regulation

The common strategy for improving specific traits of cassava is to overexpress a heterologous or native transgene using constitutive promoters such as CaMV35S and NOS. Tissue-specific promoters, such as patatin promoter isolated from potato (*Solanum tuberosum*) and p54/1.0 from cassava (Zhang et al., 2003), were also frequently used in cassava. When expressing two transgenes in one T-DNA region, different promoters were used. For example, Li et al. (2015) used different promoters to drive the expression of two vitamin B6 biosynthetic genes, *AtPDX1* and *AtPDX2*, in cassava (Li et al., 2015). Co-expression of native copper/zinc superoxide dismutase (MeCu/ZnSOD) and catalase (MeCAT1) in transgenic cassava was achieved using the native cassava vascular-specific p54/1.0 promoter and the CaMV 35S promoter, respectively (Xu et al., 2013a).

To downregulate gene expression, antisense and RNA interference (RNAi) of native genes have been used to engineer virus resistance or to modify starch composition in cassava. Using CaMV 35S, p54/1.0 or GBSSI promoter to express the hairpin DNA sequence of *MeGBSSI* was sufficient to reduce the gene expression of *MeGBSSI* significantly and result in the amylose-free (waxy) phenotype in storage roots (Zhao et al., 2011; Koehorst-van Putten et al., 2012).

With the development of genome editing technology, application of TALEN and CRISPR/Cas9 systems to mutant target genes is now in demand in cassava (Woo et al., 2015). The utilization of this tool in cassava is still in its infancy. There are no published reports involving genome editing of cassava to date, although several laboratories are working on different genes and traits. Based on draft cassava genome sequences (Prochnik et al., 2012; Wang et al., 2014; Bredeson et al., 2016), cassava breeders are at the turning point of trait improvements in important root crops. Combining the CRISPR/

Cas9 system with cross-breeding, cassava breeders, for the first time, could offer the possibility of a powerful solution towards achieving trait improvement in crops with less-perceived potential (from the general public and from policy-makers, for example) for ecological risks or hypothetical crisis using transgenic technologies. Nevertheless, gene editing can be an open question for breeders. The candidate gene identification study is still dependent on traditional methods, such as cDNA library screening under stress (Sakurai et al., 2007) or transcriptome analysis under different treatments (An et al., 2012; Utsumi et al., 2012) or development stages (Yang et al., 2011). Researchers are seeking specific ways to study massive microarray data (Yang et al., 2011; An et al., 2012; Utsumi et al., 2012), RNA-seq (Liu et al., 2014; Ma et al., 2015) and even genome sequencing (Wang et al., 2014; Bredeson et al., 2016).

4 Production of virus-resistant cassava

Cassava mosaic disease (CMD) and cassava brown streak disease (CBSD) result in great loss of cassava production in sub-Saharan Africa and the Indian subcontinent (Legg et al., 2014). The RNA-based silencing mechanism against viruses proved to be the most efficient and robust approach in engineering virus resistance (Vanderschuren et al., 2007a). Technologies such as antisense, dsRNA and miRNA have been developed to interfere with both RNA and DNA virus replication and accumulation in model plants and economically important crops. In the model cassava cultivar TMS60444, CMD-resistant transgenic cassava has been developed using both antisense and dsRNA technology (Zhang et al., 2005; Vanderschuren et al., 2007b, 2009). With the recent epidemic of CBSD in East Africa, using CMD-resistant cultivars to engineer CBSD resistance by transgenic technologies has been explored (Table 1, Fig. 1). Two groups, including one from ETH Zurich and another from Danforth Plant Science Center, reported the generation of cassava brown streak virus (CBSV)-resistant transgenic TMS60444 cassava using RNAi technology by targeting a sequence of the CBSV coat protein (CP) highly conserved between CBSV and Ugandan cassava brown streak virus (UCBSV) or the near full-length coat protein (FL-CP) of CBSUV (Yadav et al., 2011; Vanderschuren et al., 2012). Due to immunity against CBSD, the ETH group also adapted this technology to a CMD-resistant farmer-preferred Nigerian landrace TME 7 (Oko-Iyawo). All transgenic TME 7 lines showed complete protection against CBSV and UCBSV infections. The resistance was maintained when plants were co-inoculated with East African cassava mosaic virus (Vanderschuren et al., 2012). It represents a very promising approach to solve the problem of two devastating viral diseases. Unfortunately, loss of CMD resistance occurred in these transgenic plants in field due to possible epigenetic regulation of CMD2 locus (a single CMD resistance gene with a dominant effect) during somatic embryogenesis (Beyene et al., 2016). CMD2-type cultivars are widely grown by farmers in Africa and also extensively used in breeding programmes.

Using high-throughput sequencing, our group and another team from South Africa studied the interaction of cassava geminivirus and host plants (Liu et al., 2014; Allie et al., 2014). We identified several photosynthesis-related genes that are regulated in virus-infected cassava, resulting in Chl b and light harvesting complex II (LHCII) deficiencies. Furthermore, we observed fewer grana lamellae in infected leaves that demonstrated chlorosis. This study identified photosynthesis-related genes that may contribute to the development of cassava disease, and revealed the important roles of chlorophyll degradation and LHCII inhibition during ACMV attack. The results not only established the connection between gene expression and CMD phenotype for the first time but also enabled further exploration of the molecular mechanisms of viral pathogenesis. Similarly, the South African group investigated the transcriptional response of South African cassava mosaic virus-infected susceptible and tolerant landraces of cassava and highlighted the differences in resistance, basal defence and cell wall-associated genes during infection (Allie et al., 2014). It is also an important step towards understanding natural host responses to plant geminiviruses. The group also reported DNA sequences, named Sequences Enhancing Geminivirus Symptoms (SEGS-1 and SEGS-2), which enhance CMD symptoms, when occurring in the cassava genome, during South African cassava mosaic virus infection (Maredza et al., 2016). The SEGS-1 and SEGS-2 also overcame CMD2 resistance and the transmission of SEGS-2 by whiteflies indicates that cassava breeders need to look for alternative sources of CMD resistance in cassava (Ndunguru et al., 2016). Using genome-wide association mapping study, Wolfe et al. (2016) found that two possibly epistatic loci and/or multiple resistance alleles exist at the major QTL responsible for CMD resistance and two peroxidases and one thioredoxin are potential candidate genes in the major QTL region, providing a useful approach for improving resistance in breeding populations and identifying highly resistant clones as varieties.

5 Cassava biofortification for better nutrition

Biofortification of cassava has been the main target of BioCassava Plus and of Harvest Plus over the last ten years (Sayre et al., 2011; IFPRI, 2014). It is also one of the main goals of the Root, Tuber and Banana consortium targeting the underutilized potential of key vegetatively propagated starchy crops to improve food security, nutrition and livelihoods (http://www.rtb.cgiar.org). We highlight the achievements in micronutrient biofortification using transgenic approaches in the last few years (Fig. 1). Key nutritional deficiencies in areas where cassava is critical in diets are zinc, iron, and vitamins A and B6. By far, most of the research investments have targeted the carotenoid precursors of vitamin A, and lesser efforts on the other traits.

5.1 Zinc

A reported strategy for increasing zinc accumulation in the cassava storage root is to overexpress the zinc transporter genes. The research group of Schachtman has introduced *Arabidopsis* plasma membrane transporter (*AtZIP1*) and vacuolar membrane transporter (*AtMTP1*) genes into cassava plants under the control of constitutive FMV promoter or root-specific patatin promoter (Gaitán-Solís et al., 2015). The *AtZIP1* driven by FMV promoter showed stronger expression and a higher zinc level in storage roots compared to the cassava with the patatin promoter. However, co-expression of *AtZIP1* and *AtMTP1* showed no significant difference in zinc level when compared with the wild-type plant. Zinc accumulation in roots was monitored in three stages: at 4 months in a growth chamber and at 6–8 months in a greenhouse, showing that transgenic plants have 2–7-fold (24–73 mg/Kg DW) higher zinc levels compared with wild type (9 mg/kg DW). Interestingly, in the third stage, 12-month-old cassava roots grown in the field accumulate zinc up to 217 mg/kg DW, which is 10 times more than that of the wild type. Based on the zinc content in these transgenic cassava plants, the level can meet the minimum daily required amount of zinc (11 mg/day) for individuals by eating less than 400 g of fresh cassava each day. Nevertheless, the transgenic phenotype shows smaller plants and abnormal leaves in all growing conditions. The authors suggest that the overexpression of *AtZIP1* and *AtMTP1* influences zinc partitioning from leaves to roots. Reduction of zinc in the leaves by half may inhibit plant growth and development.

5.2 Iron

The algal FEA1 is an iron transporter homologous to *Arabidopsis* IRT1 that assists ferrous iron uptake from the soil. Ihemere et al. (2012) generated cassava plants containing *Patatin:FEA1*, which showed effective gene expression specific to root tissue and with normal phenotype similar to the wild-type plant. Iron concentration in the storage roots of plants from both green house and field-grown conditions showed levels as high as 36 ppm in transgenic species, which was three times higher than in the wild type. In addition, overexpression of *FEA1* transgenic cassava induced the expression of iron transporter and homeostasis genes.

Overexpression of *Arabidopsis* vacuolar transporter *AtVIT1* is regulated by patatin promoter (Narayanan et al., 2015). Young leaves of transgenic plants show 50% reduction in the iron level and exhibit yellow leaf phenotype resulting from iron deficiency. However, mature leaves show normal phenotype and accumulate iron compared with wild type. Other phenotypes including plant growth, plant size and overall biomass also showed no differences between transgenic and wild type. The iron level measured in the roots of the 4-month-old plant grown in green house was higher than in the wild type and vector control plant. Surprisingly, 462 ppm iron was detected in the stem base of transgenic plants, which was 16 times higher than in the control. Perl's Prussian blue stain clearly reveals that *AtVIT1* transgenic cassava accumulates high levels of iron in the vascular tissue of petioles, stems, fibrous roots and storage roots. Iron transport and storage genes also affected the *AtVIT1* overexpression (Narayanan et al., 2015).

FEA1 facilitates three times higher iron uptake from the soil, AtVIT1 sequesters iron into vacuoles, thus inducing 4-fold iron accumulation compared to wild type. Both approaches show similar efficiency of iron accumulation in the root. However, overexpression of *AtVIT1* shows defective iron partitioning, resulting in iron deficiency in young leaves but excess in the stem base. Co-expression of *FEA1*, *AtVIT1* and other iron transporter genes may enhance the efficiency of iron uptake and homeostasis in cassava.

5.3 Vitamin A/carotenoids

Yellow rooted cassava with high carotenoid levels exists in farmer landrace varieties. Regulation of key genes related to carotenoid biosynthesis in yellow root cassava is also desirable to quickly incorporate high carotenoid accumulation into adapted clones, or increase levels beyond what is possible with conventional crossing alone. In 2010, Beyer and his team reported that a single-nucleotide polymorphism of phytoene synthase 2 (*PSY2*) in QTL analysis of a cassava yellow root phenotype showed association with yellow colour and carotenoid content (Welsch et al., 2010). The gene *PSY2* in yellow cassava carried a mutation at position $C_{572}A$, resulting in amino acid change to $A_{191}D$, leading to higher enzyme activity. Transformation of bacterial *PSY* gene, *CrtB*, into cassava showed that overexpression enhanced carotenoid accumulation in storage roots. However, only one transgenic line exhibited strong phenotype with total carotenoids 20-fold higher than the wild type, compared with other lines showing only 2–3-fold increase (Welsch et al., 2010).

Transgenic cassava expressing 1-deoxy-xylulose-5-phosphate synthase gene from *Arabidopsis* (*AtDXS*) showed 10- to 34-fold induction in carotenoid content (Failla et al., 2012). This study also analysed the carotenoid levels after several cooking methods and found that transgenic plants have greater carotenoid retention. When cooked cassava roots were subjected to *in vitro* digestion and nutrient absorption studies, the results demonstrated that carotenoid accumulation in Caco-2 cells was highest in the *AtDXS* overexpressed root sample. These data suggest that the increase in bioavailable and bioaccessible carotenoids from *AtDXS* transgenic cassava has potential as a source of supplemental vitamin A.

5.4 Vitamin B6

Generally, cassava contains low levels of vitamin B6. A collaborative work among Swiss and Chinese scientists reported increased vitamin B6 accumulation in transgenic cassava following the co-expression of PDX1 synthase and PDX2 glutaminase. These are two enzymes required for vitamin B6 biosynthesis in plants, derived from *Arabidopsis* and driven by 35S or patatin promoter (Li et al., 2015). Field-grown cassava plants collected in China were analysed by HPLC. Compared with wild type, both transgenic lines showed 4- to 8-fold higher vitamin levels in leaves and 3- to 15-fold higher level in storage roots. The transgenic plants with CaMV 35S promoter promoted vitamin B6 synthesis in leaves, while the lines with patatin promoter showed less induction in leaves but stronger accumulation in storage roots. The overexpression of AtPDX1 and AtPDX2 resulted in an increase in non-phosphorylated forms of vitamin B6. The Caco-2 cell assay revealed increased bioavailability of vitamin B6 from leaves and roots of transgenic lines after cooking. Based on the increased concentration in transgenic lines, only 500 g of boiled roots or 50 g of leaves per day were sufficient to meet the daily vitamin B6 requirement.

6 Starch modification of cassava for industrial applications

Cassava accumulates a large amount of starch in its storage roots. However, its native starch has many shortfalls such as low solubility and retrogradation. To overcome these limitations, native starch is often modified chemically, physically or biotechnologically (Kaur et al., 2012). Modified starch is widely used in various industries, including food processing, paper manufacture, tablets, cosmetic formulations and degradable plastics. We provide an update on biotechnologically modified cassava starch (Fig. 1).

Advances in understanding starch biosynthesis and the isolation of specific genes involved have enabled genetic modification of crops to produce novel starches with improved functionality. Different amylose levels, amylopectin structure and phosphate content from various plant sources can be produced using antisense or RNAi reduction of enzyme activity of single or multiple enzymes (Sonnewald and Kossmann, 2012). Amylose-free (waxy) cassava was developed by Zhao et al. (2011) by downregulation of GBSSI expression using the RNAi approach. Two different promoters, p54/1.0 and CaMV 35S, were used to drive the expression of siRNA targeting a conserved region of the GBSSI gene. Several transgenic waxy cassava lines harvested from the field have high viscosity, high swelling index and low solubility, similar to the waxy mutant AM260-5 from CIAT, suggesting that both sources are equally effective in generating similar amylose-free starch phenotypes in cassava. Compared with native cassava starch, waxy cassava starch shows higher viscosity and different starch structure (Rolland-Sabaté et al., 2013). Parameters such as PV, gel BD and SB in transgenic waxy cassava B9 are more pronounced than in AM260-5. The partial waxy line A19 provides intermediate values in pasting behaviour analysis. Koehorst-Van Putten et al. (2012) also reported the field testing of transgenic cassava with low-amylose or amylose-free starch using RNAi technology in Indonesia. The transgenic approach, therefore, is a powerful tool for developing novel germplasm with different levels of amylose content and starches of diverse properties.

The main theoretical advantages of a transgenic approach are that the waxy trait can be engineered in a specific target variety, without changing expression of other traits, and it can be accomplished much more quickly than the transfer of the waxy gene through traditional breeding, including the long process of selecting for the wide range of other traits that would need to be incorporated into a new variety. The main disadvantage is that most countries are not yet equipped to approve transgenic waxy cassava through the regulatory process.

High-amylose starch is primarily of interest as a 'resistant starch' (slower digestibility) for use in the food industry. It can be produced in plants by downregulating the expression of branching enzymes. Zhao (2012) reported the production of high-amylose cassava using the RNAi approach (Zhao 2012). Further analysis of starch properties as well as the impact on cassava plant growth is underway. In addition, phosphorylation is the only naturally occurring covalent modification, which is catalysed by glucan water dikinase (GWD). Repression of GWD (starch phosphorylating enzyme R1) was used in the alteration of specific structural motifs of potato starch (Viksø-Nielsen et al., 2001). Overexpression of GWD was also widely used to increase starch phosphorylation and improve native starch properties in plants (Carciofi et al., 2011). Increasing the phosphate content in crop starch is therefore highly desirable because a high degree of natural phosphorylation obviates the need for expensive and environmentally harmful industrial chemical processes (Blennow et al., 2002). These studies were also conducted in cassava in several laboratories.

7 Improving storage, root production and post-harvest storage

Storage roots are the cassava plant's main commercial product, and therefore understanding storage root development is key to improving starch accumulation and root production. Genomics, transcriptomics and proteomics have been widely used to study cassava root development. Using a microarray representing 20840 genes, the differentially expressed genes and important pathways were analysed using fibrous roots, developing storage roots and mature storage roots. Among the significantly impacted pathways, glycolysis/gluconeogenesis was the most vital one (Yang et al., 2011), suggesting the importance of carbon mobilization during storage root development. Microarray analysis of storage roots at different stages shows that phytohormone-related transcription factors like MeAGL20 might play an important role in storage root initiation (Sojikul et al., 2015). Analysis of NAC transcription factors indicates the significant role of NACs in cassava root development (Hu et al., 2015). Naconsie et al. (2016) compared proteomes of roots at different development stages by 2D gel electrophoresis and LC-MS/MS, finding that secondary growth is essential for cassava root development. Comparative proteomes also show that root formation may be influenced by metabolic and regulatory processes in cassava leaves (Mitprasat et al., 2011). Through AFLP-based transcript profiling, Sojikul et al. (2010) found that four genes in the MeKD family exhibit critical expression in initiation and early stage development of the cassava storage root. Further studies are required to reveal the biological process of cassava storage root development regulated by key genes.

During storage root development, starch biosynthesis-related genes are upregulated, suggesting an intrinsic relationship between carbon partitioning and storage root growth (Yang et al., 2011; Siriwat et al. 2012). Expression of a modified *E. coli* glgC gene encoding ADP-glucose pyrophosphorylase (AGPase) by site-directed mutagenesis (G336D) to eliminate negative regulation by fructose bisphosphate in cassava showed a 2.6-fold increase of storage root biomass (Ihemere et al., 2006). In the transgenic waxy cassava (referred to in Section 1.7) showing downregulated GBSSI expression, no obvious change in storage root production was noticed (Zhao et al., 2011); however, our recent study found that downregulation of starch branching enzymes and α-glucan water dikinase (GWD) reduce the storage root production (unpublished data). Besides, many microRNA-targeted genes like MYB, APL, NAC and so on might be involved in storage root development and starch metabolism (Chen et al., 2015).

An estimated 26% of total cassava production is lost owing to post-harvest physiological deterioration (PPD, Sayre et al., 2011). Genetic improvement can potentially delay or inhibit PPD in cassava. Indeed, the PPD process differs among cassava genotypes. Sanchez et al. (2013) reported that the PPD-susceptible HMC-1 variety and PPD-tolerant experimental hybrid AM206-5 were significantly different in PPD level and in secondary metabolic synthesis (Sanchez et al., 2013). A better understanding of the mechanism of PPD processes might enable engineering of cassava storage roots with prolonged shelf life (Fig. 2).

PPD occurs mainly due to high moisture level, high respiratory rate and exposure to oxygen, especially at sites of injury of the cassava root after harvest (Salcedo et al., 2010). Recent studies based on cDNA microarray-based transcriptomes (Reilly et al., 2007) and iTRAQ proteomics (Owiti et al., 2011) showed that the onset of PPD is related to signal transduction pathways involving multiple enzymes, cellular processes including reactive oxygen turnover, signal transduction, stress response, metabolism and biosynthesis. Reactive oxygen species (ROS) production and turnover in the root is closely associated with PPD. O_2^- and H_2O_2 quickly accumulate in the root within 15 minutes after harvest, with peak levels at 4 h and 24 h, respectively (Reilly et al., 2003). Accumulation of ROS in the storage root can be observed directly by fluorescent probes (singlet oxygen sensor green, dihydrorhodamine 123), which indicates oxidative bursts during the PPD process (Xu et al., 2013a). Zidenga et al. (2012) reported that PPD is cyanide-dependent, probably because of a cyanide-dependent inhibition of respiration. ROS cannot be removed after exceeding the steady state, and the excess ROS results in PPD in the storage root (Xu et al., 2013a).

As the onset of PPD is tightly associated with ROS, PPD was delayed by overexpressing ROS turnover and scavenging-related proteins (Fig. 2). Co-expression of cytosolic MeCu/

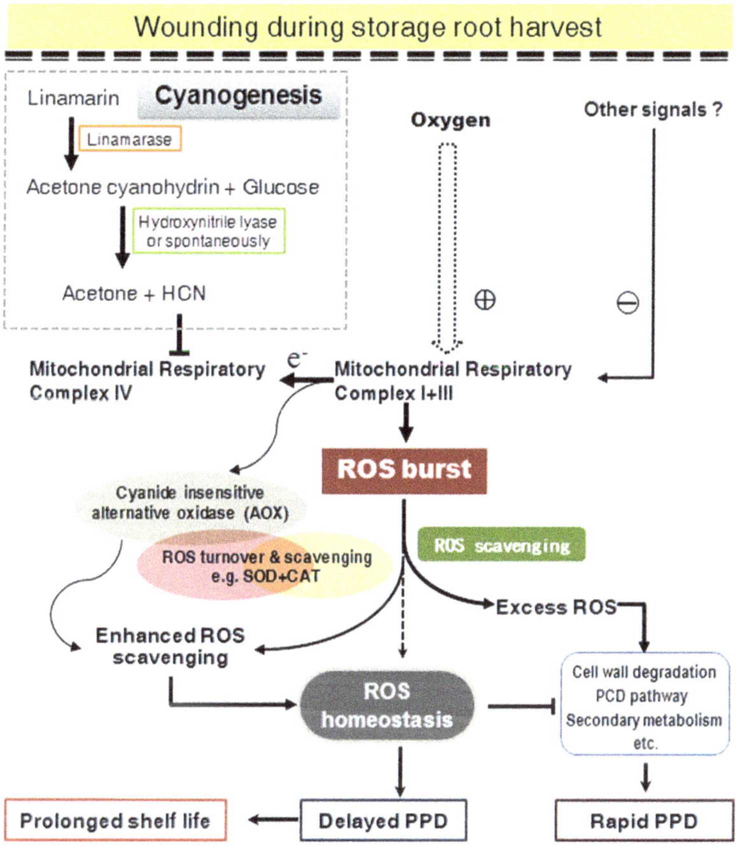

Figure 2 Regulatory mechanism of post-harvest physiological deterioration in cassava storage roots. Source: Adapted from Zhang et al., 2013.

ZnSOD (superoxide dismutase) and peroxisomal MeCAT1 (catalase) driven by the native cassava vascular-specific p54/1.0 promoter and the CaMV 35S promoter, respectively, in cassava, improved the capacity of ROS scavenging and delayed PPD occurrence over 10 days (Xu et al., 2013a). This gene stacking approach synergistically increased ROS turnover, thereby enhancing improved tolerance of transgenic cassava cells to oxidative stress caused by H_2O_2 and MV. The increased SOD and CAT enzyme activities enabled rapid scavenging of ROS at the site of generation, as well as prevented the formation of hydroxyl radicals. Therefore, this study convincingly demonstrates that the co-expression of antioxidant enzymes SOD and CAT in transgenic cassava leads to a synergistic effect that not only reduces ROS levels but also delays cassava PPD (Xu et al., 2013a). Using the transgenic cassava plants, Liao et al. (2016) also proved that ROS regulate leaf pulvinus abscission zone cell separation in response to water-deficit stress in cassava. Expression of a cyanide-insensitive *AOX*, codon-optimized *Arabidopsis* mitochondrial alternative oxidase gene driven by patatin promoter in transgenic cassava showed increased resistance to PPD, but also resulting in decreased storage root yield (Zidenga et al., 2012). Transgenic cassava plants overexpressing an *Arabidopsis* cytosolic *GPX* (glutathione peroxidase) in storage roots also showed reduced PPD symptoms in root slices until 48 hours after harvest (Vanderschuren et al., 2014). Controversially, since PPD symptoms typically begin to occur at about 48 hours after harvest, the root slice method used in the study cannot properly reflect the real situation of PPD occurrence in these storage roots.

Blue/black discolouration is the main initial visual symptom of PPD. The hydroxycoumarins play a key role in this discolouration, including scopoletin, scopolin, esculin and esculetin (Bayoumi et al., 2008). Discolouration of the root is mediated by various enzymes and pathways such as the phenylpropanoid metabolic pathway and antioxidant enzymes. Research has focused on the metabolism and regulation of these hydroxycoumarins. Histochemical analysis indicated that acidic polysaccharides may delay PPD based on starch catabolism (Uarrota and Maraschin 2015). Therefore, inhibition of hydroxycoumarin pathways might be useful in reducing the symptoms of PPD in storage roots. Such studies are ongoing in several laboratories, according to the presentations at the World Congress of Root and Tuber Crops in Nanning in 2016.

8 Future trends and conclusion

Cassava breeding is at a transition phase from sole use of traditional hybridization to the application of genetic engineering as an additional tool to generate new diversity. In the last 5 years, cassava genetic improvement using transgenic approaches has yielded significant progress worldwide. Farmer-preferred cultivars have been explored to develop cassava with increased resistance to viruses and abiotic stress, enhanced nutritional value, improved starch yield and quality and prolonged shelf life. Furthermore, genome editing technologies, such as TALLEN and CRISPR/Cas9, provide new strategies for genetic manipulation of targeting genes and produce marker-free plants, which have been used in several advanced laboratories. Nonetheless, as of yet, there are no commercially planted transgenic cassava varieties available, although several field trials of transgenic cassava have been reported in the last few years. The impact to date from cassava genetic improvement has been exclusively from traditional cross-breeding approaches. As in other key crops, this is likely to change in the near future as regulatory agencies develop rules and policy on transgenic cassava, and the traits available begin to meet farmer and consumer demands. Therefore, it is essential to understand and prioritize the specific economic traits and biological features of cassava in order to effectively improve the crop using biotechnology tools. Increasing our understanding of cassava response to low temperature (e.g. winter in the subtropics, or in the tropical highlands) and drought conditions is essential to develop useful strategies in plant tolerance to abiotic stress. Cassava storage roots also provide a unique model to understand the sink/source regulation of photosynthetic transport and partitioning in cassava. As the phenomena underlying cassava storage roots differ from those of other tuberous crops that cause huge production losses, understanding the regulatory mechanisms of PPD is critical. Advances in molecular biology have drawn increased public and private research attention globally. Molecular breeding of cassava is greatly promoted by the development of novel approaches derived from intensive basic research.

9 Where to look for further information

Cassava genetic transformation protocols are available in the following references:

1 Production of transgenic cassava (*Manihot esculenta* Crantz), http://dx.doi.org/10.1007/978-1-4020-2333-0_23
2 Regeneration of transgenic cassava from transformed embryogenic tissues, http://dx.doi.org/10.1385/1-59259-827-7:165
3 *Agrobacterium*-mediated transformation of friable embryogenic calli and regeneration of transgenic cassava, http://dx.doi.org/10.1038/nprot.2009.208

Cassava reviews related to genetic modification are available in the following references:

1 Development and application of transgenic technologies in cassava, http://dx.doi.org/10.1007/s11103-004-4872-x
2 Cassava genetic transformation and its application in breeding, http://dx.doi.org/10.1111/j.1744-7909.2011.01048.x
3 The potential of using biotechnology to improve cassava: a review, http://dx.doi.org/10.1007/s11627-016-9776-3

10 Acknowledgements

This work was supported by grants from the National Key Technology Research and Development Program of China (2015BAD15B01), the National Natural Science Foundation of China (31271775, 31571729) and the Earmarked Fund for China Agriculture Research System (CARS-12-shzp).

11 References

Allie, F., Pierce, E. J., Okoniewski, M. J. and Rey, C. (2014), Transcriptional analysis of South African cassava mosaic virus-infected susceptible and tolerant landraces of cassava highlights differences in resistance, basal defense and cell wall associated genes during infection. *BMC Genomics* 15(1): 1–30.

An, D., Yang, J. and Zhang, P. (2012), Transcriptome profiling of low temperature-treated cassava apical shoots showed dynamic responses of tropical plant to cold stress. *BMC Genomics* 13: 64.

Bayoumi, S. A., Rowan, M. G., Blagbrough, I. S. and Beeching, J. R. (2008), Biosynthesis of scopoletin and scopolin in cassava roots during post-harvest physiological deterioration: The *E-Z*-isomerisation stage. *Phytochemistry* 69(17): 2928–36.

Beyene, G., Chauhan, R. D., Wagaba, H., Moll, T., Alicai, T., Miano, D., Carrington, J. and Taylor, N. J. (2016), Loss of CMD2-mediated resistance to cassava mosaic disease in plants regenerated through somatic embryogenesis. *Molecular Plant Pathology* 17: 1095–110.

Blennow, A., Nielsen, T. H., Baunsgaard, L., Mikkelsen, R. and Engelsen, S. B. (2002), Starch phosphorylation: a new front line in starch research. *Trends in Plant Science* 7(10): 445–50.

Bredeson, J. V., Lyons, J. B., Prochnik, S. E., Wu, G. A., Ha, C. M., Edsinger-Gonzales, E., Grimwood, J., Schmutz, J., Rabbi, I. Y., Egesi, C., Nauluvula, P., Lebot, V., Ndunguru, J., Mkamilo, G., Bart, R. S.,

Setter, T. L., Gleadow, R. M., Kulakow, P., Ferguson, M. E., Rounsley, S. and Rokhsar, D. S. (2016), Sequencing wild and cultivated cassava and related species reveals extensive interspecific hybridization and genetic diversity. *Nature Biotechnology* 34(5): 562–70.

Bull, S. E., Owiti, J. A., Niklaus, M., Beeching, J. R., Gruissem, W. and Vanderschuren, H. (2009), *Agrobacterium*-mediated transformation of friable embryogenic calli and regeneration of transgenic cassava. *Nature Protocols* 4(12): 1845–54.

Carciofi, M., Shaik, S. S., Jensen, S. L., Blennow, A., Svensson, J. T., Vincze, É. and Hebelstrup, K. H. (2011), Hyperphosphorylation of cereal starch. *Journal of Cereal Science* 54(3): 339–46.

Ceballos, H., Kulakow, P. and Hershey, C. (2012), Cassava breeding: current status, bottlenecks and the potential of biotechnology tools. *Tropical Plant Biology* 5(1): 73–87.

Chauhan, R. D., Beyene, G., Kalyaeva, M., Fauquet, C. M. and Taylor, N. (2015), Improvements in *Agrobacterium*-mediated transformation of cassava (*Manihot esculenta* Crantz) for large-scale production of transgenic plants. *Plant Cell, Tissue & Organ Culture* 121: 591–603.

Chen, X., Xia, J., Xia, Z., Zhang, H., Zeng, C., Lu, C., Zhang, W. and Wang, W. (2015), Potential functions of microRNAs in starch metabolism and development revealed by miRNA transcriptome profiling of cassava cultivars and their wild progenitor. *BMC Plant Biology* 15: 33.

Chetty, C. C., Rossin, C. B., Gruissem, W., Vanderschuren, H. and Rey, M. E. C. (2013), Empowering biotechnology in southern Africa: establishment of a robust transformation platform for the production of transgenic industry-preferred cassava. *New Biotechnology* 30(2): 136–43.

Duan, X., Xu, J., Ling, E. and Zhang, P. (2013), Expression of Cry1Aa in cassava improves its insect resistance against *Helicoverpa armigera. Plant Molecular Biology* 83(1): 131–41.

Failla, M. L., Chitchumroonchokchai, C., Siritunga, D., De Moura, F. F., Fregene, M., Manary, M. J. and Sayre, R. T. (2012), Retention during processing and bioaccessibility of β-carotene in high β-carotene transgenic cassava root. *Journal of Agricultural Food Chemistry* 60(15): 3861–6.

Gaitán-Solís, E., Taylor, N. J., Siritunga, D., Stevens, W. and Schachtman, D. P. (2015), Overexpression of the transporters AtZIP1 and AtMTP1 in cassava changes zinc accumulation and partitioning. *Frontiers in Plant Science* 6: 492.

Hu, W., Wei, Y., Xia, Z., Yan, Y., Hou, X., Zou, M., Lu, C., Wang, W. and Peng, M. (2015), Genome-wide identification and expression analysis of the NAC transcription factor family in cassava. *PLoS ONE* 10(8): e0136993.

IFPRI (2014), *Biofortification Progress Briefs*, HarvestPlus, pp. 1–82.

Ihemere, U., Arias-Garzon, D., Lawrence, S. and Sayre, R. (2006), Genetic modification of cassava for enhanced starch production. *Plant Biotechnology Journal* 4: 453–65.

Ihemere, U. E, Narayanan, N. N. and Sayre, R. T. (2012), Iron biofortification and homeostasis in transgenic cassava roots expressing the algal iron assimilatory gene, FEA1. *Frontiers in Plant Science* 3(3): 171.

Kaur, B., Ariffin, F., Bhat, R. and Karim, A. A. (2012), Progress in starch modification in the last decade. *Food Hydrocolloids* 26(2): 398–404.

Koehorst-Van Putten, H., Sudarmonowati, E., Herman, M., Pereira-Bertram, I., Wolters, A., Meima, H., De Vetten, N., Raemakers, C. and Visser, R. (2012), Field testing and exploitation of genetically modified cassava with low-amylose or amylose-free starch in Indonesia. *Transgenic Research* 21: 39–50.

Legg, J., Lava-Kumar, P., Makeshkumar, T., Tripathi, L., Ferguson, M., Kanju, E., Ntawuruhunga, P. and Cuellar, W. (2014), Cassava virus diseases: biology, epidemiology, and management. In Gad, L. and Nikolaos, I. K. (eds), *Advances in Virus Research* 91: 85–142.

Li, K. T., Moulin, M., Mangel, N., Albersen, M., Verhoeven-Duif, N. M., Ma, Q., Zhang, P., Fitzpatrick, T. B., Gruissem, W. and Vanderschuren, H. (2015), Increased bioavailable vitamin B6 in field-grown transgenic cassava for dietary sufficiency. *Nature Biotechnology* 33(10): 1029–32.

Liao, W. B, Wang, G., Li, Y. Y., Wang, B., Zhang, P. and Peng, M. (2016), Reactive oxygen species regulate leaf pulvinus abscission zone cell separation in response to water-deficit stress in cassava. *Scientific Reports* 6: 21542.

Liu, J., Yang, J., Bi, H. and Zhang, P. (2014), Why mosaic? Gene expression profiling of african cassava mosaic virus-infected cassava reveals the effect of chlorophyll degradation on symptom development. *Journal of Integrative Plant Biology* 56(2): 122–32.

Liu, J., Zheng, Q., Ma, Q., Gadidasu, K. K. and Zhang, P. (2011), Cassava genetic transformation and its application in breeding. *Journal of Integrative Plant Biology* 53: 552–69.

Lu, F., Liang, X., Lu, H., Li, Q., Chen, Q., Zhang, P., Li, K., Liu, G., Yan, W., Song, J., Duan, C. and Zhang, L. (2016), Overproduction of superoxide dismutase and catalase confers cassava resistance to Tetranychus cinnabarinus. *Scientific Reports* (in press).

Ma, Q., Zhou, W. and Zhang, P. (2015), Transition from somatic embryo to friable embryogenic callus in cassava: dynamic changes in cellular structure, physiological status, and gene expression profiles. *Frontiers in Plant Science* 6: 824.

Maredza, A. T., Allie, F., Plata, G. and Rey, M. E. (2016), Sequences enhancing cassava mosaic disease symptoms occur in the cassava genome and are associated with South African cassava mosaic virus infection. *Molecular General Genetics* 291(3): 1467–85.

Mitprasat, M., Roytrakul, S., Jiemsup, S., Boonseng, O. and Yokthongwattana, K. (2011), Leaf proteomic analysis in cassava (*Manihot esculenta* Crantz) during plant development, from planting of stem cutting to storage root formation. *Planta* 233: 1209–21.

Naconsie, M., Lertpanyasampatha, M., Viboonjun, U., Netrphan, S., Kuwano, M., Ogasawara, N. and Narangajavana, J. (2016), Cassava root membrane proteome reveals activities during storage root maturation. *Journal of Plant Research* 129: 51–65.

Narayanan, N., Beyene, G., Chauhan, R. D., Gaitán-Solis, E., Grusak, M. A. and Taylor, N. (2015), Overexpression of Arabidopsis VIT1 increases accumulation of iron in cassava roots and stems. *Plant Science* 240: 170–81.

Narayanan, N. N., Uzoma, I., Claire, E. and Sayre, R. T. (2011), Overexpression of hydroxynitrile lyase in cassava roots elevates protein and free amino acids while reducing residual cyanogen levels. *PLoS ONE* 6(7): e21996.

Nassar, N. and Ortiz, R. (2010), Breeding cassava to feed the poor. *Scientific American* 302(5): 78–82, 84.

Ndunguru, J., De León, L., Doyle, C. D., Sseruwagi, P., Plata, G., Legg, J. P., Thompson, G., Tohme, J., Aveling, T., Ascencio-Ibáñez, J. T. and Hanley-Bowdoin, L. (2016), Two novel DNAs that enhance symptoms and overcome CMD2 resistance to cassava mosaic disease. *Journal of Virology* 90(8): 4160–73.

Niklaus, M., Gruissem, W. and Vanderschuren, H. (2011), Efficient transformation and regeneration of transgenic cassava using the neomycin phosphotransferase gene as aminoglycoside resistance marker gene. *GM Crops* 2(3): 193–200.

Ntui, V. O., Kong, K., Khan, R. S., Igawa, T., Janavi, G. J., Rabindran, R., Nakamura, I. and Mii, M. (2015), Resistance to Sri Lankan Cassava Mosaic Virus (SLCMV) in genetically engineered cassava cv. KU50 through RNA silencing. *PLoS ONE* 10: e0120551.

Nyaboga, E., Njiru, J., Nguu, E., Gruissem, W., Vanderschuren, H. and Tripathi, L. (2013), Unlocking the potential of tropical root crop biotechnology in east Africa by establishing a genetic transformation platform for local farmer-preferred cassava cultivars. *Frontiers in Plant Science* 4: 526.

Nyaboga, E. N., Njiru, J. M. and Tripathi, L. (2015), Factors influencing somatic embryogenesis, regeneration, and *Agrobacterium*-mediated transformation of cassava (*Manihot esculenta* Crantz) cultivar TME14. *Frontiers in Plant Science* 6:411.

Ogwok, E., Odipio, J., Halsey, M., Gaitán-Solís, E., Bua, A., Taylor, N. J., Fauquet, C. M. and Alicai, T. (2012), Transgenic RNA interference (RNAi)-derived field resistance to cassava brown streak disease. *Molecular Plant Pathology* 13: 1019–31.

Okwuonu, I. C., Achi, O. K., Egesi, C. N. and Taylor, N. J. (2015), Evaluation of red fluorescent protein (dsRed) as alternative visual marker of genetic transformation in cassava (*Manihot esculenta* crantz). *In Vitro Cellular and Developmental Biology – Plant* 51: 1–9.

Owiti, J., Grossmann, J., Gehrig, P., Dessimoz, C., Laloi, C., Hansen, M. B., Gruissem, W. and Vanderschuren, H. (2011), iTRAQ-based analysis of changes in the cassava root proteome reveals pathways associated with post-harvest physiological deterioration. *Plant Journal* 67(1): 145–56.

Prochnik, S., Reddy, M. P., Desany, B., Rabinowicz, P. D., Kodira, C., Mohiuddin, M., Rodriguez, F., Fauquet, C., Tohme, J., Harkins, T., Rokhsar, D. S. and Rounsley, S. (2012), The cassava genome: current progress, future directions. *Tropical Plant Biology* 5(1): 88–94.

Reilly, K., Bernal, D., Cortés, D. F., Gómez-Vásquez, R., Tohme, J. and Beeching, J. R. (2007), Towards identifying the full set of genes expressed during cassava post-harvest physiological deterioration. *Plant Molecular Biology* 64: 187–203.

Reilly, K., Gomez-Vasquez, R., Buschmann, H., Tohme, J. and Beeching, J. R. (2003), Oxidative stress responses during cassava post-harvest physiological deterioration. *Plant Molecular Biology* 53(5): 669–85.

Rolland-Sabaté, A., Sanchez, T., Buléon, A., Colonna, P., Ceballos, H., Zhao, S. S., Zhang, P. and Dufour, D. (2013), Molecular and supra-molecular structure of waxy starches developed from cassava (*Manihot esculenta* Crantz). *Carbohydrate Polymers* 92(2): 1451–62.

Rossin, C. B. and Rey, M. E. C. (2011), Effect of explant source and auxins on somatic embryogenesis of selected cassava (*Manihot esculenta* Crantz) cultivars. *South African Journal of Botany* 77: 59–65.

Sakurai, T., Plata, G., Rodríguez-Zapata, F., Seki, M., Salcedo, A., Toyoda, A., Ishiwata, A., Tohme, J., Sakaki, Y., Shinozaki, K. and Ishitani, M. (2007), Sequencing analysis of 20,000 full length cDNA clones from cassava reveals lineage specific expansions in gene families related to stress response. *BMC Plant Biology* 7: 66.

Salcedo, A., Valle, A. D., Sanchez, B., Ocasio, V., Ortiz, A., Marquez, P. and Siritunga, D. (2009), Comparative evaluation of physiological post-harvest root deterioration of 25 cassava (*Manihot esculenta*) accessions: Visual vs. hydroxycoumarins fluorescent accumulation analysis. *African Journal of Agricultural Research* 5(22): 3138–44.

Sanchez, T., Dufour, D., Moreno, J. L., Pizarro, M., Aragon, I. J., Dominguez, M. and Ceballos, H. (2013), Changes in extended shelf life of cassava roots during storage in ambient conditions. *Postharvest Biology and Technology* 86: 520–8.

Sayre, R., Beeching, J. R., Cahoon, E. B., Egesi, C., Fauquet, C., Fellman, J., Fregene, M., Gruissem, W., Mallowa, S., Manary, M., Maziya-Dixon, B., Mbanaso, A., Schachtman, D. P., Siritunga, D., Taylor, N., Vanderschuren, H. and Zhang, P. (2011), The biocassava plus program: biofortification of cassava for sub-Saharan Africa. *Annual Review of Plant Biology* 62(1): 251–72.

Siriwat, W., Kalapanulak, S., Suksangpanomrung, M., Netrphan, S., Meechai, A. and Saithong, T. (2012), Transcriptomic data integration inferring the dominance of starch biosynthesis in carbon utilization of developing cassava roots. In Chan, J. H., Meechai, A. and Kwoh, C. K. (eds), *Proceedings of the 3rd International Conference on Computational Systems-Biology and Bioinformatics*, pp. 96–106.

Sojikul, P., Kongsawadworakul, P., Viboonjun, U., Thaiprasit, J., Intawong, B., Narangajavana, J. and Svasti, M. R. J. (2010), AFLP-based transcript profiling for cassava genome-wide expression analysis in the onset of storage root formation. *Physiologia Plantarum* 140: 189–98.

Sojikul, P., Saithong, T., Kalapanulak, S., Pisuttinusart, N., Limsirichaikul, S., Tanaka, M., Utsumi, Y., Sakurai, T., Seki, M. and Narangajavana, J. (2015), Genome-wide analysis reveals phytohormone action during cassava storage root initiation. *Plant Molecular Biology* 88: 531–43.

Sonnewald, U. and Kossmann, J. (2012), Starches-from current models to genetic engineering. *Plant Biotechnology Journal* 11(2): 223–32.

Taylor, N., Gaitán-Solís, E., Moll, T., Trauterman, B., Jones, T., Pranjal, A., Trembley, C., Abernathy, V., Corbin, D. and Fauquet, C. (2012), A high-throughput platform for the production and analysis of transgenic cassava (*Manihot esculenta*) plants. *Tropical Plant Biology* 5(1): 127–39.

Uarrota, V. G. and Maraschin, M. (2015), Metabolomic, enzymatic, and histochemical analyzes of cassava roots during postharvest physiological deterioration. *BMC Research Notes* 8(1): 648.

Utsumi, Y., Tanaka, M., Morosawa, T., Kurotani, A., Yoshida, T., Mochida, K., Matsui, A., Umemura, Y., Ishitani, M., Shinozaki, K., Sakurai, T. and Seki, M. (2012), Transcriptome analysis using a high-density oligomicroarray under drought stress in various genotypes of cassava: an important tropical crop. *DNA Research* 19(4): 335–45.

Vanderschuren, H., Akbergenov, R., Pooggin, M., Hohn, T., Gruissem, W. and Zhang, P. (2007b), Transgenic cassava resistance to African cassava mosaic virus is enhanced by viral DNA-A bidirectional promoter-derived siRNAs. *Plant Molecular Biology* 64: 549–57.

Vanderschuren, H., Alder, A., Zhang, P. and Gruissem, W. (2009), Dose-dependent RNAi-mediated geminivirus resistance in the tropical root crop cassava. *Plant Molecular Biology* 70: 265–72.

Vanderschuren, H., Moreno, I., Anjanappa, R. B., Zainuddin, I. M. and Gruissem, W. (2012), Exploiting the combination of natural and genetically engineered resistance to cassava mosaic and cassava brown streak viruses impacting cassava production in Africa. *PLoS ONE* 7(9): e45277.

Vanderschuren, H., Nyaboga, E., Poon, J. S., Baerenfaller, K., Grossmann, J., Hirsch-Hoffmann, M., Kirchgessner, N., Nanni, P. and Gruissem, W. (2014), Large-scale proteomics of the cassava storage root and identification of a target gene to reduce postharvest deterioration. *Plant Cell* 26(5): 1913–24.

Vanderschuren, H., Stupak, M., Fütterer, J., Gruissem, W. and Zhang, P. (2007a), Engineering resistance to geminiviruses – Review and perspectives. *Plant Biotechnology Journal* 5: 207–20.

Viksø-Nielsen, A., Blennow, A., Jørgensen, K., Kristensen, K. H., Jensen, A. and Møller, B. L. (2001), Structural, physicochemical, and pasting properties of starches from potato plants with repressed r1-gene. *Biomacromolecules* 2(3): 836–43.

Wang, W., Feng, B., Xiao, J., Xia, Z., Zhou, X., Li, P., et al. (2014), Cassava genome from a wild ancestor to cultivated varieties. *Nature Communications* 5: 5110.

Welsch, R., Arango, J., Bär, C., Salazar, B., Al-Babili, S., Beltrán, J., Chavarriaga, P., Ceballos, H., Tohme, J. and Beyer, P. (2010), Provitamin A accumulation in cassava (*Manihot esculenta*) roots driven by a single nucleotide polymorphism in a phytoene synthase gene. *Plant Cell* 22(10): 3348–56.

Wolfe, M. D., Rabbi, I. Y., Egesi, C., Hamblin, M., Kawuki, R., Kulakow, P., Lozano, R., Carpio, D. P. D., Ramu, P. and Jannink, J. (2016), Genome-wide association and prediction reveals genetic architecture of Cassava Mosaic Disease resistance and prospects for rapid genetic improvement. *Plant Genome* 9. doi:10.3835/plantgenome2015.11.0118.

Woo, J. W., Kim, J., Kwon, S. I., Corvalán, C., Cho, S. W., Kim, H., Kim, S. G., Kim, S. T., Choe, S. and Kim, J. S. (2015), DNA-free genome editing in plants with preassembled CRISPR-Cas9 ribonucleoproteins. *Nature Biotechnology* 33(11): 1162–4.

Xu, J., Duan, X. G., Yang, J., Beeching, J. R. and Zhang, P. (2013a), Enhanced reactive oxygen species scavenging by overproduction of superoxide dismutase and catalase delays post-harvest physiological deterioration of cassava storage roots. *Plant Physiology* 161(3): 1517–28.

Xu, J., Duan, X., Yang, J., Beeching, J. R. and Zhang, P. (2013b), Coupled expression of Cu/Zn-superoxide dismutase and catalase in cassava improves tolerance against cold and drought stresses. *Plant Signaling and Behavior* 8: e24525.

Xu, J., Yang, J., Duan, X., Jiang, Y. and Zhang, P. (2014), Increased expression of native cytosolic Cu/Zn superoxide dismutase and ascorbate peroxidase improves tolerance to oxidative and chilling stresses in cassava (*Manihot esculenta* Crantz). *BMC Plant Biology* 14: 208.

Yadav, J., Ogwok, E., Wagaba, H., Patil, B., Bagewadi, B., Alicai, T., Gaitan- Solis, E., Taylor, N. and Fauquet, C. (2011), RNAi mediated resistance to cassava brown streak Uganda virus in transgenic cassava. *Molecular Plant Pathology* 12: 677–87.

Yang, J., An, D. and Zhang, P. (2011), Expression profiling of cassava storage roots reveals an active process of glycolysis/gluconeogenesis. *Journal of Integrative Plant Biology* 53(3): 193–211.

Zainuddin, I. M., Schlegel, K., Gruissem, W. and Vanderschuren, H. (2012), Robust transformation procedure for the production of transgenic farmer-preferred cassava landraces. *Plant Methods* 8: 24.

Zhang, P., An, D., Ma, Q., Wang, H., Wang, L., Xu, J., Yang, J. and Zhou, W. (2013), Key scientific questions and recent advances in cassava molecular breeding. *Scientia Sinica Vitae* 43: 1082–9.

Zhang, P., Bohl-Zenger, S., Puonti-Kaerlas, J., Potrykus, I. and Gruissem, W. (2003), Two cassava promoters related to vascular expression and storage root formation. *Planta* 218(2): 192–203.

Zhang, P., Phansiri, S. and Puonti-Kaerlas, J. (2001), Improvement of cassava shoot organogenesis by the use of silver nitrate *in vitro*. *Plant Cell Tissue and Organ Culture* 67(1): 47–54.

Zhang, P., Vanderschuren, H., Fütterer, J. and Gruissem, W. (2005), Resistance to cassava mosaic disease in transgenic cassava expressing antisense RNAs targeting virus replication genes. *Plant Biotechnology Journal* 3: 385–97.

Zhao, S., Dufour, D., Sánchez, T., Ceballos, H. and Zhang, P. (2011), Development of waxy cassava with different biological and physico-chemical characteristics of starches for industrial applications. *Biotechnology and Bioengineering* 108(8): 1925–35.

Zhao, S. S. (2012), Quality improvement of cassava starch by the expressional regulation of GBSSI and SBE genes. Dissertation of Graduate University of Chinese Academy of Sciences, Institute of Plant Physiology and Ecology, Shanghai Institutes for Biological Sciences, Chinese Academy of Sciences.

Zidenga, T., Leyva-Guerrero, E., Moon, H., Siritunga, D. and Sayre, R. (2012), Extending cassava root shelf life via reduction of reactive oxygen species production. *Plant Physiology* 159(4): 1396–407.

Part 2

Breeding improved cassava varieties

Chapter 7

Breeding cassava for higher yield

Piya Kittipadakul, Pasajee Kongsil and Chalermpol Phumichai, Kasetsart University, Thailand; and Shelley H. Jansky, USDA-ARS Vegetable Crops Research Unit and University of Wisconsin-Madison, USA

1 Introduction

This chapter reviews the strategies and factors for success in cassava breeding for higher yield, including the importance of genetic diversity and breeding schemes. There is a particular focus on Thailand, which has some of the most successful breeding programmes toward improving cassava varieties over a period of more than 40 years. The programmes have increased root yield and starch content, which have provided great benefits to many stakeholders through raising farmers' income and increasing starch production efficiency in the industry.

Cassava is a root crop that adapts well to a wide range of both high productivity and stressful environments (Ceballos et al., 2012). Farmers mostly grow cassava in poor soil and drought-stressed areas, where the crop typically has comparative adaptive advantages

http://dx.doi.org/10.19103/AS.2016.0014.18

over other crops such as cereal grains and grain legumes. African farmers grow cassava mainly for human consumption but in Asia industrial starch and animal feed predominate (Kawano, 2003). Particularly, a key factor for grower decisions to continuously plant cassava or not is the expectation of getting a benefit from planting the crop in a volatile price environment. It is increasingly true that only cassava growers with high productivity can survive. Skillful growers achieve high productivity by using improved high-yielding cassava varieties along with good cultural practices. Kittipadakul et al. (2014) reported examples of the combined effects of improved high-yielding varieties and good cultural practices (including drip irrigation in one-rai (0.16 ha) experimental units at Nakhon Ratchsima Province). The breeding line CMR 34-08-89 (CMR89) yielded 67.5 tonnes per hectare with 18.7% starch content and the variety Huay Bong 60 (HB60) yielded 51.9 tonnes per hectare with 30.2% starch content.

Nowadays, cassava is grown not only in Latin America, but it is also popular in tropical and subtropical regions of Asia and Africa, where its genetic base is narrower than in Latin America. It is an important source of raw material for food, feed, fuel and starch-based industries.

Higher yield is a trait of near-universal interest to cassava growers. Cultivated cassava is a functional diploid crop with 18 pairs of chromosomes (Nassar and Ortiz, 2008). It is a heterozygous species, with high inbreeding depression. This means that breeding cassava for higher root yield is one of the most challenging goals for breeders.

The variety Kasetsart 50 (KU50) is the best example of the success of breeding for high cassava yield in Thailand. The variety has made a huge contribution to economic development in Thailand, and appears to be the most successful cassava variety of all time, anywhere in the world. Compared to the yield of Rayong 1, the popular variety in the past, it was estimated that the accumulated producer benefits from KU50 adoption (Net Present Value, NPV) were about $1629.7 million in Thailand (based on exchange rate $1 = 30 Baht) (Praneetvatakul et al., 2015). Now, KU50 has expanded all over Thailand and Southeast Asia. By 2013, it was grown on over 1 million hectares in Thailand and Vietnam (as KM94) and was also being grown in Indonesia, Cambodia, Myanmar, Philippines and probably China. The breeding of KU50 and other varieties from Thai programmes provide valuable lessons more broadly to breeding cassava for high yield and high starch.

2 Genetic diversity for cassava breeding

The foundation of successful breeding is access to appropriate genetic diversity, that is, sources of genes that control the traits of interest and which can be combined to confer high levels of expression. Centres of origin and of diversification are always of key consideration for breeders in determining sources of genetic diversity. The genus *Manihot*, with some 100 current-day species, originated in South America (Nassar and Ortiz, 2008) but the evolutionary pathway of cassava (*Manihot esculenta*) is unclear. It has been hypothesized as a natural hybrid among wild *Manihot* subspecies such as *M. pilosa* or developed from *M. esculenta* ssp. *flabellifolia* (Pohl) Ciferri (Ceballos et al., 2012). There have been several germplasm collections established in a number of countries in the past half century. Systematic phenotyping has been carried out for priority traits such as high productivity, pest and disease resistance, and starch quality, among others (Ceballos et al., 2012). The combination of new technologies for producing high-quality vegetative planting materials and for minimizing the risk of transfer of pests and pathogens through *in vitro* or seed-based germplasm exchange means that global cassava breeding programmes have access to a rich genetic resource (Ceballos et al., 2012).

The largest cassava collection is at CIAT (International Center for Tropical Agriculture) in Cali, Colombia, which now includes a global collection of around 6600 cassava accessions, wild *Manihot* species and advanced breeding lines. This collection was initially established in 1969 and held solely as a field collection until 1978, when an *in vitro* collection was initiated. Evaluation of the CIAT germplasm collection as part of a new breeding effort started in 1975. Base genetic materials were mostly landraces from farmers' fields in Latin America, including Colombia, Venezuela, Ecuador, Mexico, Panama, Brazil, Costa Rica, Puerto Rico and Peru (Kawano, 2003). Germplasm improvement efforts at CIAT have focused on productivity, root quality and responses to biotic and abiotic stresses in diverse agro-ecosystems. A core collection of about 10% of the whole collection is based on sampling the diversity of collection origin, morphological and performance traits, and molecular markers (Ceballos et al., 2004). The core collection is currently under re-evaluation based on diversity of new-generation molecular markers, such as SNPs, which were not part of the original criteria.

An additional 5000 field and *in vitro* accessions of cassava are also maintained at the Cruz das Almas Experimental Station of Empresa Brasileira de Pesquisa Agropecuária (EMBRAPA) in the Bahia state, Brazil, along with some wild *Manihot* species maintained at the Universidade de Brasília (Ceballos et al., 2012).

In Nigeria, the International Institute of Tropical Agriculture (IITA) has a germplasm collection that contains mostly West African cassava landraces, including over 2000 accessions, along with some wild species. In Asia, the Central Tuber Crops Research Institute of India has a cassava collection of around 1700 accessions and the National Bureau of Plant Genetic Resources also has around 184 landraces. Moreover, there are over 500 accessions of cassava and eight *Manihot* species maintained in state universities in India (Ceballos et al., 2012). In 2000, CIAT transferred tissue culture plantlets of 600 cassava clones, which were included in the core collection, to the Rayong Field Crops Research Center (RFCRC), Thailand (Rojanaridpiched et al., 1993).

Kawano (2003) concluded that the success of cassava breeding programmes in Thailand and other countries in Asia in collaboration with CIAT was due to two main factors:

- an inclusion of a broad base of genetic diversity from centres of origin and diversification
- the evaluation of germplasm in a range of environmental conditions.

One method to broaden the genetic base is through crossing with wild relatives to generate interspecific hybrids having new traits. Some interspecific hybrids have been reported to have high protein content, high resistance to pests and diseases or high carotenoid content; however, root productivity was low when compared to hybrids among the cultivated cassava clones (Nassar, 2007). Wild *Manihot* species do not typically have significant production of storage roots, although two species, *M. brachyandra* and *M. epruinosa*, have been reported to exhibit considerable root production.

Other non-conventional methods to create or reveal variation in cassava are mutagenesis and creating doubled-haploid lines for true cassava hybrid production (Ceballos et al., 2004). Partially inbred line production has revealed hidden positive recessive traits such as waxy cassava (Ceballos et al., 2007) and reduced genetic load (Ceballos et al., 2004) which can occur frequently in vegetatively propagated plants which have less recombination than seed-propagated species. Ramu et al. (2016) have reported a HapMap from deep sequencing of cassava elite accessions and wild relatives, with an attempt to get rid of deleterious mutations as an alternative way of yield improvement through inbreeding rather than hiding deleterious mutations through crossing to make heterozygotes. Moreover, with inbred lines, near-isogenic lines can be formed to facilitate conventional and molecular genetic studies in the future (Ceballos et al., 2004).

In India, triploid cassava lines were generated from crosses between a diploid cassava with a colchicine-induced tetraploid. Two lines were reported to have high root yield, ranging from 23.4 to 58.3 tonnes per hectare in five locations in India with dry matter ranging from 44.3 to 46.9% and starch content ranging from 36.5 to 36.8%.

Mutation-breeding programmes in India also resulted in some mutants having significantly higher root yields (20–30% higher than the wild type M-4) and higher starch content (Khumaida et al., 2015). A higher photosynthesis rate and lower cyanide content were also reported in mutant lines (Edison et al., 2006).

Kawano (2003) emphasized that a comprehensive field evaluation is a critical methodology for successfully assessing genetic diversity. Soil types and climatic conditions are important factors to be considered in choosing field trial sites. Three CIAT germplasm evaluation sites in Colombia are good examples of diverse environmental conditions, ranging from highly productive to highly stressful (both biotic and abiotic) in the lowland tropics. The location at the CIAT headquarters represents favourable environmental conditions, including fertile soil.

The classification of cassava cultivation zones in Latin America is a good example of breeding cassava for a defined zone. These zones, as defined by CIAT, are called edapho-climatic zones (ECZ), and include sub-humid tropics (ECZ1), acid soil savannas (ECZ2), humid tropics (ECZ3), mid-altitude tropics (ECZ4), high-altitude tropics (ECZ5), sub-tropics (ECZ6) and semi-arid regions (ECZ7). Top genotypes selected from each ECZ are included in the elite clone collection.

Early evaluation of the CIAT genebank was conducted in three key environments of Colombia: the seasonally dry north coastal eastern region (ECZ1), the eastern lowland plains (ECZ2) and the highly productive CIAT experiment station of Palmira, Valle (ECZ4). Regional selection trials were conducted in six locations, ranging from dry to semi-arid lowland tropics (Kawano, 2003). The key selection criteria for evaluating cassava genetic diversity were harvest index, biomass and root dry matter. Root yield per plant was given lower priority due to the unreliability of inter-genotypic competition between rows. Moreover, in the selection criteria, diverse phenotypes were selected in spite of low yield in order to maintain wide genetic variability.

Current germplasm collections in Thailand can be traced back to the 1930s (Komkris, 1937). Three varieties were imported from the Philippines and 17 varieties from Malaysia to a southern agricultural experiment station (which is now known as a rubber tree research station in the Had-Yai district in the Songkla province). At that time, there was no report about the number of cassava varieties cultivated in Thailand. Then, in 1956, there was a collection of cassava landraces from the north-eastern part of Thailand selected at Huay Bong agricultural station in the Rayong province (which is now known as the Rayong Field Crops Research Station). The Department of Agriculture (DOA) imported seven varieties from Indonesia in 1963. Two years later, the Thai Tapioca Trade Association imported 43 varieties from the Virgin Islands for the DOA. In 1970, there was the first clonal cassava introduction from CIAT of five varieties (CMC9, CMC36, CMC72, CMC76 and CMC84) followed by a total of 20 varieties and 132 756 seeds introduced to Thailand from 1975 to 1999. Moreover, in 2001, the DOA signed an MOU with CIAT for introduction of

600 accessions to Thailand in the form of tissue culture, and 30 more accessions which are resistant to virus introduced to the Thailand germplasm collection in 2007. Recently in May 2013, the Department of Agronomy, Kasetsart University has introduced 21 clones of cassava germplasm from CIAT which are yellow root; good cooking; and resistant to cassava mosaic disease (CMD), bacterial blight and super-elongation disease to provide new diversity in a proactive manner for Thai breeding programmes. Crosses among cassava varieties were started in 1974. The first successful cross in Thailand was a cross between Rayong 1 and Jek-Kui. A large amount of cassava germplasm (943 accessions) was imported to Thailand from CIAT from 2001 to 2005, including the core collection referred to above, at the Rayong Field Crop Research Station.

When the breeding programme in Thailand began in the late 1970s, there was no accession that had better yield than the Thai landraces, especially when compared to Rayong 1, the local variety (Rojanaridpiched et al., 1993). Rayong 1, grown in most of Thailand by the late1950s, has high root yield across Thai field conditions, good sprouting, vigorous vegetative growth, moderate harvest index and good plant type. Rayong 1 was first officially named in 1975. Before that there were many landrace varieties grown in Thailand, but they had not been recorded officially. None of the early introductions, for example from Indonesia, the Virgin Islands and CIAT, could outperform Rayong 1 at that time (Rodjanaridpiched et al., 1993). Rayong 1 continued to be the dominant variety until products from the Thai breeding programmes began to have widespread impact by the early 1990s.

The importance of Rayong 1 as a parent can be seen in subsequent varieties that include it in their pedigree. For example, the high-yielding Rayong 60 came from a cross between MCol1684 and Rayong 1. Sriracha 1, from a cross between (MCol113 × MCol22) and Rayong 1, and had high root dry matter and good plant type. In 1992, Kasetsart 50 (KU50), which came from a cross between Rayong 1 and Rayong 90, had high root dry matter and high yield. Currently KU50 has become the most popular variety in Thailand and as well as several other countries in Asia because of good adaptation to a broad range of environments. Since the release of new varieties starting in the late 1980s, the area planted to Rayong 1 has been decreasing continuously. Although nowadays Rayong 1 is no longer grown in Thai farm fields, its legacy continues as one of the parents of Rayong 60, Rayong 72 and KU50 (Rodjanaridpiched et al., 1993).

3 Breeding programmes: key objectives and selection stages

At the establishment of the breeding programme at CIAT in the mid-1970s, the main goal was to help small farmers increase their income and food security by growing more productive, higher-quality cassava. To accomplish this goal, the focus was on covering an agro-ecological range from low-input or marginal environments to high-productivity conditions. More recently, CIAT added other goals such as nutritional value by breeding for high beta-carotene content, and to provide options for new, specialized markets such as waxy cassava (Ceballos et al., 2012).

The cassava breeding programme in Thailand is an excellent example of success in a centralized national breeding system to increase productivity to meet the demand for domestic cassava root processing as well as export of starches and dried root chips

(Kawano, 2003; Ceballos et al., 2012). For cassava breeding in Thailand, where the end uses are mostly for industry, pest and disease pressure was not a significant factor for many years in determining yield. Nonetheless, in 2009 the cassava mealy bug was introduced and briefly caused devastating yield losses until a biological control programme was successfully established.

Breeding objectives concentrate on agronomic traits such as fresh root yield, root dry matter content, harvest index, root starch content, leaf and stem weight, total plant weight and plant type. Moreover, new cassava varieties bred for Thailand should have specific starch quality properties for starch industry demand, in addition to high root starch content. These varieties should have very white roots, and starch gels should be clear in colour, with high viscosity. Important agronomic traits include good sprouting, vigorous growth, erect plant type (late branching), ease of harvest and long storability of roots (delayed onset of post-harvest deterioration) (Rojanaridpiched et al., 1993). Even though Thailand does not grow cassava as a major crop for human consumption, this breeding effort still continues to add value to cassava in the niche market as a human consumption product. This will help to promote food security in Thailand in the future because cassava is a staple crop that tolerates severe environmental changes better than other economic crops of Thailand, such as rice. Moreover, this breeding objective also maintains genetic diversity in the cassava gene pool for the breeding programme, not only in Thailand, but also in other countries where cassava is regarded as a staple crop.

Breeders have also used parents with a high net photosynthesis combined with other yield-determinant traits, such as high leaf area index and harvest index to improve root yield of breeding lines. Physiological traits related to photosynthesis can be used for selection, but they are not practical in early generations where precise measurement is difficult due to large numbers of genotypes that need to be managed.

The main stages in a cassava breeding programme are:

1. Germplasm collection of local landraces, introduction from other institutions and evaluation
2. Parent selection for crossing
3. Crossing to create new hybrids (clones)
4. Seedling selection trial (SST) from botanical seeds
5. Clonal selection trials (CST)
6. Preliminary selection trials (PST)
7. Advanced selection trials (AST)
8. Regional selection trials (RST)
9. Release of variety

The first steps of developing and choosing parental material to produce novel breeding lines and high seed numbers are a key component of success in cassava breeding. The more seeds there are, the higher the probability of developing high-yielding lines. For instance, in the beginning of the breeding programme at CIAT (in 1975), between 40 000 and 60 000 F1 seeds from 200 to 400 crosses were made each year. Approximately 30 000 seeds were sown in seedling selection trials with a 75% survival rate. Around 3 000 clones were selected in clonal selection trials (single-row trials) and 500 clones were selected for replicated selection trials (Kawano, 2003).

Between 20000 and 40000 F1 hybrid seeds were produced annually by controlled pollination in the Thai-CIAT breeding programme in Thailand, during 1982–1998. Between 12000 and 25000 genotypes were grown in seedling trials at the Rayong research station in eastern Thailand. Between 1200 and 2400 clones were then selected to be used in clonal selection trials at the same location. Anywhere from 100 to 160 clones were planted in preliminary selection trials with two replications, again at the same location. In the next step, between 16 and 20 clones were selected for advanced selection trials with four replications in each of the three locations. In the last stage, seven to nine elite clones were selected for regional selection trials at six locations (Kawano, 2003). After the variety improvement programme started in Thailand, a number of new varieties were soon released by the RFCRC, DOA, as well as by Kasetsart University, and this process of varietal release has continued to the present.

4 Selection schemes for breeding

Cassava is a highly heterozygous plant. Fortunately, it can be asexually propagated to facilitate estimating genetic effects by evaluation in multiple environments. However, due to non-additive effects and high within-family variation for quantitative and complex traits such as fresh root yield, phenotypic recurrent selection from full-sib or half-sib progenies to exploit heterosis has been done in large part without benefit of good genetic information. Therefore, the success of selection of cassava lines is mostly based on precise evaluation and selection of individual genotypes without considering family structure (Ceballos et al., 2015). For more precise selection, stratified phenotypic selection conducted within each replication has been used to reduce the environmental effect on phenotypic expression (Ceballos et al., 2004). Due to high genetic variation within families, and the fact that the genetic integrity of heterozygous individuals is maintained through clonal propagation, single outstanding clones are selected *per se*, rather than by average family performance.

Diallel crosses in cassava have been developed to evaluate general combining ability (GCA) and specific combining ability (SCA) effects of each cross. As an example, dry matter content has a higher heritability than fresh root yield. Within-family variation is higher for fresh root yield than for dry matter content (Ceballos et al., 2015). Since within-family variation of fresh root yield is high, prediction of breeding value of parent lines based on SCA is more precise than prediction based upon GCA (Ceballos et al., 2015).

GCA is used for estimating average performance of half-sib progenies for traits which have high additive effects, such as pest and disease resistance, for example, CMD resistance, cassava brown streak disease (CBSD), dry matter content, harvest index and plant type. SCA is used for estimating average performance of full-sib progenies in complex traits with high non-additive effects such as fresh root yield (Ceballos et al., 2015). Mean squares for GCA and SCA are used to estimate the additive and non-additive effects in different types of traits (Ceballos et al., 2015). For fresh root yield, the non-additive effect was predominant, so that SCA was large and the GCA/SCA ratio was around 1. For plant architecture-related traits and pest and disease resistance, the additive effect is the major effect on phenotype so the ratio of GCA/SCA in these traits is larger than 1 (Ceballos et al., 2015).

More recently, marker-assisted selection (MAS) has been used in plant breeding given the following advantages (Ceballos et al., 2015):

- For traits that are expensive or time consuming to measure conventionally, MAS is helpful for the reduction of cost and time.
- For pest and disease-resistance traits, MAS can allow countries in which those pests or diseases have not occurred yet to screen for resistant lines without exposing plants to actual pest and disease pressure.
- In species for which selfing has been successful, MAS can help identify recessive alleles during backcrossing and help pyramid traits into elite lines. However, selfing in cassava can cause inbreeding depression. Therefore, this advantage of MAS will not be applied in cassava breeding until inbred cassava lines can be made.

However, the success of MAS depends on simple traits with relatively high heritability. For complex traits, environmental interactions can cause the misinterpretation of genetic effects, especially in the case of non-additive effects.

For this reason, for cassava productivity traits, quantitative trait loci (QTLs) related to early bulking, dry matter content and plant architecture (Okogbenin and Fregene, 2002, 2003; Kizito et al., 2007; Okogbenin et al., 2008; Boonchanawiwat et al., 2011) were identified. However, QTL directly linked to fresh root yield have not been found because experimental errors in phenotyping can be high due to variation in maturity, biotic and abiotic stresses, a low number of replications, small plot size and a reduced number of testing locations (Ceballos et al., 2004, 2015).

To overcome the disadvantages of MAS, genomic selection (GS) is now being applied in some crops. GS that uses a random-effects statistical modelling approach to jointly estimate all marker effects does not require significance testing and has the goal of capturing small-effect QTL that are excluded by significance thresholds used in conventional MAS. This procedure involves the use of dense genome-wide markers covering a whole plant genome of a training population to determine genomic estimated breeding value (GEBV) (Heffner et al., 2009). A GS 'prediction model' is subsequently created that integrates these two sets of data. The model produces a GEBV for lines within the breeding population for which genotypic information is available. GEBVs represent the overall predicted quality, or value, of a line as a potential parent for crossing. Essentially, it is this GEBV that tells a breeder how a plant will perform in the field, before it is field-tested

The GS model will effectively estimate the breeder's choice of multiple traits in a selection index based on genome-wide markers, including markers linked to loci with small effects. Through recurrent selection, GS can maximize genetic gains by the reduction of the cycle time (de Oliviera et al., 2012). However, GS is still not efficient to estimate complex traits with non-additive effects, especially epistatic effects, which require more information and computational resources for identification than dominance effects (Nakaya and Isobe, 2012). Therefore, the use of GS for predicting key yield components such as cassava dry matter content or even fresh root yield is still an ongoing project of the NextGen Cassava Project, which is mainly developing proof of concept for GS and towards solving cassava issues for Africa.

Before the prediction model can be applied to the breeding population, the accuracy of the model must be tested. For this, the majority of the training population is used to create a prediction model, which is then used to predict the GEBVs of the remaining individuals in the training population, using genotypic data only. This enables researchers to 'test' and refine the prediction model to ensure the prediction accuracy is high enough that future predictions can be relied upon. Once validated, the model can be applied to a breeding population to calculate GEBVs of lines for which genotypic, but not phenotypic, information is available.

The significant advantage of GS is that phenotyping for lines for which GEBVs have been calculated can be done after selection and crossing, while the breeding cycle advances on the basis of the GEBVs alone. This advantage allows for more cycles and increased genetic gain per unit time using GS, compared to phenotypic selection. Of course, GS does not eliminate the need for phenotyping. On the contrary, the training population must be regularly updated with phenotypic and genotypic data of new lines in the breeding population, to maintain or increase accuracy as the breeding population evolves.

In conventional cassava breeding, new breeding material resulting from a cross is assessed by breeders over several growing seasons and in multiple locations before they select superior clones for the next cycle of germplasm improvement. Cassava is vegetatively propagated, and limited numbers (5–10) of propagules can be obtained from one plant. To obtain sufficient propagules to perform replicated and multi-locational trials for meaningful phenotypic characterization, several cycles of propagation are thus required.

Cassava breeding cycles, from seedling to multi-location field trials, can therefore take more than five years, limiting the rate of variety improvement and breeders' ability to respond to new challenges. Shifting from phenotype-based selection to genotype-based selection has the potential to overcome lengthy cassava breeding cycles. By using GS in cassava breeding programme, breeders can select superior clones after only two years if GEBVs are accurate enough.

In addition to shortening the cassava breeding cycle, GS also promises to increase the accuracy of selection in cassava breeding. While feasible, it is not necessarily efficient to advance many thousands of cassava seedlings through to adult clonal evaluation. In conventional cassava breeding schemes, breeders thus discard many clones at the seedling stage, a point at which important cassava traits can only be poorly predicted. Thus, valuable clones may be lost. GS allows breeders to use GEBVs to predict adult performance at very early growth stages. This significantly increases the number of seedlings that can be accurately evaluated.

GS promises to revolutionize cassava breeding for the future. Giving breeders the ability to select based on predictions rather than observations will result in much improved genetic gains and efficiency in cassava breeding. This dramatic increase in efficiency will allow cassava breeders to meet the unprecedented demands of crop improvement.

Cassava breeding without GS could still make effective use of inbreeding. Assuming that inbred lines of cassava can be created to be parental lines for true hybrid production, as in maize breeding, reciprocal recurrent selection of two heterotic populations can truly improve cassava genetic gain even in root yield traits that are difficult to improve upon using highly heterozygous parents. The use of homozygous parents can exploit heterosis and allow expression of non-additive effects (Ceballos et al., 2015). Moreover, backcrossing to introgress many traits for yield stability into elite inbred lines can improve those lines to be elite parental lines. Another way to improve the inbred lines is to recombine them to form a new population in the same heterotic group (Ceballos et al., 2015).

In India, progress in capitalizing on hybrid vigour has been made by selfing cassava lines for four generations. Six of the selfed lines were crossed in a diallel system. Root yields of hybrids were higher than the better parent and better than standard varieties such as Sree Sahya, Sree Visakham and Sree Prakash by 36.8%, 26.8% and 13.0%, respectively. Seven elite top-cross hybrids had high root yield ranging from 37 to 44 tonnes per hectare with high harvest index (61.6–74.9%) and high starch content (26.7–34.3%).

To sum up, high root yield is a quantitative trait that cannot be predicted as easily as most qualitative traits such as waxy starch trait, high provitamin A and several disease-resistance traits. Therefore, breeding for high root yield is challenging because it is difficult to bring all yield-related genes together in a single genotype, but also the effects of environment are high. The concept of inbreeding cassava to create inbred lines can lead the way to real molecular breeding of cassava as molecular markers can be studied more thoroughly without ambiguous interpretations of gene interactions. Near-isogenic line populations generated from inbreeding avoid the challenges of dealing with variation within families.

Until molecular tools are better developed and more widely applied, conventional breeding of cassava still relies on the good judgement of cassava breeders in the field to select parents to improve the chances of obtaining superior hybrid genotypes that can be indefinitely cloned. Breeders must select lines that not only perform well in specific areas, such as in experiment stations where soil fertility typically is high, but are also adapted to farmers' fields, which may have significant environmental challenges. Therefore, yield stability is as important as root yield potential. If multiple selection trials can be done near the beginning of a breeding programme, then the probability of combining high yield with good stability increases. If specific varieties are developed to fit certain environments, then dissemination of new cassava varieties to farmers should be done with care to provide specific information on agronomic requirements.

In addition, breeding can only increase the *yield potential* based on genetic factors, but the transfer of knowledge to farmers so they can adopt effective and sustainable agricultural production practices is also important for elevating cassava root yield in the long run. Most cassava lines from breeding programmes have been selected from trials with good agricultural practices applied. Farmers also need to know how to implement good agricultural practices in their own fields together with understanding of these following issues:

1 Morphological characteristics of varieties: Many farmers adopt commercial varieties knowing only the name of these varieties and some basic traits such as yield and starch content, without knowing morphological identification such as young leaf colour, stem colour, leaf shape or stipule colour, which are important to identify the true varieties. Without proper identification, farmers could buy stems of the wrong varieties and not obtain optimum yield. Field mixtures can be continued from one season to the next unless farmers know how to accurately identify specific varieties.
2 Selection trials of varieties in their own field: farmers also need to try 3–4 high-yielding varieties in their own field before choosing one to plant in larger areas. Even though registered varieties were tested in many trials across various environments, they will still have some advantage in some kinds of environments over others. For this reason, comparing commercial varieties in a farmer's own field will help them to choose the best-adapted variety for their own soil type and bio-physical environments.
3 Optimizing for the best spacing and fertilizer application: Other than selecting the most suitable variety for their field, farmers also need to test for the best spacing and

fertilizer application, which will help them find the cost–benefit management for their own location.

These are general guidelines for farmer practices for adopting new varieties in their field. Some practices are still not common in standard cassava production, such as irrigation during the dry season to increase yield. Therefore, in the Thai breeding programme, there has not been varietal selection for good response to irrigation. Some varieties, for example, may produce excess foliage in response to irrigation, with little change in root yield compared to non-irrigated. In cases where farmers have access to irrigation, it is very important to select a suitable variety for irrigated conditions, based on responses in farmers' own fields, together with other good agricultural practices.

In addition to farmer demands, the starch industry requirements are also important factors to determine the direction of cassava breeding. If starch plants pay a differential for high starch content, that will motivate breeding programmes to focus on high starch content and high yield at the same time. However, to date, many starch factories still buy raw materials from farmers with less incentive on high per cent starch. This means that farmers do not focus much on varieties with high starch content and it also has a feedback to breeding programmes to release varieties with high yield regardless of per cent starch, to best suit farmers' needs. This will also reduce competitiveness of Thai starch industry in long run.

5 Breeding for higher yield: Thailand as a case study

During the decade from 2004 to 2014, Thailand annually produced between 21.4 and 30.0 million tonnes of fresh cassava roots. The estimate for fresh cassava production in 2015 is 31.1 million tonnes, making Thailand the second largest cassava producer in the world after Nigeria at 55.0 million tonnes. Thailand is the world's largest exporter of cassava products with a value of $2790 million, or 76% of the world total of $3688 million (Thai Office of Agricultural Economics, 2015).

Cassava in Thailand is mostly grown in the Northeast, which is characterized by highlands and plateaus. The other main planted areas are the western and the eastern parts of the country. The western part also consists of highlands, while the eastern part is lowland with sandy soils. Cassava in Thailand can be grown in various kinds of soil ranging from sandy to loamy clay soil. However, higher clay percentage in the soil will cause waterlogging problems during the rainy season and may also cause severe drought effects during the dry season. For more sandy soils, cassava yield is relatively low because nutrients may be rapidly lost, and because of low water retention capacity. The best locations for cassava production have adequate rainfall, with a rainy season from May to June through October to November. Most farmers in Thailand plant early in the rainy season, which is optimal for cassava vegetative growth. However, some farmers also plant late in the rainy season, especially if they also grow rotation crops in the same field, such as maize or rice. These different practices can impact variety traits, and breeders need to take into account the common practices of farmers in their target regions.

In Thailand, two organizations have operated cassava breeding programmes for more than 40 years. One is the RFCRC, DOA, Ministry of Agriculture and Cooperatives. The RFCRC initiated cassava breeding by collecting and evaluating landrace varieties more than a decade before Rayong 1 (R1) was released. The other programme is the Department of Agronomy, Faculty of Agriculture, Kasetsart University. This programme was initiated in 1976 by Professor Dr. Chareinsuk Rojanaridpiched, who introduced seeds from CIAT for selection at Kasetsart University's research station. Until 2015, RFCRC has released 11 cassava varieties. The names of varieties released by that centre was named Rayong followed by a number and abbreviation with the letter 'R'. The following new varieties developed in Thailand were released from RFCRC, DOA, Thailand (see also Table 1): Rayong 3 (R3, in 1983), Rayong 2 (R2, in 1984), Rayong 60 (R60, in 1987), Rayong 90 (R90, in 1991), Rayong 5 (R5, in 1994), Rayong 72 (R72, in 1999), Rayong 7 (R7, in 2005), Rayong 9 (R9, in 2006), Rayong 11 (R11, in 2010) and Rayong 86-13 (R86-13, in 2013). Until 2015, Kasetsart University has released five cassava varieties: Sriracha 1 (SRC1, in 1990), Kasetsart 50 (KU50, in 1992), Huay Bong 60 (HB60, in 2003), Huay Bong 80 (HB80, in 2008) and Kasetsart 72 (KU72, in 2015). The National Science and Technology Development Agency (NSTDA) together with Mahidol University also released Pirun 1 and 2 (PR1, in 2014; PR2, in 2015). Currently, according to the Thai Office of Agricultural Economics report, Kasetsart 50 (KU50) makes up 52% of the area in cassava planted in Thailand. The second most widely grown cassava variety is R5, grown on 21% of cassava planted area (data are averaged from 2003 to 2014).

6 Measuring the success of the Thai breeding programme

Cassava breeding programmes in Thailand have been successful at continuously increasing yield. Average cassava yields in Thailand from 1957 to 2015 are available from national routine surveys of the Thai Office of Agricultural Economics. They can be classified into four time periods (Table 1) according to the releasing of the varieties R1 and KU50 as follows:

1. From 1957 to 1974 (17 years): throughout this period, Thailand had not released any cassava varieties. The average cassava yield of this period was only 14.8 t/ha. It should be noted here that the reported average yield in this period may have been from a different survey methods (it is a very early stage of data recording by surveying in Thailand) since yields should not be higher than the second (1975–1982) and the third (1983–1991) periods.
2. From 1975 to 1982 (7 years): the period in which Thailand had the first Thai commercial variety. This variety, R1, while a landrace variety, was officially promoted to farmers by the DOA. Average cassava yield in this period did not change, and remained at 14.0 t/ha.
3. From 1983 to 1991 (8 years): the period in which five Thai cassava varieties were released (R3, R2, R60, SRC1 and R90), but adoption and impact were not yet significant. The average cassava yield in this period remained stagnant at 13.7 t/ha.
4. From 1992 to 2015 (23 years): the period in which Thailand had modern cultivated Thai varieties released from breeding programmes and widely adopted by farmers. The average cassava yield for the period was 17.7 t/ha. By the end of this period, average national yields (in 2015) had climbed to over 22 t/ha, among the world's highest.

Table 1 Periods of release of Thai varieties, national average yield of cassava in Thailand and yield increase due to variety

Year	Year no. in each period	Released variety	Institution releasing variety*	National average yield (t/ha)	Yield compared to 1957–1974 (%)
1957–1974	17	No released varieties		14.8	
1975		R1	RFCRC		
1975–1982	7			14.0	95.0
1983		R3	RFCRC		
1984		R2	RFCRC		
1987		R60	RFCRC		
1990		SRC1	KU		
1991		R90	RFCRC		
1983–1991	8			13.7	92.7
1992		KU50	KU		
1994		R5	RFCRC		
1999		R72	RFCRC		
2003		HB60	KU		
2005		R7	RFCRC		
2006		R9	RFCRC		
2008		HB80	KU		
2010		R11	RFCRC		
2013		R86-13	RFCRC		
2014		PR1	NSTDA		
2015		KU72	KU		
1992–2015	23			17.7	119.7

* RFCRC stands for Rayong Field Crops Research Center;
KU stands for Kasetsart University;
NSTDA stands for National Science and Technology Development Agency.

There was no increase in average cassava yield in periods 1 through 3, before the variety Kasetsart 50 (KU50), released by the Department of Agronomy of Kasetsart University, and other potentially high-yielding varieties were released from both the Department of Agronomy and the DOA from 1992 to 2015 (Table 1). The average yield in the fourth period increased by 19.7% compared to the first period. Based on the 2015 production area of 1 491 155 ha, production in Thailand would be 26 348 709 tonnes based on the yields of fourth period varieties, while production based on first period varieties would be only 22 009 448 tonnes. The difference between the potential of the fourth and the first period is approximately 4.3 million tonnes and is based on the use of high-yielding varieties.

Rojanaridpiched et al. (1993) analyzed the yield response of commercial varieties at that time in different environments and across years and then found that R1 had the lowest harvest index compared to others. R3 had a high harvest index, but low total biological yield. This variety is only adapted to high-yield environments. R60 and R90 have a high harvest index and the same total biological yield as R1. KU50 has a high harvest index and high total biological yield. The variety is adapted to both low- and high-yielding environments due to its canopy abundance. This broad adaptation is confirmed by its very broad adoption by farmers throughout Southeast Asia. Regression of fresh root yields of cassava variety on the location mean (average fresh root yield of all varieties at each location) of all different yielding environments showed that R1 has a lower harvest index than others, but high upper ground yield. Therefore, the regression coefficient of fresh root yield of R1 on the location mean of both low- and high-yielding environments was lower than others (b = 0.84). R3 has a higher yield than R1 only in high-yielding conditions (higher than the location mean at the interception point) even though it has a higher regression coefficient than Rayong 1 (b = 1.02) (Fig. 1). KU50 has a higher yield than R1 in all conditions, indicated by the regression line of KU50 being higher than that of R1 at all locations (Fig. 1).

R5 was released in 1994; it has a high harvest index and high total biological yield. It was selected from a cross between 27-77-10 and R3. The breeding programme at Kasetsart University focused on crosses between R5 and KU50 since these parents are unrelated and they both have high root yield, harvest index, starch content and total biological yield. This resulted the variety HB60 and HB80, the hybrids from the cross R5 x KU50, was released in 2003 and 2008.

The period of time from crossing to releasing varieties can be considered as one of the indicators of the efficiency of a breeding project (Table 2). R11 has the largest time interval – 18 years. R60 has the shortest time interval – 6 years. On average Thai commercial varieties take 10.5 years to develop. There does not appear to be any relationship between the period of time to develop a variety and the level of success for the variety in terms of farmer adoption. While this relationship has not been closely studied, the variations in period of varietal development appear to just depend on how each breeder manages their breeding programme. For example, some breeders are able to dedicate their full time to breeding while others have multiple responsibilities in other research and teaching areas as well.

7 Relationships among Thai cassava varieties

A diagram of relationships between cultivated Thai varieties (highlighted in green) from the DOA and the Department of Agronomy are shown in Fig. 2.

This information is important for breeders in order to best select parents for the creation of new hybrids. For example, in Fig. 2, HB60 and HB80 are varieties from the Department of Agronomy, Kasetsart University breeding programme. They are full siblings, both have R5 as the maternal parent and KU50 as the paternal parent. R5 has clone 27-77-10 as a maternal parent and R3 as a paternal parent. KU50 has R1 as a maternal parent and R90 as a paternal parent. Therefore, clones 27-77-10, R3, R1 and R90 can be regarded as ancestors of HB60 and HB80. There are many Thai commercial varieties with R1 or

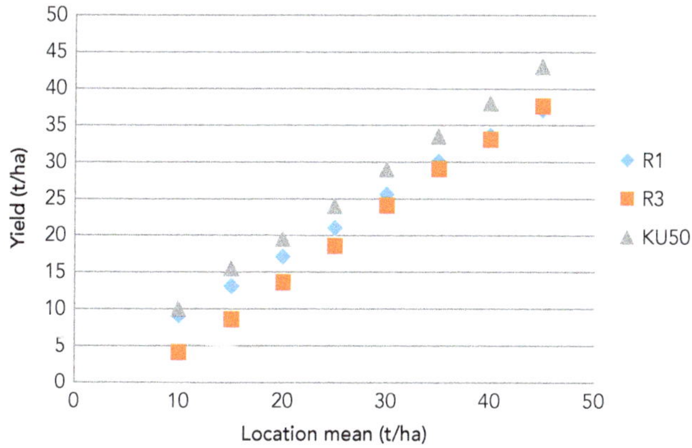

Figure 1 Regression of fresh root yields of three cassava varieties, R1, R3 and KU50 on the location mean of all different yielding environments.

Table 2 Length of time required to develop new cassava varieties

Variety	Abbreviation	Crossing year	Release year	Breeding period
Rayong 1	R1	NA	1975	NA
Rayong 3	R3	1975	1983	8
Rayong 2	R2	1976	1984	8
Rayong 60	R60	1981	1987	6
Sriracha 1	SRC1	1983	1990	7
Rayong 90	R90	1978	1991	13
Kasetsart 50	KU50	1985	1992	7
Rayong 5	R5	1982	1994	12
Rayong 72	R72	1990	1999	9
Huay Bong 60	HB60	1991	2003	12
Rayong 7	R7	1992	2005	13
Rayong 9	R9	1992	2006	14
Huay Bong 80	HB80	1991	2008	17
Rayong 11	R11	1992	2010	18
Rayong 86-13	R86-13	2003	2013	10
Pirun 1	PR1	2006	2014	8
Pirun 2	PR2	2006	2015	9
Kasetsart 72	KU72	2008	2015	7
			Average	10.5

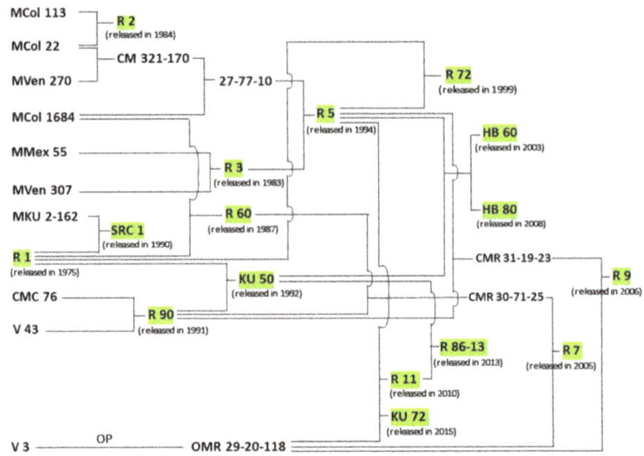

Figure 2 Pedigrees of Thai cassava varieties (highlighted in green), including the year of release and relationships to other commercial varieties.

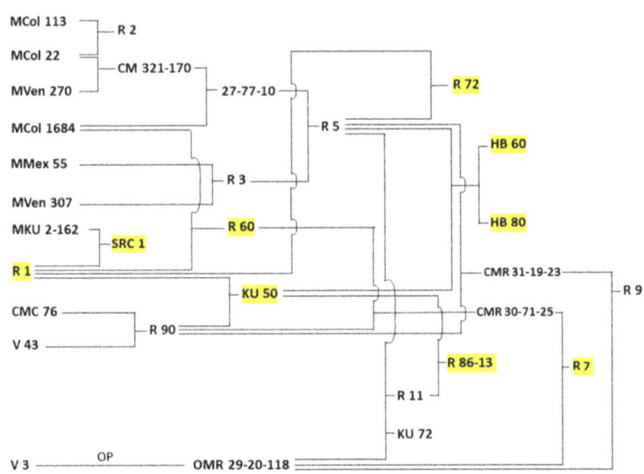

Figure 3 Pedigrees of Thai cassava varieties that have the variety R1 in their genetic background (highlighted in yellow).

R5 as one of their parents, as shown in many connections at R1 and R5. Varieties having R5 as an ancestor are R72, R9, HB60, HB80, R11, KU72 and R86-13. Nine commercial varieties have R1 in their pedigree (R60, SRC1, KU50, R72, HB60, HB80, R7 and R86-13; highlighted in yellow in Fig. 3) from a total of 16 released varieties. In other words, 56% of Thai commercial varieties have genes of R1 in their background. This example illustrates a somewhat narrow germplasm base in Thai cassava breeding programmes in the past. This is one possible reason there has not been much genetic gain in yield since KU50 was released. Kittipadakul et al. (2011) indicated that through the process of inter-varietal crossing among Thai commercial varieties followed by conventional selection, it will be difficult to improve yield over KU50.

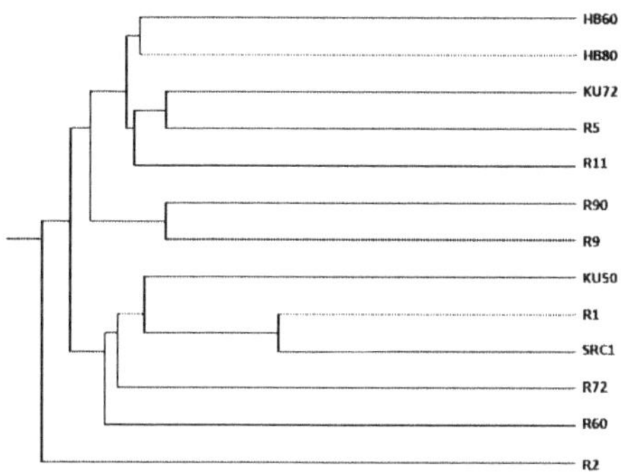

Figure 4 Cladogram from cluster analysis of 13 Thai cultivated varieties using 53 078 SNP markers.

Pedigree information is also important for utilizing DNA markers in genotyping. For example, dendrograms based on SNP markers can be used to classify groups of commercial varieties. In addition, data on yield and other agronomic traits is also more informative if genotypes are grouped based on relatedness. For example, using the Trait Analysis by Association Evolution and Linkage (TASSEL) programme (Bradbury et al., 2007) with genotypic data of 53 078 SNP markers in 13 Thai cultivated cassava varieties generated the cladogram in Fig. 4 using a distance matrix based on a neighbour-joining algorithm. The cladogram corresponds with the expected grouping based on pedigrees of Thai commercial varieties (Fig. 2 and 3). For example, in the cladogram, HB60 and HB80 are in the same group, which is in accordance with the pedigree (they have the same parents) (Vichukit et al., 2003, 2008). KU72 was grouped with R5 in the cladogram, which is logical because R5 is a parent of KU72 (Rayong Field Crops Research Center, 1999). Another example is the close relationship between R9 and R90 shown in the cladogram: R9 has CMR31-19-23 as a maternal parent and CMR31-19-23 has R90 as a maternal parent (Rayong Field Crops Research Center, 2005). As a basis for the close relationship between SRC1 and R1 in the cladogram, the pedigree shows that SRC1 has R1 as a paternal parent (Rodjanaridpiched et al., 1990). Cluster analysis was done using only SNP genotype data (Fig. 4). Different algorithms could show a different cladogram. If the breeding programme does not have other information such as the pedigree data or phenotype data, varietal grouping done using only genotype data would be less useful for the breeding programme. Therefore, three components are required for optimized cluster analysis in a breeding programme: 1) genotype data, 2) pedigree data and 3) phenotype data.

8 Progress in the current Thai breeding programme

Five Thai cassava commercial varieties (HB60, KU50, R1, R5 and Pirun1) and five Thai breeding lines (CMR89, KUC3, KUC19, KUC99 and KUC132) were evaluated in regional trials planted in the early rainy season at 19 locations in six provinces (Kanchanaburi, Chachoengsao, Chonburi, Chaiyaphum, Nakhon Ratchsima and Rayoung) in Thailand. Among these, KU50 is the most widely grown variety in Thailand, followed by R5. R1 was included in the trial because it was the first Thai cassava variety in Thailand and the most popular variety back in 1975. When considering the progress of fresh root yield improvement, it was noted that all the varieties had higher fresh root yield than R1, except for KUC19, which had fresh root yield that was not significantly different from R1. Commercial varieties ranged from 13.8% (R5) to 20.6% (HB60) higher in yield than R1 and other entries ranged from 23.0% (KUC132) to 40.2% (CMR89) in breeding lines. CMR89 was the only variety that had higher fresh root yield than KU50 (Table 3). Breeding line CMR89 (CMR 34-08-89) is a high-yielding line bred from the Rayong Field Crop Research Station. It was produced from a cross made in 1991, but was not selected due to low starch content. However, it was popular among farmers in Nakhon Ratchsima because of high yield without regard to starch content. This is an example where a breeding line does not meet the standards for varietal release or recommendation, but may still be adopted by farmers in specific agro-ecological or market conditions where the variety performs well. When experimental lines are tested on farmers' fields, those farmers usually have the option to select and continue growing materials they find desirable, even if a broader release is not warranted.

The promising lines of the KUC series mentioned above, from the breeding programme of the Department of Agronomy, Kasetsart University, are in the process of further evaluation in farmer field trials. The starch contents of Thai commercial varieties were higher than R1, ranging from 15.8% (Pirun1) to 53.8% (HB60) higher. Starch content of promising new lines ranged from 35.3% (KUC3) to 65.9% (KUC19) higher than R1. When compared to KU50, only KUC19 had higher starch content (15.6%), among the Thai commercial varieties, none had higher starch contents than KU50 (Table 4). Based on data from regional trials in recent years (2014 to 2016), Thai released cassava varieties are improved relative to R1 for both fresh root yield and starch content. HB60 has the highest improvement, by 20.6% for fresh root yield and 53.8% for starch content (Tables 3 and 4, respectively). However,

Table 3 Average root yield of five Thai cassava varieties and five Thai cassava lines combined over 17 selection trials in farmers' fields between 2014 and 2015

Variety/Line[1]	Yield (ton/ha)[2]		Relative to R1 (%)	Relative to KU50 (%)
CMR89	29.5	a	140	127
KUC99	25.0	b	124	108
KUC3	24.9	b	123	107
KUC132	24.9	b	123	107
HB60	24.4	bc	121	105
KU50	23.2	bc	115	100
Pirun1	23.2	bc	115	100
R5	22.9	c	114	99
R1	21.0	d	104	90
KUC19	20.2	d	100	87

[1] The five Thai cassava varieties were R1, R5, KU50, HB60 and Pirun1 (the letter 'R' stands for Rayong and 'HB' stands for Huay Bong) and the five Thai cassava lines were CMR89, KUC 3, KUC19, KUC99 and KUC132 (the letter 'CMR' stands for Cross *Manihot* Rayong and 'KUC' stands for Kasetsart University Cross).

[2] Within a column, means followed by different letters are different at $\alpha = 0.05$, based on Duncan's multiple range test (DMRT).

Table 4 The average starch content of five Thai cassava varieties and five Thai cassava lines combined over 17 selection trials in farmers' fields between 2014 and 2015

Varity/Line[1]	Starch (%)[2]		Relative to R1 (%)	Relative to KU50 (%)
KUC19	21.4	a	166	116
KUC132	20.2	ab	157	109
HB60	19.9	abc	154	107
R5	18.8	bcd	146	102
KU50	18.5	bcd	143	100
KUC99	18.5	bcd	143	100
CMR89	17.7	cd	137	96
KUC3	17.5	d	135	94
Pirun1	15.0	e	116	81
R1	12.9	f	100	70

[1] The five Thai cassava varieties were R1, R5, KU50, HB60 and Pirun1 (the letter 'R' stands for Rayong and 'HB' stands for Huay Bong) and the five Thai cassava lines were CMR89, KUC 3, KUC19, KUC99 and KUC132 (the letter 'CMR' stands for Cross *Manihot* Rayong and 'KUC' stands for Kasetsart University Cross).
[2]Within a column, means followed by different letters are different at $\alpha = 0.05$, based on DMRT.

compared to KU50, there was no other commercial variety that had higher fresh root yield or starch content.

Kittipadakul et al. (2011) used data from the Thai Office of Agricultural Economics for an analysis of Thai commercial cassava fresh root yield per hectare. Based on this analysis, KU50 had higher yield than R1, R3, R60 and SRC1, except for R90, all of which were released before KU50. For varieties released after KU50, there were three varieties that had higher yield than KU50. They were R5 (4% higher yielding), R7 (8%) and HB60 (12%). However, since these data were based on surveys rather than on managed yield trials, they could not be analysed statistically (Table 5). In fact, based on 17 selection trials where HB60 was planted along with KU50, there was no significant difference in yield. Thus, it can be concluded that the commercial Thai cassava varieties released after R1 have significantly higher yield and starch content than R1, but after KU50 was released, there was not much progress in variety improvement in either yield or starch content.

Regional selection trials in many locations, together with stability evaluations of varieties before release, are mandatory in order to determine adaptability of varieties to specific planting zones. Different varieties frequently perform differently in different environments, known as genotype by environment interactions (G×E). G×E can be affected by location (L), season (S) and year (Y). As well, there can be higher level interactions, for example of G×L×S or G×S×Y and so on. Thus, the phenotypic expression of genotypes changes across environments. Yan and Kang (2003): classified GxE into three types 1) no interaction, 2) no crossover interaction and 3) crossover interaction. From stability evaluations of KU50, HB60, R72 and R5 in 11 trials (Kittipadakul et al., 2004), KU50 and HB60 had high stability for starch content, fresh root yield and dry root yield, as well as higher starch content and

Table 5 Fresh root yield (tonnes per ha) of Thai cassava varieties planted between 2002 and 2008 and average yield across years

Variety	Release year	Fresh root yield (t/ha)								Percentage relative to KU50[1]
		2002	2003	2004	2005	2006	2007	2008	Average	
R1	1975	12.9	19.8	19.7	14.6	17.3	17.9	16.8	17.0	86
R3	1983	13.6	18.8	22.9	16.5	20.3	20.2	18.1	18.6	94
R60	1987	13.6	17.5	20.6	16.8	20.3	22.8	21.2	19.0	96
R90	1991	18.9	21.2	22.0	16.4	21.7	24.0	21.4	20.8	105
SRC1	1990	NA	19.2	18.6	14.9	19.0	NA	NA	17.9	91
KU50	1992	17.4	18.7	19.8	17.4	21.1	22.9	21.3	19.8	100
R5	1994	16.8	20.7	21.1	17.9	21.8	23.6	22.0	20.6	104
R72	1999	NA	NA	NA	NA	NA	NA	18.6	18.6	94
HB60	2003	NA	NA	NA	NA	19.6	24.1	23.0	22.2	112
R7	2005	NA	NA	NA	NA	NA	NA	21.4	21.4	108
R9	2005	NA	NA	NA	NA	NA	NA	19.4	19.4	98

[1]The percentage of fresh root yield per unit area relative to Kasetsart 50 (KU50).
Source: Kittipadakul et al. (2011).

dry root yield than the grand mean and R72 and R5. R72 had the lowest stability for starch content, fresh root yield and dry root yield. Also, the starch content of R72 was lower than the mean of the starch content of all four varieties. R5 had a lower fresh root yield and dry root yield than the mean of the four varieties even though it had a high stability for these three traits.

9 Adaptability of varieties

Determining stability of breeding lines (e.g. across locations and years) in a breeding programme is important before releasing them as new varieties. Varieties can have specific or wide adaptability for key traits. Breeders can determine stability by conducting multi-location or multi-season selection trials to identify G×E effect. Analysis of variance (ANOVA) for each trial has been done to determine the best lines in each trial in terms of high starch yield per area and other characteristics of importance. Total starch yield take into consideration both fresh root yield and starch content for variety selection. Combined ANOVA from all data trials provides the average performance of each line. However, data from many locations or seasons would have different experimental variances, so it is good to do homogeneity test of variances, for example, using the Bartlett test (Bartlett, 1937) among trials to get an unbiased comparison of breeding lines. If G×E interaction is significant from combined ANOVA of all trials, genotype stability and the best genotypes for environments can be identified by GGE biplot (Yan et al., 2000; Yan and Kang, 2003). To illustrate how the KU cassava breeding programme identifies which breeding line is selected for specific or broad adaptation, the 17 location trials in the early rainy season from 2014 to 2015 were used as a case study. From the Bartlett test, only nine of seventeen trials (Table 6) had high homogeneity of variance, and these nine were used for analysis with combined ANOVA and GGE analysis. Each regional selection trial consisted of four breeding lines (KUC50-13-132, KUC50-17-3, KUC51-18-19 and KUC51-13-99) and five check varieties (Kasetsart 50, Huay Bong 60, Rayong 1, Rayong 5 and Pirun 1). The combined analysis of the nine lines (or varieties) in nine environments shows that the main effect of lines and environments and their interaction effects were significant for starch

Table 6 Symbol, location and year of nine regional selection trials

Symbol	City/Place	Province	Year
BP_2014	Bo Phloi	Kanchanaburi	2014
HB4_2014	Huay Bong	Nakhon Ratchasima	2014
HB7_2014	Huay Bong	Nakhon Ratchasima	2014
SWF_2014	Suwan Farm	Nakhon Ratchasima	2014
RY_2014	Mueang Rayong	Rayong	2014
SK_2014	Sikhio	Nakhon Ratchasima	2014
SK1_2015	Sikhio	Nakhon Ratchasima	2014
BH4_2015	Huay Bong	Nakhon Ratchasima	2015
RY_2015	Mueang Rayong	Rayong	2015

Table 7 Combined ANOVA of starch yield of breeding lines and check varieties in nine regional selection trials at early raining season

Source of variation	df	SS	MS	*F*-value	*P*-value
Environment	8	28545112	3568139	21.13	9.734e-10
Block (Environment)	27	4558601	168837	2.38	0.0003094
Line	8	4663193	582899	5.60	0.0000221
Line × Environment	64	6663384	104115	1.47	0.0222713
Error	216	15303900	70851		

Table 8 Average starch yield of five Thai cassava varieties and four Thai cassava lines combined over 9 selection trials in farmers' fields between 2014 and 2015

Line/Variety[1]	Starch yield (t/ha)[1]	
Kuc132	6.6	a
Hb60	6.2	ab
Kuc99	6.1	ab
R5	5.7	abc
Kuc19	5.6	abc
Ku50	5.4	bc
Kuc3	5.2	bc
Pirun1	4.7	cd
R1	4.0	d

[1] The code of lines or varieties: Hb60 = Huay Bong 60, Ku50 = Kasetsart 50, Kuc3 = KUC50-17-3, Kuc19 = KUC51-18-19, Kuc99 = KUC51-13-99, Kuc132 = KUC50-13-132, Pirun1 = Pirun 1, R1 = Rayong 1 and R5 = Rayong 5. Within a column, means followed by different letters are different at $\alpha = 0.05$, based on DMRT.

yield ($P < 0.05$) (Table 7). From the mean comparison, it was found that the promising lines KUC50-13-132 had higher starch yield than the varieties Kasetsart 50, Pirun 1 and Rayoug 1, and the line KUC50-17-3 (Table 8), on average. However, the mean comparison of combined ANOVA did not provide enough detail for breeders to select the best line for specific environments. GGE biplot is a tool to assist breeders to determine the stability or specificity of lines for testing environments.

Since the result from combined ANOVA gives only the overall mean of starch yield for selection trials without the information of stability, there will not be enough information to make a decision to pick the winner. To provide more information for doing effective selection, stability analysis (e.g. GGE biplot) is another step worth conducting for genotype evaluation. The GGE biplot in Fig. 5 allows breeders to examine environmental effects on starch yield. In the case of GGE biplot, estimation of starch yield and genotype stability can be identified by considering the average environment axis (AEA). This axis is the line drawn

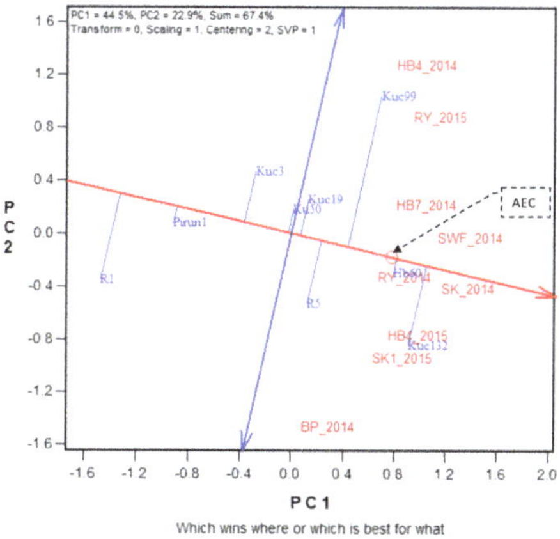

Figure 5 Identification of the stability of genotypes for starch yield by GGE biplot.

through the origin of the biplot and average environment coordinate (AEC) point. The AEC (the red circle in Fig. 5) is the average environment defined by the average values of PC1 and PC2 of all environments. The value of starch yield (or other traits of interest) as *high* or *low* is determined by the average ordinate environment (AOE). The AOE is the line that is perpendicular to the AEA line and passes through the origin. The AOE separates genotypes that have starch yields higher than the grand mean on right of the line, and genotypes that have a starch yield lower than the grand mean on the left of the line. Figure 5 shows that the line KUC50-13-132 has the highest starch yield and Rayoug 1 has lowest starch yield, on average. There are five lines/varieties (KUC50-13-132, Huay Bong 60, KUC51-13-99, Rayong 5 and KUC51-18-19) with starch yield higher than the grand mean. The genotype stability is indicated by the distance between the AEA line and the genotype position. The genotypes closer to the AEA line are more stable than others. In the overall genotype stability for starch yield, Fig. 5 shows that the variety Huay Bong 60 is the most stable variety for high starch yield and the variety Pirun 1 is the second most stable variety, but stability in term of having low starch yield. The promising line KUC50-13-132 has the highest starch yield with moderate starch yield stability and the line KUC51-13-99 is the third highest in starch yield, but this line fluctuates in starch yield across environments. The GGE biplot has another option to identify the best genotype for environments (Fig. 6 and 7). This kind of biplot will provide a polygon to reveal the genotype that is the best in environments of interest. The polygon derives from drawing the genotype that is located farthest from the biplot origin (0, 0) that served as corners of polygon. Therefore, all genotypes are located in the polygon. All environments were divided into several sections by drawing a perpendicular line to each side of the polygon, passing through the biplot origin. The best genotypes in environments located in each section is the one located at the top (corner) of polygon (Yan et al., 2000; Yan and Kang, 2003). Figure 6 shows that the line KUC51-13-99 (Kuc99) gave high starch yield in three environments (HB4_2014, RY_2015 and HB7_2014) and the line KUC50-13-132 (Kuc132) gave high starch yield in six environments (SWF_2014, RY_2014, SK_2014, HB4_2015, SK1_2015 and BP_2014). This information allows breeders to recommend these two lines as new varieties for specific locations. Figure 7 shows that the genotype KUC51-18-19 (Kuc19) is an excellent breeding line with the best performance

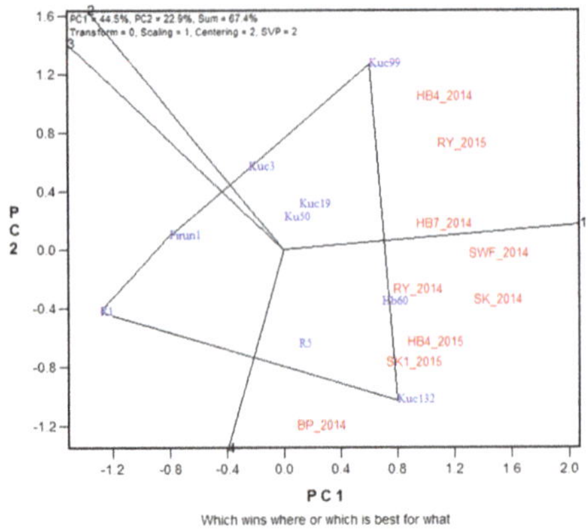

Figure 6 Identification of the best genotypes for starch yield for each environment by GGE biplot.

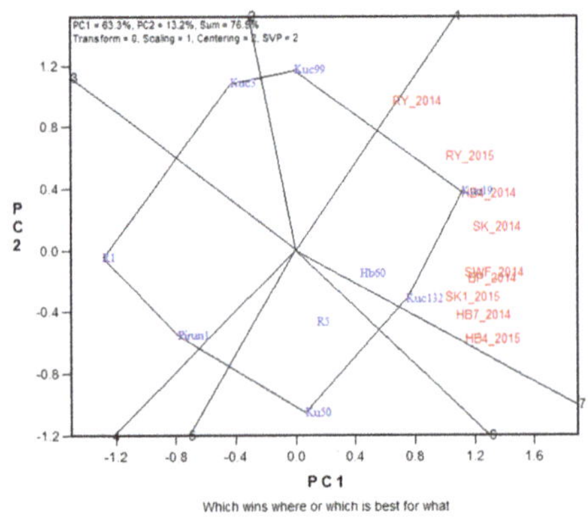

Figure 7 Identification of the best genotypes for starch content for each environment by GGE biplot.

of high starch content in all nine trials. These analyses show the advantages of the use of combined ANOVA and stability analysis to provide more information for making decisions on variety selection. In the recent selections of the KU Cassava Breeding Program, more than 48 regional selection trials between 2014 and 2017 have been planted and evaluated to make sure the KU breeding programme obtains enough data to make a decision on best selections before releasing new varieties.

10 Combining ability in Thai varieties

The evaluation of potential parents is key for the developing of successful hybrids in a cassava breeding programme. Understanding the combining ability of parents can help breeders predict good combinations to maximize heterosis and identify high-yielding hybrids with the appropriate balance of traits required by farmers and by the marketplace.

General combining ability (GCA) is one indication of which genotypes should be chosen as parents. In addition, specific combining ability (SCA) can indicate which cross combinations have a high potential for generating good breeding lines. However, conducting an experiment using a combining ability mating design in cassava is challenging in practice because it is difficult to get enough seedlings from all required cross combinations. This explains the scarcity of combining ability analysis in Thai and other cassava breeding programmes in the past.

Kittipadakul et al. (2011) indicated GCA and SCA using a modified model and simplified calculations, based on Ceballos et al. (2007). The method involves calculation of the percentages of selected breeding lines (seedling selection from botanical seeds) relative to all breeding lines developed from the parent. In the case of Thailand, the trials were repeated over three years – 2006-2008. For example in 2007, the value for KU50 ranged from 13.1 to 21.2, for the percentage of progeny that were selected from among all progeny where KU50 was the female parent. The GCA was examined on the percentages of selected seedlings from the male parent, regardless of female parent, and vice versa from the total number of seedlings of each cross whether a clone was used as male or female parent. With this method, Kittipadakul et al. (2011) found that the range and average percentage of progeny selected from female and male parents can be used for GCA and SCA estimates. A wide range between maximum and minimum percentage of progeny selected from all lines of a parent was used as the male or female progenitor showed that a parent has a good SCA in a specific cross. For instance, in Table 9, HB60, as a female progenitor in 2007, had a range of 0.6–30.9. This means that using HB60 as a female parent in crosses gave progeny with as low as a 0.6% chance of being selected. However, when it was crossed with another male parent, the progeny had a 30.9% chance of being selected. This information shows that HB60 can give good progeny when crossed with some male parents, but not all. This is an example of SCA. Parents with high GCA who had narrow range between maximum and minimum percentage of progeny selected from all lines of a parent were used as the male or female progenitor. An example is R9. While this method can give some very preliminary information on GCA and SCA, ideally there should be a full set of reciprocal crosses. If the crossing is very unbalanced, the estimates will have limited validity.

Khongsomyut et al. (2015) determined GCA and SCA effects for six cassava varieties (KU50, HB80, R90, R1, R5 and Hanatee) for fresh root yield, to identify superior parents in the KU breeding programme. A complete 6 × 6 diallel mating design among the six varieties was used in the study to produce a total of 1080 segregating genotypes. The results from seedling evaluations showed that R90 had the best GCA for fresh root yield. R90 × R1, R90 × R5, R90 × Hanatee, R5 × Hanatee, KU50 × Hanatee, Hanatee × R5 and Hanatee × HB80 showed good SCA for the same trait.

There have been several studies to determine GCA and SCA for many agronomic traits in cassava, mostly conducted in Africa, using diallel crossing and North Carolina II methods (Bueno, 1991; Chipeta et al., 2013; Easwari Amma et al., 1995; Kamau et al., 2010; Lokko et al., 2006; Njenga et al., 2014; Owolade et al., 2006; Parkes et al., 2013; Were et al., 2012; Zacarias and Labuschagne, 2010). Ceballos et al. (2015) concluded that SCA is important for selecting complex traits such as cassava fresh root yield because the ratio of GCA/SCA ranges from less than 1 to slightly more than 1. Harvest index, plant type, CMD resistance and CBSD resistance mostly have a high GCA/SCA ratio.

Regarding the potential of Thai varieties in other countries, Ceballos et al. (2007) reported that among half-sib families in Colombia, more than 40% of the breeding lines

Table 9 The range and average of percentage of progeny selected from female and male progenitors in 2007

Female progenitor	Range (%)	Average (%)	Male progenitor	Range (%)	Average (%)
R9	17.3–19.7	18.3	KM140	16.7–30.9	22.7
KU50	13.1–21.2	17.8	CMR196	7.0–19.0	12.0
Rayong7	4.1–27.0	12.7	R72	1.6–22.7	10.9
MKUC235	8.1–16.7	11.5	KM98-5	1.6–17.3	7.0
MKUC245	1.6–27.4	10.7	KU50	0.0–13.0	6.5
MKUC106	0.0–18.7	8.9	R90	0.0–11.3	5.6
HB60	0.6–30.9	8.2	Hanatee	1.1–5.7	4.2
R90	0.0–22.7	6.6	R1	3.5–3.6	3.6
R5	0.0–20.0	5.5	CM1223-11	0.0–1.7	0.9
KM98-5	0.0–7.1	3.6	MKUC106	0.0–0.6	0.3

Source: Kittipadakul et al. (2011).

from R90, KU50, R60 and R5 were selected for the next cycle. Therefore, some key Thai commercial varieties can be regarded as good parents with the potential to produce high-yielding offspring, even outside the Thai conditions where they were selected. The use of Thai commercial varieties may result in higher heterosis than cross combinations with Colombian germplasm. Hence, the continuous inclusion of novel germplasm in breeding programmes is an important factor for success.

11 Exploitation of homozygosity and heterosis in cassava

One of the major factors affecting success in a breeding programme is selecting the best germplasm for use as parents (Kawano et al., 1978). The report of Kittipadakul et al. (2011) states that parents in crosses in the breeding programme of the Department of Agronomy, Kasetsart University, from 2006 to 2008 had high SCA, thus contributing to highly successful progeny selection. With regard to the potential of Thai commercial varieties as parents for the cassava breeding programme at CIAT, Ceballos et al. (2007) concluded that that crossing between parents from different heterotic groups can potentially be a good strategy to produce good hybrids. In maize, for example, crosses between heterotic groups have been highly successful, and the mainstay of commercial breeding advances. Therefore, heterotic patterns in maize have been determined in temperate, semi-temperate and tropical regions (Hallauer and Miranda Fo, 1981; Pandey and Gardner, 1992; Melchinger and Gumber, 1998; Parentoni et al., 2001). However, heterotic patterns in cassava are not as well established. Currently, the cassava breeding programme of the Department of Agronomy at KU is constructing putative heterotic groups by using one group from the RFCRC and another group from the Department of Agronomy at Kasetsart University. The hypothesis is that the two groups are based on distinct pedigrees, therefore have a low degree of genetic relatedness, and therefore a potential to show heterotic effects when crossed. The efficacy of this approach is yet to be determined.

Inbreeding is another important strategy that can potentially contribute to breeding progress. First, selfing allows more opportunities to discover recessive traits. Secondly, the process can fix genotypes as homozygous parents, which is the only viable option for backcross breeding schemes where recovery of the background genotype is a goal. This is not possible with heterozygous parents. Thirdly, another advantage of using homozygous parents is that mutation breeding can be implemented more easily. Moreover, homozygous cassava can be propagated true to type by botanical seed. Fourthly, by selfing cassava, recessive genes that reduce yield (commonly described as 'genetic load') can be identified and eliminated. Finally, inbred lines would allow breeders to fully exploit both GCA and SCA, unlike the crossing of heterozygous parents.

However, inbreeding depression, a common consequence of successive selfing, is a concern to breeders. Breeders could expect a similar process to what occurred in maize in the transformation from open-pollinated (heterozygous parents) varieties to hybrids (homozygous parents). Inbreeding depression was a challenge in the development of early maize inbred lines. Nowadays, though, inbred lines of maize can have yields of up to four tonnes per hectare, so inbreeding depression is no longer a problem in maize, in the sense of providing relatively vigorous plants and good seed production (Duvick, 1999). Attempting to develop highly homozygous lines, the IITA has self-pollinated cassava for five cycles (Devries and Toenniessen, 2001). There is currently a new initiative, as well, to inbreed cassava at IITA as part of a multi-institutional project on doubled-haploid development.

The efficacy of trait discovery through partial inbreeding by self-pollination was demonstrated for both the amylose-free (waxy) mutation (Ceballos et al., 2007) and the discovery of a line with small starch granules in 2006 in the breeding programme at CIAT. The waxy gene was a natural mutation found in several landrace varieties, while the small granule mutation was the result of mutagenesis. Recently, the breeding programme of the Department of Agronomy, Kasetsart University, has selfed Thai commercial varieties and a landrace, including R90, R1, KU50, HB80, R5 and Hanatee, in a project aimed at developing parent lines in an inbred-hybrid strategy. All 133 S_1 lines showed good agronomic traits and high yield, except for the progeny from R90. Twelve S_1 lines were selected for S_2 production.

12 Conclusion: how to improve cassava breeding programmes

The techniques used in cassava breeding are often one step behind those used in the breeding programmes of other major crops such as maize, rice and wheat. This is largely because cassava is a highly heterozygous crop and inbred line development is difficult. Cassava breeders also generally work in more limited funding situations because of the low priority often given to the crop in agricultural policy, and the virtual absence of the private sector in breeding. Many cassava breeding programmes still choose parental materials to create new progeny without considering well-planned experimental designs. Moreover, new crosses are commonly carried out regardless of the combining ability of parents or heterotic groupings. Field plot techniques are important for reducing experimental error due to heterogeneity in field trials. Based on the experiences of the authors, working in the cassava breeding programme since 1999, the following guidance is suggested for improving efficiency of cassava breeding:

1 Continue to import cassava germplasm to add new sources of diversity. Focus on germplasm that is not available in the programme, such as accessions that carry novel disease and pest resistance genes required to combat the relevant problems of the region, or advanced lines or varieties with known pedigrees from other countries.
2 Systematically assess the value of germplasm in breeding with proven methodologies, such as augmented designs (Federer, 1956; Federer and Raghavarao, 1975).
3 Examine kinship among varieties and breeding lines in the programme to avoid inbreeding depression in selected lines, or, conversely, to select for highest level of homozygosity in the case of inbred line development.
4 Parental selection should be based on evaluation trials in large area plots to reduce error and increase accuracy of data for selection or association analysis between genotype and phenotype data.
5 As much as possible, maintain uniform trial sites for all selection cycles (e.g. for soil type and soil nutrients, rainfall patterns, temperature). Variation in the experimental unit can provide misleading information for breeders when making selection decisions. Therefore, choosing an appropriate experimental design, applying best field plot techniques and blocking of experimental units is important to reduce variation within the experimental site.
6 In the process of parental line development for the breeding programme, inter-varietal crossing of current commercial varieties should not be the only, or necessarily the main strategy. Selfing or full-sib matings of elite lines or high performance varieties may be a good option to create parents with good agronomic traits because undesirable recessive alleles are revealed and can be discarded.
7 Heterotic groupings of parental materials should be done for important traits, such as high yield and starch content. However, heterotic groupings are difficult to assign as long as cassava is heterozygous. Developing partial inbred lines, combined with progeny testing, is an option as a starting point in the case of cassava.
8 Determine the combining ability of parents through mating design analysis to increase the probability of systematically generating high-yielding breeding lines and reducing the number of unsuccessful crosses.
9 Develop a comprehensive database of all breeding lines and parents, including performance data, to help breeders estimate combining abilities and classify clones into heterotic groups.
10 Maintain a germplasm enhancement effort to discover and introgress novel traits such as high nutrient levels, high amylose starch, high starch viscosity and other physico-chemical properties of starch for value-added products, both for varieties for human consumption and for industry.
11 Move towards molecular MAS. If high precision and high accuracy markers are identified for important agronomic traits, they will be useful for accelerating genetic gains and increasing efficiency in the breeding programme.

13 Where to look for further information

Ceballos, H., Iglesias, C. A., Pérez, J. C. and Dixon, A. G. O. 2004. Cassava breeding: opportunities and Challenges. *Plant Mol. Biol.* 56: 503–16.

Ceballos, H., M. Fregene, J. C. Pérez, N. Morante and F. Calle. 2007. Chapter 12 cassava genetic improvement, pp. 365–91. In M. S. Kang and P. M. Priyadarshan (Eds) *Breeding Major Food Staples*. Blackwell Publishing Ltd, Oxford, UK.

Ceballos, H., R. S. Kawuki, V. E. Gracen, G. C. Yencho and C. H. Hershey. 2015. Conventional breeding, marker-assisted selection, genomic selection and inbreeding in clonally propagated crops: a case study for cassava. *Theor. Appl. Genet.* 128: 1647–67.

de Oliveira E. J., M. D. Vilela de Resende, V. da Silva Santos, C. Fortes Ferreira, G. Alvarenga Fachardo Oliveira, M. Suzarte da Silva, L. Alves de Oliveira and C. I. Aguilar-Vildoso. 2012. Genome-wide selection in cassava. *Euphytica* 187: 263–76.

Federer, W. T. 1956. Augmented (or hoonuiaku) design. *Hawaii. planters' Rec.* 55: 191–208.

Heffner, E. L., M. E. Sorrells and J. L. Jannink. 2009. Genomic selection for crop improvement. *Crop Sci.* 49: 1–12.

Kawano, K. 2003. Thirty years of cassava breeding for productivity- biological and social factors for success. *Crop Sci.* 43: 1325–35.

Melchinger, A. E. and R. K. Gumber. 1998. Overview of heterosis and heterotic groups in agronomic crops, pp. 29–56. In Lamkey, K. R. and Staub, J. E. (Eds) *Concepts and Breeding of Heterosis in Crop Plants*. Crop Science Society of America, Madison.

Okogbenin, E. and M. Fregene. 2003. Genetic mapping of QTLs affecting productivity and plant architecture in a full-sib cross from non-inbred parents in cassava (*Manihot esculenta* Crantz). *Theor. Appl. Genet.* 107: 1452–62.

14 References

Bartlett, M. S. 1937. Some examples of statistical methods or research in agriculture and applied biology. *J. R. Stat. Soc.* 4(2): 137–83.

Boonchanawiwat, A., S. Sraphet, O. Boonseng, D. A. Lightfoot and K. Triwitayakorn. 2011. QTL underlying plant and first branch height in cassava (*Manihot esculenta* Crantz). *Field Crop Res.* 121: 343–9.

Bradbury, P. J., Z. Zhang, D. E. Kroon and et al. 2007. TASSEL: software for association mapping of complex traits in diverse samples. *Bioinformatics Oxford England* 23: 2633–5.

Bueno, A. 1991. Estimación de los parámetros genéticos en la yuca, pp. 197–220. In C. H. Hershey (Ed.) *Mejoramiento genético de la yuca en América Latina*. CIAT Publication, Cali.

Ceballos, H., C. A. Iglesias, J. C. Pérez and A. G. O. Dixon. 2004. Cassava breeding: opportunities and Challenges. *Plant Mol. Biol.* 56: 503–16.

Ceballos, H., M. Fregene, J. C. Pérez, N. Morante and F. Calle. 2007. Chapter 12 cassava genetic improvement, pp. 365–91. In M. S. Kang and P. M. Priyadarshan (Eds) *Breeding Major Food Staples*. Blackwell Publishing Ltd, Oxford, UK.

Ceballos, H., R. S. Kawuki, V. E. Gracen, G. C. Yencho and C. H. Hershey. 2015. Conventional breeding, marker-assisted selection, genomic selection and inbreeding in clonally propagated crops: a case study for cassava. *Theor. Appl. Genet.* 128: 1647–67.

Ceballos, H., T. Sánchez, N. Morante, M. Fregene, D. Dufour, A. M. Smith, K. Denyer, J. C. Pérez, F. Calle and C. Mestres. 2007. Discovery of an amylose-free starch mutant in cassava (*Manihot esculenta* Crantz). *J. Agric. Food Chem.* 55(18): 7469–76.

Chipeta, M. M., J. M. Bokosi, V. W. Saka and I. R.M. Benesi. 2013. Combining ability and mode of gene action in cassava for resistance to cassava green mite and cassava mealy bug in Malawi. *J. Plant Breed. Crop Sci.* 5(9): 195–202.

de Carvalho, R. and M. Guerra. 2002. Cytogenetics of *Manihot esculenta* Crantz (cassava) and eight related species. *Hereditas* 136: 159–68.

de Oliveira, E. J., M. D. Vilela de Resende, V. da Silva Santos, C. Fortes Ferreira, G. Alvarenga Fachardo Oliveira, M. Suzarte da Silva, L. Alves de Oliveira and C. I. Aguilar-Vildoso. 2012. Genome-wide selection in cassava. *Euphytica* 187: 263–76.

Department of Agriculture. 2007. *Cassava Breeding Techniques*. Agricultural Cooperation Association, Bangkok, Thailand. 67p. (in Thai).

Devries, J. and G. Toenniessen. 2001. Chapter 13: Cassava, pp. 147–56. In J. Devries and G. Toenniessen (Eds) *Securing the Harvest: Biotechnology, Breeding and Seed Systems for African Crops*. CABI Publishing Oxon, UK.

Duvick, D. N. 1999. Heterosis: feeding people and protecting natural resources, pp. 19–29. In J. G. Coors and S. Pandey (Eds) *Genetics and Exploitation of Heterosis in Crops*. American Society of Agronomy, Inc., Crop Science Society of America, Inc. Madison, WI, USA.

Easwari Amma, C. S., N. Sheela and P. K. Thankamma Pillai. 1995. Combining ability analysis in cassava. *J. Root Crops* 21(2): 65–71.

Edison, S., M. Unnikrishnan, B. Vimala, S. V. Pillai, M. N. Sheela, M. T. Sreekumari and K. Abraham. 2006. *Biodiversity of Tropical Tuber Crops in India*. National Biodiversity Authority, Chennai, TamilNadu, India.

Federer, W. T. 1956. Augmented (or hoonuiaku) design. *Hawaii. planters' Rec.* 55: 191–208.

Federer, W. T. and D. Raghavarao. 1975. On augmented designs. *Biometrics* 31: 29–35.

Hallauer, A. R. and J. B. Miranda Fo. 1981. *Quantitative Genetics in Maize Breeding*, 2nd edn. Iowa State University Press, Ames.

Heffner, E. L., M. E. Sorrells and J. L. Jannink. 2009. Genomic selection for crop improvement. *Crop Sci.* 49: 1–12.

Kamau, J., R. Melis, M. Laing, J. Derera, P. Shanahan, E. Ngugi. 2010. Combining the yield ability and secondary traits of selected cassava genotypes in the semi-arid areas of Eastern Kenya. *J. Plant Breed. Crop Sci.* 2(7): 181–91.

Kawano, K., A. Amaya, P. Paza and M. Rios. 1978. Factors affecting efficiency hybridization and selection in cassava. *Crop Sci.* 18: 373–37.

Kawano, K. 2003. Thirty years of cassava breeding for productivity- biological and social factors for success. *Crop Sci.* 43: 1325–35.

Khongsomyut, S., V. Vichukit, R. Kaveeta and P. Kittipadakul. 2016. Combining ability in selected hybrids of six improved cassava varieties in Thailand. *In* the First World Congress on Root and Tuber Crops, Nanning, China.

Khumaida, N., S. W. Ardie, M. Dianasari and M. Syukur. 2015. Cassava (*Manihot esculenta* Crantz) improvement through gamma irradiation. *Procedia Food Sci.* 3: 27–34.

Kittipadakul P., C. Rojanaridpiched, V. Vichukit and J. Chantaworn. 2011. Efficiency of combining ability selection of parents in cassava breeding program. *In* The 49th Kasetsart University Annual Conference, February 2011, Kasetsart University, Bangkok. (in Thai).

Kittipadakul P., C. Rojanaridpiched, V. Vichukit, P. Changlek, C. Malumpong and K. Kaewtrakulpong. 2004. Stability of Thai cassava varieties, pp. 191–201. *In* The 42nd Kasetsart University Annual Conference, Kasetsart University, Bangkok. (in Thai).

Kittipadakul, P., V. Vichukit and S. Pongpungjamit. 2014. Increasing cassava yield with drip irrigation system (I). *Kehakaset* 38: 229–32. (in Thai)

Kizito, E., A. C. Rönnberg-Wästljung, T. Egwang, U. Gullgerg, M. Fregene and A. Westerberg. 2007. Quantitative trait loci controlling cyanogenic glucoside and dry matter content in cassava (*Manihot esculenta* Crantz) roots. *Hereditas* 144: 129–36.

Komkris, T. 1937. Cassava. *Kasikorn* 10: 495–506. (in Thai).

Lokko, Y., A. G. O. Dixon, S. K. Offei and E. Y. Danquah. 2006. Combinning ability analysis of resistance to mosaic virus disease in cassava. *Afr. Crop Sci. J.* 14(3): 221–30.

Magoon, M. L., R. Krishnan and K. Vijayabai. 1969. Morphology of pachytene chromosomes and meiosis in *Manihot esculenta* Crantz. *Cytologia* 34: 612–18.

Melchinger, A. E. and R. K. Gumber. 1998. Overview of heterosis and heterotic groups in agronomic crops, pp. 29–56. In Lamkey, K. R. and Staub, J. E. (Eds) *Concepts and Breeding of Heterosis in Crop Plants*. Crop Science Society of America, Madison.

Nakaya, A. and S. N. Isobe. 2012. Will genomic selection be a practical method for plant breeding? *Ann. Bot.* 110(6): 1303–16.

Nassar, N. M. A. 2007. Cassava genetic resources and their utilization for breeding of the crop. *Genet. Mol. Res.* 6(4): 1151–68.

Njenga, P., R. Edema and J. Kamay. 2014. Combining ability for beta-carotene and important quantitative traits in a cassava F1 population. *J. Plant Breed. Crop Sci.* 6(2): 24–30.

Okogbenin, E. and M. Fregene. 2002. Genetic analysis and QTL mapping of early root bulking in an F1 population from non-inbred parents in cassava (*Manihot esculenta* Crantz). *Theor. Appl. Genet.* 106: 58–66.

Okogbenin, E. and M. Fregene. 2003. Genetic mapping of QTLs affecting productivity and plant architecture in a full-sib cross from non-inbred parents in cassava (*Manihot esculenta* Crantz). *Theor. Appl. Genet.* 107: 1452–62.

Okogbenin, E., J. Marín and M. Fregene. 2008. QTL analysis for early yield in a pseudo F2 population of cassava. *Afr. J. Biotechnol.* 7: 131–8.

Owolade, O. F., A. G. O. Dixon and A. Y. A. Adeoti. 2006. Diallel analysis of cassava genotypes to anthracnose disease. *World J. Agric. Sci.* 2(1): 98–104.

Pandey, S. and Gardner, C. O. 1992. Recurrent selection for population, variety and hybrid improvement in tropcial maize. *Adv. Agron.* 48: 2–79.

Parentoni, S. N., J. V. Magalhães, C. A. P. Pacheco, M. X. Santos, T. Abadie, E. E. G. Gama, P. E. O. Guimarães, W. F. Merielles, M. A. Lopes, M. J. V. Vasconcelos and E. Paiva. 2001. Heterotic groups based on yieldspecific combining ability data and phylogenetic relationship determined by RAPD markers for 28 tropical maize open pollinated varieties. *Euphytica* 121: 197–208.

Parkes, E. Y., M. Fregene, A. Dixon, B. B. Peprah and M. T. Labuschagne. 2013. Combining ability of cassava genotypes for cassava mosaic disease and cassava bacterial blight, yield and its related components in two ecological zones in Ghana. *Euphytica* 194: 13–24.

Ramu, P., W. Esuma, R. Kawuki, I. Y. Rabbi, C. Egesi, J. V. Bredeson, R. S. Bart, J. Verma, E. S. Buckler and F. Lu. 2016. Cassava hapmap: Masking deleterious mutations in a clonal crop species. *bioRxiv*. doi: http://dx.doi.org/10.1101/077123.

Rayong Field Crop Research Center. 1994. Cassava. Rayong Field Crop Research Institute, Department of Agriculture, Ministry of Agriculture and Cooperation, Bangkok, Thailand. 210 p. (in Thai).

Rayong Field Crop Research Center. 1999. Cassava Variety: Rayong 72. Rayong Field Crop Research Institute, Department of Agriculture, Ministry of Agriculture and Cooperation, Bangkok, Thailand. 34p. (in Thai).

Rayong Field Crop Research Center. 2005. *Registration Request Form of Cassava Variety: Rayong 9.* Rayong Field Crop Research Institute, Department of Agriculture, Ministry of Agriculture and Cooperation, Bangkok, Thailand. 47p. (in Thai).

Rodanardpiched, C., A. Limsila, D. Supraharn, O. Boonseng, P. Poolsanguan, C. Tiraporn and K. Kawano. 1993. Recent progress in cassava varietal improvement in Thailand. In *Cassava Breeding, Agronomy Research and Technology Transfer in Asia*, Proceedings of the Fourth Regional Workshop, Trivandrum, Kerela, India, 2–6 November 1993.

Rojanaridpiched, C. and P. Srinives. 1986. *Economic Crop Breeding in Thailand.* Agriculture Book Group, Bangkok, Thailand. 381p. (in Thai).

Rojanaridpiched, C., P. Poonsanguan, J. Jiamjumnanja, S. Bhuddacharein, T. Kasirawat, V. Vichukit, P. Duangpattra and N. Taweechai. 1990. New cassava variety: Sriracha 1. *Wittayasarn Kasetsart.* 24: 145–60. (in Thai).

Thai Office of Agricultural Economics. 2015. The Situation and Trend of the Key Agricultural Products for 2016, Thai Office of Agricultural Economics, Bangkok, Thailand. 239p. (in Thai).

Vichukit, V., P. Poonsanguan, C. Rojanaridpiched, E. Sarobol, C. Jiamjumnanja, P. Kittipadakul, P. Changlek, K. Sriroth and K. Piyachomkwan. 2003. New Cassava Variety: MKUC34-114-206, pp. 264–73. *In* The 41st Kasetsart University Annual Conference, Kasetsart University, Bangkok, Thailand. (in Thai).

Vichukit, V., C. Rojanaridpched, E. Sarobol, P. Changlek, C. Jiamjumnanja, K. Sriroth, K. Piyachomkwan, W. Lertmongkol, C. Sripunya and S. Boonma. 2008. New Cassava Variety Registering Form: Huay Bong 80. Department of Agronomy, Faculty of Agriculture, Kasetsart University, Bangkok, Thailand. (in Thai).

Were, W. V, Shanahan, P., Melis, R. and Omari, O. O. 2012. Gene action controlling farmer preferred traits in cassava varieties adapted to mid-altitude tropical climatic conditions of western Kenya. *Field Crop Res.* 133: 113–18.

Yan, W., Hunt, L. A., Sheng, Q. and Szlavnics, Z. 2000. Cultivar evaluation and mega-environment investigation based on GGE biplot. *Crop Sci.* 40: 597–605.

Yan, W. and M. S. Kang. 2003. *GGE Biplot Analysis: A Graphical Tool for Breeders, Geneticists, and Agronomists.* CRC Press, Boca Raton, FL, USA, 271p.

Zacarias, A. M. and M. T. Labuschagne. 2010. Diallel analysis of cassava brown streak disease, yield and yield related characteristics in Mozambique. *Euphytica* 176: 309–20.

Chapter 8

Breeding, delivery, use and benefits of bio-fortified cassava

Elizabeth Parkes and Olufemi Aina, International Institute of Tropical Agriculture (IITA), Nigeria

1 Introduction

Cassava (*Manihot esculenta* Crantz) is a perennial root crop largely grown and consumed in the tropics and sub-tropical regions, feeding over one billion people around the globe. Cassava's high caloric value, as well as its ability to tolerate dry conditions and poor soils, makes it a key food security crop in developing countries, particularly in Africa. The ability to transform roots into many different food and non-food products and derivatives (livestock feed and industrial non-food products) also makes cassava a valuable multi-purpose crop contributing significantly to food and nutrition security (Apea-Bah et al., 2009).

Although cassava originated from South America, the crop is now cultivated in over 100 countries around the world. Africa accounts for about 57% of global production, followed by Asia at 32% and Latin America, at 11% (FAOSTAT 2014). The leading cassava-producing countries include: Nigeria (54MMT), Thailand (29.3MMT), Indonesia (23.9MMT), Brazil (21.2MMT), Democratic Republic of the Congo (16.5MMT), Angola (16.4MMT) and Ghana (14.6MMT) (FAOSTAT, 2014).

http://dx.doi.org/10.19103/AS.2016.0014.19

Cassava is the third most important source of calories in the tropics after rice and maize, serving as a rich source of carbohydrates with appreciable amounts of calcium and vitamin C. Storage roots form the basic carbohydrate part of the diet of many West African countries such as Nigeria, Cameroon, Togo and Côte d'Ivoire with an estimated 70 million people obtaining over 500 calories per day from the roots (Bokanga, 2001). Cassava has also been identified as a low glycaemic index food controlling the release of glucose in the blood with potential benefits in ameliorating conditions such as diabetes (Okafor, 2017). Nweke (2001) has classified the utilization of cassava products in Africa into five main groups. They include fresh roots, dried roots, pasty products, granulated products and cassava leaves. Utilization of cassava as food takes a large proportion of over 90% of the production in Nigeria and Africa at large with very little left for industrial processing (Akinpelu, et al., 2011). In most African countries, cassava is eaten in processed form, mostly as *gari* and *fufu*. Other processed products include *abacha and lafun* in Nigeria, *kokonte* and *agbelima* in Ghana and *atieke in* Côte d'Ivoire (Apea-Bah et al., 2009). In Ghana, cassava is also popularly consumed in its boiled form.

Apart from biotic, abiotic and agronomic constraints, utilization of cassava also faces some nutritional challenges, mainly insufficient amounts of micronutrients (pro-vitamin A, zinc and iron). Micronutrient malnutrition, also known as hidden hunger, reduces learning and cognitive ability; impairs growth; reduces immunity; decreases working capacity; causes several pregnancy complications, blindness, goitre; and raises the risk of mortality (Darnton-Hill et al., 2005). The commonly known and widely consumed cassava varieties in Africa are white-fleshed varieties with lower carotenoid content even though yellow-fleshed varieties contain higher carotenoid content. It has been estimated that white-fleshed varieties contain 0.05–0.35 ppm of carotenoids whilst yellow-fleshed varieties contain 5–25 ppm (Ceballos et al., 2013). The main carotenoid present in cassava is β-carotene though other carotenoids are present in smaller amounts such as lycopene, lutein and β-cryptoxanthin (Rodriguez-Amaya et al., 2004; Chavez et al., 2005). The main condition for pro-vitamin A activity is that the carotenoid must possess at least one β-ionone ring residue (Fig. 1). β-Carotene has two β-ionone rings in its structure and is the most widespread of

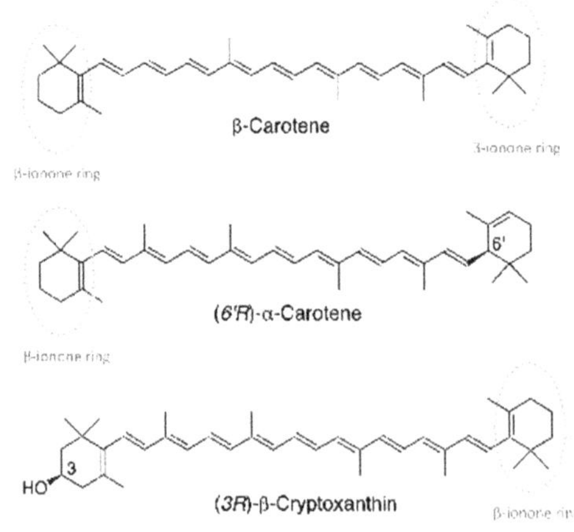

Figure 1 Chemical structures of pro-vitamin A carotenoids. Source: Fabiana F. De Moura HarvestPlus Reviews.

all carotenoids in food. Structurally, β-carotene can generate two molecules of vitamin A (retinol) when centrally cleaved, while α-carotene and β-cryptoxanthin can generate only one molecule. This suggests the main focus in breeding should be on improving β-carotene levels in cassava. It is also important to note that carotenoids are vulnerable to oxidation and start to degrade after harvest, further reducing the carotenoid content of the roots on consumption (Nassar and Ortiz, 2007; Akinwale et al., 2010; Njenga et al., 2014). A key issue therefore is bioavailability of carotenoids in cassava after processing.

Heavy reliance on cassava as the main source of food without additional intake of protein and vitamin-mineral rich food sources can lead to nutrient deficiencies. Reports by Rice et al. (2004) and Gichuki et al. (2010) pointed out vitamin A deficiency as the main nutritional problem in areas with heavy reliance on cassava as food. The deficiency of vitamin A is of major public health concern in sub-Saharan Africa, contributing significantly to widespread mortality and morbidity, and impaired growth and mental development in children (Black et al., 2008). The World Health Organization (2009) estimated that 21% of 250 million children were at risk of vitamin A deficiency, with most of them are in tropical sub-Saharan Africa. Groups at highest risk include children less than five years of age, adolescents, women of childbearing age (particularly the pregnant and lactating), refugees and victims of famine. Nutritional surveys conducted in Nigeria from 2001 to 2003 showed that about 20% of pregnant women and 30% of children under the age of five are deficient in vitamin A (Maziya-Dixon et al. 2006). In Kenya, it is estimated that 23 500 children's deaths annually are linked to micronutrient malnutrition and 70% of the children under six have subclinical vitamin A deficiency (UNICEF, 2005). A recent study on school children at Telemu, Nigeria, by the HarvestPlus project and a team from Wageningen in 2016 is yet to be published by Alida Melse, Wageningen University and team.

Increasing the carotenoid content of cassava would help to address the problem of vitamin A deficiency.

2 The HarvestPlus breeding programme for pro-vitamin A cassava

Various strategies have been developed to enhance the cultivation and nutrient density of cassava. A major strategy is through breeding (Okogbenin et al., 2013). Research has shown that relatively few major genes are involved in the determination of carotenoid accumulation in cassava roots (Iglesias et al., 1997; Chavez et al., 2005). Akinwale et al. (2010) further confirmed that two or more genes are involved in the accumulation of yellow root colour in cassava. Chavez et al. (2005) confirmed a positive strong correlation between the total carotenoids content (TCC) and the intensity of yellow flesh colour of cassava roots while Akinwale et al (2010) also confirmed that the deeper the flesh root colour, the higher the beta carotene content. Thus, the level of beta carotene in cassava varieties can be improved through simple recurrent selection that allows the concentration of the genes through selection and recombination. Although most relevant target traits for cassava breeding are quantitatively inherited and strongly affected by the environment, by contrast, heritability of carotenoid content in roots is relatively high (Morillo, 2009).

Of the two major genes controlling carotenoid content, one appears to control the transport of the precursor to the roots, and the other is responsible for the accumulation process (Chavez et al., 2000). Welsh et al. (2010) working on carotenoid synthesis and accumulation in cassava roots revealed that *phytoene synthase* (PSY) regulation in cassava roots is dominant and affects carotene synthesis by different enzymatic activity efficiencies. The intensity of the root colour has been shown to be highly correlated with carotene content, making it feasible to use visual selection for colour intensity in improvement of carotene content (Iglesias et al., 1997). Very importantly for improvement through breeding, genes with minor effects are also involved in the wide variability observed in the carotenoid content of the cassava roots, and allow for the possibility of continual improvement through recurrent selection even if the major genes are fixed in a population (Njenga et al., 2014).

As a starting point, and in order to enrich the cassava gene pool for carotenoid improvement in Africa, germplasm has been introduced from Ghana, Democratic Republic of Congo (DRC), Uganda and Cameroon, and from Colombia in South America. The breeding starts with the set of objectives and plan for a crossing block. An appropriate set of parents with required traits is assembled in a crossing block and managed with a desired mating design. A typical breeding cycle (Fig. 1) begins with creation of recombinants (progeny) by crossing of desirable parents, taking into account the full range of target genes or traits, including high carotenoid content in at least one of the parents, followed by selection based on progeny performance. Selected seedlings from each segregating family are advanced to the clonal stage for further evaluation. Desirable individuals are then advanced through sequential stages of screening and selection, focusing on key traits such as high yield, disease and pest resistance, among others, with selection made at each trial stage. Effective phenotypic selection relies on the evaluation of progeny over several clonal generations, in a series of trials where precision of measurement increases each year, through a combination of larger plot size, more replications and more locations.

In the International Institute of Tropical Agriculture (IITA) Ibadan, Nigeria, genetic materials that perform well in advanced multi-clonal yield trials are usually passed on to national programmes or to other cassava programmes managed by IITA (Zundel et al., 2010). The complete cassava breeding cycle takes about 7–8 years. Further selection cycles may be initiated in order to continue to accumulate a higher frequency of desirable alleles within the population at each stage of selection (Fig. 2). Genomic selection presents an opportunity to reduce the breeding cycle time and increase genetic gain in cassava for important traits. Institutional collaboration with partners involving germplasm exchange and sharing of technical expertise has been very productive towards attaining the desired breeding objectives (Fig. 3).

Over the last decade, the HarvestPlus Programme (www.harvestplus.org) has partnered with the International Centre for Tropical Agriculture (CIAT), IITA and the National Agriculture Research Institutes in Nigeria and in the DRC to develop and deliver bio-fortified cassava with increased TCCs. The HarvestPlus Programme aims to develop genotypes with total carotenoid (TC) target levels of 15 µg/g fresh weight basis and dry matter content above 30%, combined with acceptable culinary qualities of high mealiness/poundability. Mealiness is considered a key textural and processing quality in many African countries such as Ghana, Sierra Leone, Gambia, Nigeria, Cameroon and the DRC.

Full-scale breeding programmes for pro-vitamin A cassava at CIAT and IITA started in 2011. CIAT generates high pro-vitamin A clones that serve as sources for genes that confer high TCC . These clones are shared with partners as in-vitro plantlets and seed. Such high TCC clones when introduced to Africa by IITA and other partners tend to be susceptible to CMD. Pollens from such parents are collected for hybridization. Hybridization work in IITA involves land races, introduction from Latin America and other African countries. IITA together with NRCRI, in Nigeria have released six bio fortified cassava varieties in nigeria. The HarvestPlus Programme has been divided into two phases:

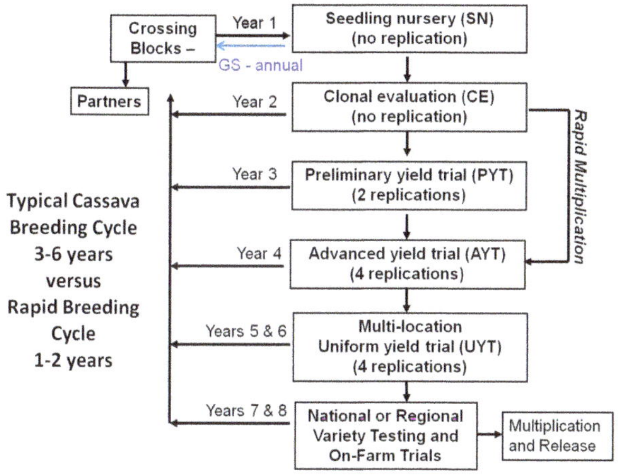

Figure 2 Schematic representation of the Cassava breeding cycle. Source: IITA.

Figure 3 Breeders identifying cassava genotypes with desirable traits.

- Phase I (Discovery, 2003–2008)
- Phase II (Development, 2009–2013)
- Phase III (Final product, Dissemination and Delivery, 2014–2018)

During Phase I, initial screening of germplasm accessions found ranges of 0–19 ppm pro-vitamin A in existing cassava varieties. Studies on genotype-by-environment (GxE) interaction for carotenoid content found high consistency in the relative ranking of genotypes across sites, that is, low GxE interaction. Ssemakula et al. (2007) found relatively high heritability of carotenoid content in cassava roots. The degradation rate of carotenoids was also investigated. Sun drying was more detrimental to the pro-vitamin A levels (44–73% degradation) than shade (41%) or oven drying (10–45%).

During Phase II HarvestPlus and its partners developed analytical methods for carotenoid content screening, demonstrating that spectrophotometric screening overestimated levels relative to high-performance liquid chromatography (HPLC) values in yellow-fleshed cassava (Kimura et al., 2007). Rapid-cycling recurrent selection was applied as proof of concept to demonstrate the efficacy of achieving increased carotenoid levels. In this system, individual plants from the seedling stage were tested for carotenoid content and selected clones incorporated directly into crossing nurseries. The rapid-cycling recurrent selection procedure has been experimented to shorten the breeding cycle from the typical eight years to a cycle of two to three years (Ceballos et al., 2013).

Phase III involves continuous research to meet up with the release of genotypes with the final target range of 15 μg/g fresh weight and continued dissemination and delivery of the already released materials.

The research centres involved in the HarvestPlus Programme have succeeded in developing elite yellow varieties with higher concentrations of carotenoids in comparison with their progenitors (De Moura et al., 2015). Bio-fortified yellow cassava genotypes with increased carotenoid content ranging from 6 to 11 μg/g on fresh root basis, have been developed. Six varieties (Table 1 and 2) have been released in Nigeria, one in DRC and one in Sierra Leone. Ghana, Malawi and Cameroon are testing elite varieties for release.

Table 1 First-wave pro-vitamin A cassava varieties released in Nigeria (2011). Source: (HarvestPlus, 2014a)

Variety name	Fresh root yield (t/ha)	Dry matter (%)	Total carotene content (TCC) (μg/g FWT basis)
IBA011371 *(IBA940561 x IBA940263)	20.1	30.7	8
IBA011412 *(IBA950971 x IBA940561)	29.8	30.1	7
IBA011368 (IBA940561 x IBA940263)	26.7	33.4	7

*Pedigree of genotypes (Female x male)
Pro-vitamin A content is approximately 80% of total carotenoid content
FWT = fresh weight
ppm = parts per million

Table 2 Second-wave pro-vitamin A cassava varieties released in Nigeria (2014). Source: (HarvestPlus, 2014a)

Clone	Fresh root yield (t/ha)	Dry matter (%)	Total carotene content (TCC) (μg/g FRWT)
NR070220 (TMS01 x 1368 OP)	25.0	32.0	12.0
IBA070593 (IBA011277 x IBA990067)	27.0	33.0	12.0
IBA070539 (IBA011610 x IBA011442)	25.0	30.0	11.0

*Pedigree of genotypes (Female x male)
Pro-vitamin A content is approximately 80% of total carotenoid content
FWT = fresh weight
ppm = parts per million

In 2017 over 400 clones will be harvested with TC greater than the benchmark variety IBA070593. However, despite the significant strides achieved, major challenges still exist in obtaining pro-vitamin A varieties that combine high carotenoid content with the right physico-chemical, biochemical, functional and sensory properties that are required to meet consumer expectations. There is a need for further integration of breeding approaches with post-harvest and processing requirements, through more interdisciplinary and collaborative research.

More than 50 pro-vitamin A cassava varieties are now at different stages of evaluation to identify those that are agronomically competitive for third-wave release. The top five clones have up to 15 ppm (100% of target level). These varieties will be put in tissue culture for international distribution, particularly targeting potential expansion countries. IITA plans to distribute elite pro-vitamin A clones to numerous countries in the region. Local genotype by environment testing of the deployed clones provides information on pro-vitamin A levels and agronomic performance from multiple sites per country and allows high-precision identification of fast-track candidates for release and parent clones for breeding. Pro-vitamin A cassava varieties are also being tested in Brazil, Central African Republic, Colombia, Ethiopia, Ghana, Côte d'Ivoire, Kenya, Malawi, Mozambique, Sierra Leone, Tanzania and Uganda.

The ongoing bio-fortified cassava research funded by HarvestPlus is recently using SNPs to identify quantitative trait loci (QTLs) associated with the key traits of interest. The QTLs found on a single chromosome – Chromosome 1 – were associated with the synthesis of carotenoid content. The same Chromosome 1 also had QTLs that influence dry matter content in cassava – a key quality root trait that is also a key target for genetic improvement. However, Rabbi et al. (2014) suggested a negative association between the TCs and dry matter. In other words, there would be a tendency for one trait to decrease if selection were made to increase the other. Work is in progress to deploy these markers in breeding programmes and overcome the negative association between these two key traits.

3 Delivering pro-vitamin A cassava varieties to farmers: the HarvestPlus Programme in Nigeria

The delivery of pro-vitamin A cassava in Nigeria has been implemented by HarvestPlus in partnership with stakeholders in the public and private sectors. The Federal Ministries of Agriculture, Health, Education as well as NGOs, including international and national research institutions, and the media, are the strongest partners in the public sector. Food processing companies, marketers and the entertainment industry are the most important private sector partners. Support from public sector partners comes in the form of policy direction, crop variety release, extension, quality control and licensing while support from private sector partners focuses on food processing, packaging and marketing.

The delivery strategy is illustrated in Fig. 4. Cultivation started in 2010 with only 500 bundles (i.e. 50 stakes of 1 m length) of stems of the three first-wave pro-vitamin A cassava varieties planted in 4 ha of land. Through on-station rapid multiplication of elite genotypes, the land area under cultivation was increased to 66 ha in 2011. A decentralized community-based seed production scheme was used to increase stem availability to

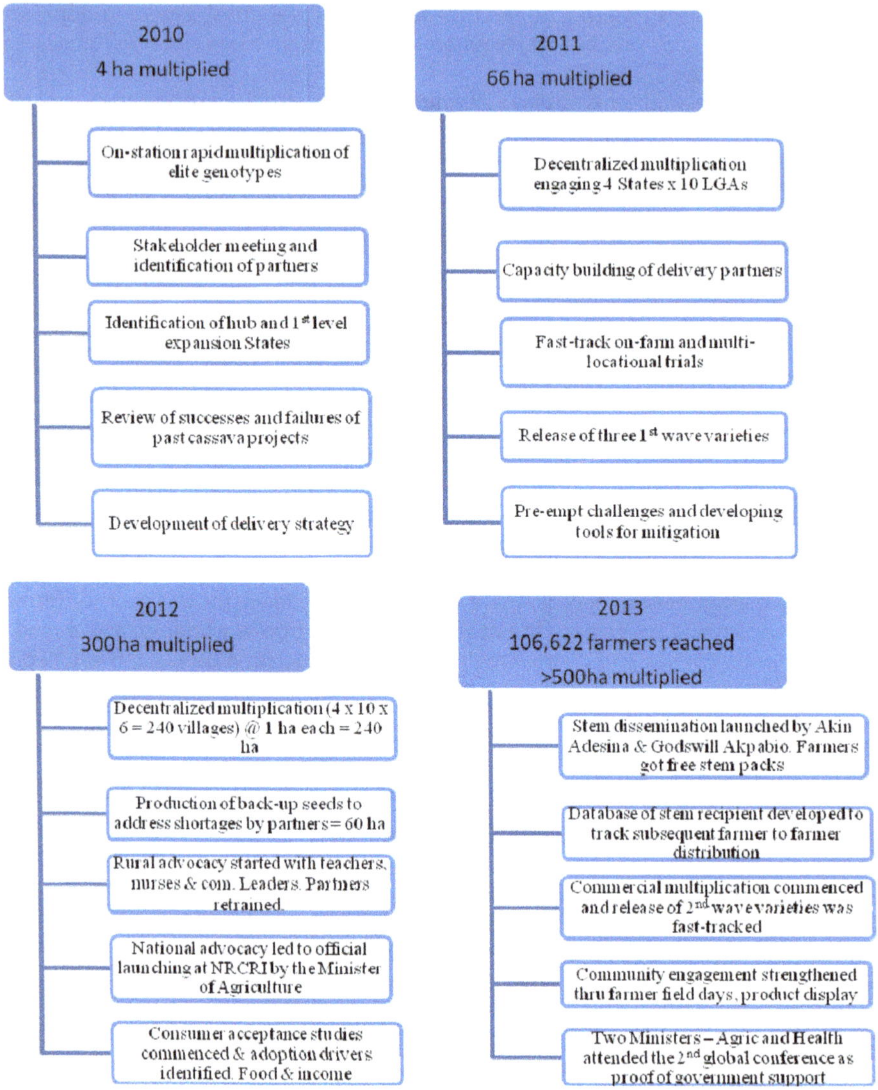

Figure 4 Schematic description of pro-vitamin A cassava delivery strategy in Nigeria. Source: Ilona, 2017.

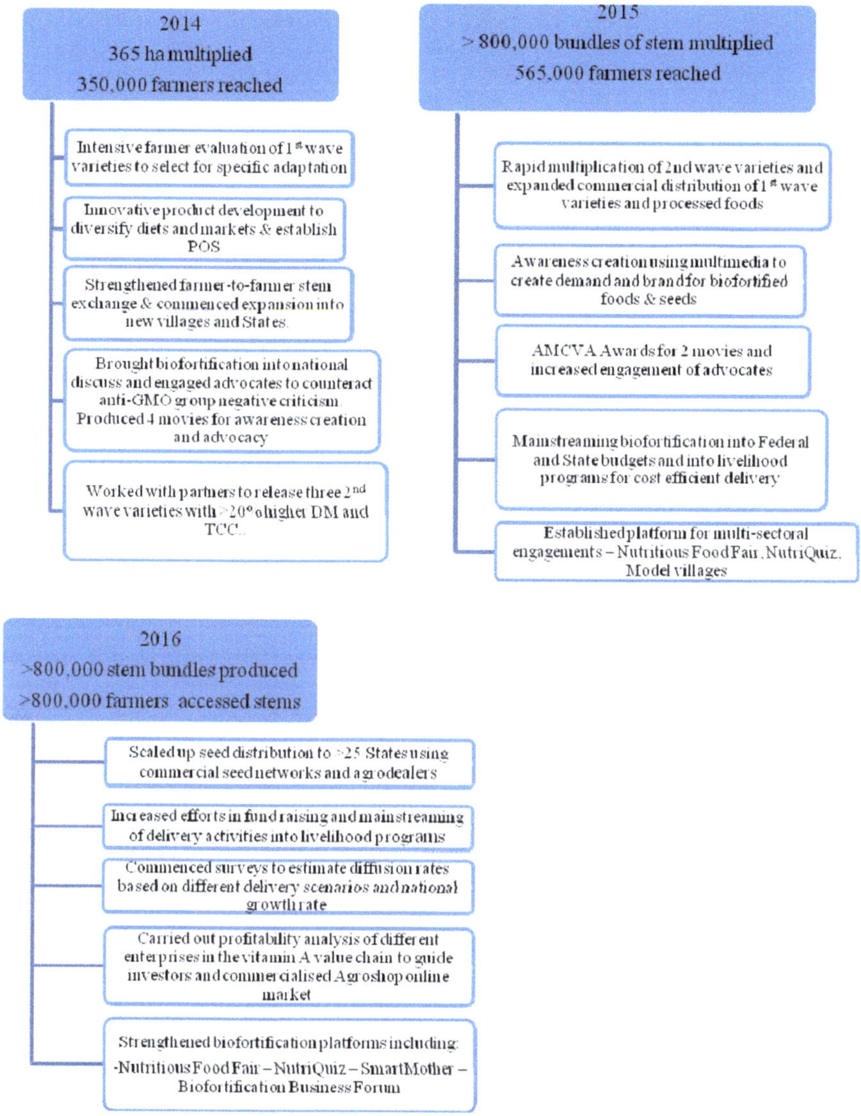

Figure 4 (*Continued*).

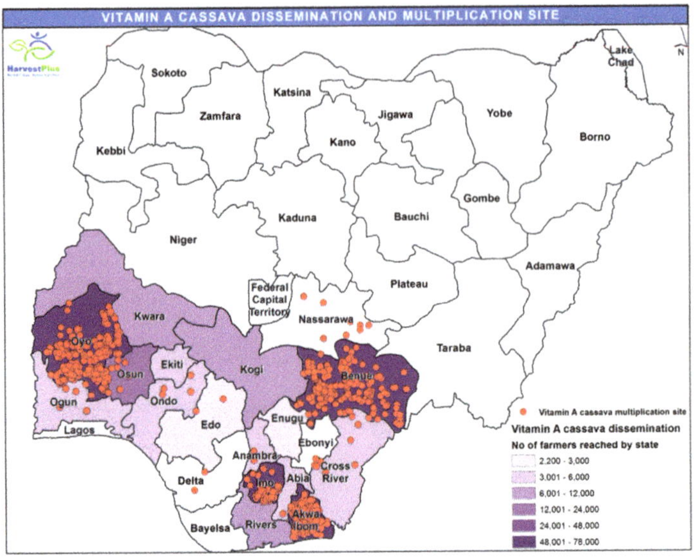

Figure 5 Pro-vitamin A cassava dissemination and multiplication sites. Source: HarvestPlus (2014b).

250000 bundles by 2012, thus demonstrating that multiplication ratio in cassava can be increased from the conventional 1:5 to 1:>30 even in on-farm situations, using good agronomic practices.

In 2011, the pro-vitamin A cassava crop delivery programme started with seed multiplication in 10 Local Government Areas (LGAs) in four states across Nigeria. The states included: Oyo in the South West, Imo in the South East, Akwa Ibom in the South-South and Benue in the North (Fig. 5). The four states were selected on the basis of being the highest producers and consumers of cassava and cassava products across the region. In 2012, the programme expanded to six villages in each LGA, making a total of 60 villages per state and 240 villages in the four targeted states. To ensure profitability in seed production, farmers were trained to adopt crop management practices that double stem yield from the conventional 300 bundles to 600 bundles per hectare. Over 1000 extension agents and facilitators have been trained to train farmers and investors in Oyo, Benue, Akwa Ibom and Imo States.

The programme rolled out to 18 more states between 2013 and 2015, thus covering over 60% of all the states in the country and over 80% of the major cassava-producing states. The level of coverage differs from one state to the other. This exceeded the initial multiplication target of 150000 bundles by over 40% for this period. As a result, and with adjustments to the size of seed packs, the project increased the number of households targeted to receive free stem packs from 25000 to 100000 in 2013. Multiplication by the public sector dominated the early years while commercial stem multiplication started very slowly in 2011 and picked up rapidly from 2014. The land area multiplied commercially grew from 66 ha in 2011 to 300 ha in 2012 to 500 ha in 2013. 350000 farmers were reached in 2014, 565000 farmers in 2015 and 800000 farmers in 2016. By 2015 about 25 states in Nigeria were already cultivating the pro-vitamin A cassava varieties (Fig. 6). Commercial seed production is expected to account for over 90% of seed supply for bio-fortified varieties by 2018.

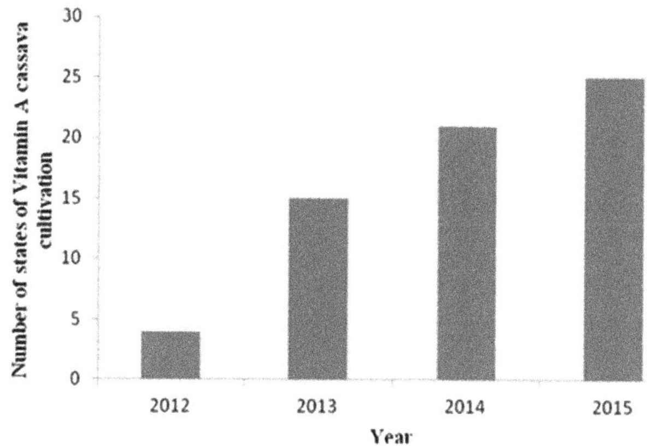

Figure 6 Graph showing number of states involved in the production of pro-vitamin A cassava (2012–2015) in Nigeria. Source: HarvestPlus (2015).

HarvestPlus employs three major channels for seed distribution to farmers:

- direct distribution
- indirect distribution
- commercial distribution

In the direct channel the project and partners give free seed packs (100 stem cuttings/pack) to each farmer in rural areas, targeting the poorest since they are the most vulnerable and may not be able to purchase stems. The indirect channel, which follows up on the direct channel, requires that each farmer who received free stem packs gives a pack of free stems to another farmer in return, usually at the time of harvest. In the commercial channel, seed investors multiply and supply stems to farmers on request. Sixty bundles are required to plant one hectare of cassava.

4 Encouraging use of pro-vitamin A cassava by consumers

Successful introduction of pro-vitamin A cassava varieties depends on their acceptability and consumption by target populations. There is therefore the need to create awareness of the varieties and their benefits. Strategies adopted for the awareness creation included decentralized dissemination; farmers' field days; training of partners on innovative extension, networking and partnership; engaging more extension agents, cooperatives and champions; improved communication among partners, engaging village facilitators and project staff; and improved data capturing. The establishment of Agro-shops in locations that are accessible to consumers has also increased awareness (Fig. 7).

A programme to encourage take-up of pro-vitamin A cassava by consumers was initiated in 2014. This necessitated a quick study to understand the constraints and opportunities for using market-based approaches to accelerate the widespread adoption and integration

Figure 7 A cassava farmer purchasing pro-vitamin A cassava stem cutting from Agro-shop. Source: HarvestPlus, 2014a.

Table 3 Consumption pattern by zone and cassava products. Source: HarvestPlus, 2015

Region	Cassava products
West Africa	Boiled root, *gari*, *lafun*, *fufu/akpu abacha*, *kokonte*, *agbelima*, *atieke* and flour
East Africa	Cassava leaves, local drinks and cassava flour
Central Africa	Cassava leaves, local drinks and cassava flour

Notes:
Gari – Toasted cassava granule
Lafun – Dried cassava flour
Fufu/Akpu –Fermented cassava dough
Abacha – Boiled cassava chips (dried/wet)

of bio-fortified cassava in Nigeria. The Global Alliance for Improved Nutrition (GAIN) contracted Sahel Capital to conduct this study in 2014.

Consumption pattern of cassava varies among countries (Table 3). The 2014 GAIN study indicated that over 90% of the 53 million tons of fresh cassava roots (both white and yellow fleshed) harvested annually, for all varieties cultivated in Nigeria, is processed into foods. Gari alone accounts for approximately 60% of processed cassava food while *fufu*, and to a lesser extent lafun, abacha and flour, account for >30% (GAIN, 2014). Less than 10% of total cassava production goes to the industrial sector. Currently, there is no apparent opportunity to use bio-fortified cassava for products such as starch, ethanol and sweeteners as these production processes destroy micronutrients. The GAIN study therefore suggested that vitamin A cassava was best targeted at the *gari* and *fufu* markets. The project prioritized *gari* and *fufu* enterprises and emphasized two key areas: awareness creation and product development to strengthen both demand and supply. The project also built capacities for increased efficiency, quality control and branding to improve the competitive ability of these products, in markets that are sometimes already saturated. In addition to conventional food products like *gari* and *fufu*, the project also developed 20

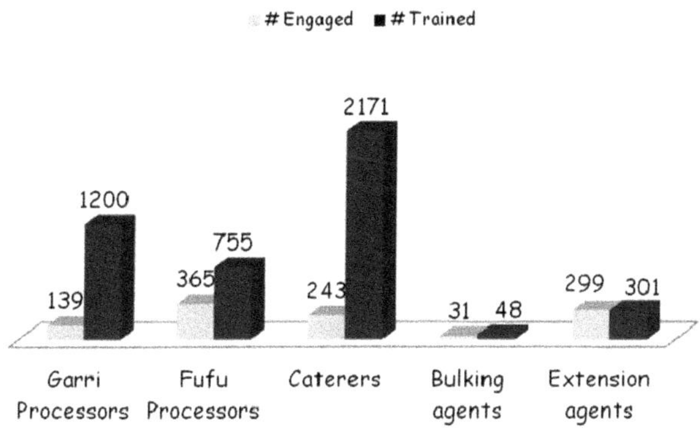

Figure 8 Number of trained and engaged entrepreneurs in the bio-fortified cassava value chain. Source: (HarvestPlus, 2014a).

other food products for children and adults, targeting rural and urban consumers. A recipe book was produced and over 4000 potential investors have been trained. Among them, 778 have invested, thus making vitamin A cassava products available in the market (Fig. 8).

A consumer acceptance study survey conducted in 2011 in two Nigerian states suggested a good initial response to the new varieties. If delivered together with nutrition information, bio-fortified varieties were generally preferred both after sensory evaluations of taste, appearance and texture and in terms of consumers' willingness to pay. This willingness was stimulated through an experimental auction technique known as the Becker-DeGroot-Marschak mechanism (Oparinde et al., 2016).

Market data show that *gari* is the most traded bio-fortified cassava product, accounting for 58% of total sales. This is closely followed by *fufu*, which accounted for 30% of the total sales. The high demand for flour (12%) was the result of the high demand for pro-vitamin A cassava-based snacks and confectioneries, such as queen cake, combo-bits and combo-strips. Other products like abacha and pudding traded in very low quantities. Household consumption is estimated to have accounted for more than 75% of the all harvested roots while the remaining 25% is traded in local markets.

The increase in the consumption of yellow flesh cassava has led to the development of other new food products or snacks. Obasi and Chukwuma (2015) developed cassava crackers from yellow cassava. In their preparation, cassava starch was mixed with water, salt and mixed spices, and crayfish into a pasty product which was moulded in rolls and steam-cooked. They were allowed to cool, then sliced, dried at 60°C for 3 hours, and deep fried into cassava crackers. Sensory tests using three clones of yellow cassava were conducted with 30 panellists. All three samples were generally accepted by the panellists with no statistically significant difference ($p \leq 0.05$) in the acceptability scores, nor in the taste and crispiness.

The gross income in 2015 for medium and small-scale investors in the bio-fortified cassava value chain is tracked to be 265 million naira, equivalent to US$1.3 million. The sales data show that *gari* has the highest gross sale of 135.5 million naira accounting for 52% of the income from sales. Flour, *fufu* and stems contributed 26%, 13% and 9%, respectively (Fig. 9).

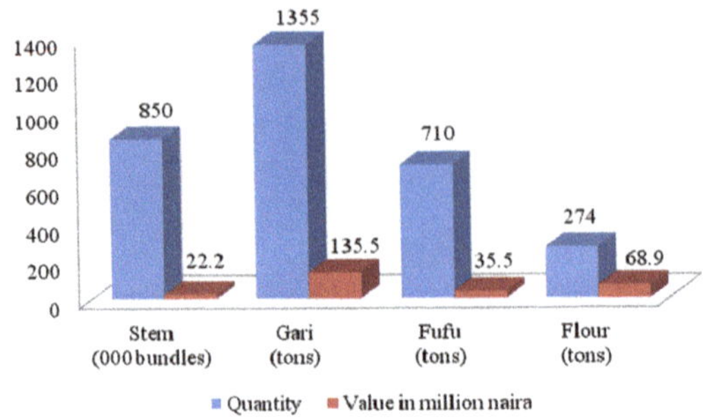

Figure 9 Volume of sales and gross income from bio-fortified cassava products in 2015. Source: (HarvestPlus, 2014a).

5 Retention of carotenoids and bioavailability after processing

The increase in nutrition awareness of people living in sub-Saharan Africa has led to an increase in the consumer acceptability and utilization of yellow flesh cassava. However, of equal or greater importance for nutrition impact is the retention of carotenoids and their bioavailability in the body after processing into the products. During processing, the product is exposed to some heat treatment, which can affect carotene concentration and availability. Considerable losses in carotene that occur during and/or after processing arise from thermally mediated isomerization and oxidation (Chandler and Schwartz, 1987; 1988; Schwartz, 1989; Van der pol et al., 1988). Differences in sequestration and intracellular localization of carotenoids in the tissue may be crucial factors in the susceptibility of these pigments to trans-cis isomerization and oxidation (Borsarelli and Mercadante 2009). Additionally, when the crop tissues are disrupted by cutting, chopping, shredding, cooking or natural ageing these physical barriers are affected, thus rendering the carotenoids open to exposure to oxygen and oxidizing enzymes (Britton and Khachik, 2009).

De Moura et al. (2015) reviewed research on the effects of methods of product preparation on the retention of TCs in *gari*, *fufu* and flour. Research results revealed that during processing, boiling retained the highest TCs followed by *gari*, followed by raw *fufu* and the lowest in cooked *fufu* (Fig. 10). On average, boiling reduced carotene content the least (34% reduction), followed by oven-dried flour with a 44% reduction. Sun-dried flour reduced the carotene concentration to the lowest level (73% reduction), which seems to imply that carotene is photo-labile. In spite of the processing steps involved in producing dry cassava flour, considerable amounts of carotene were conserved, particularly when oven drying was carried out (McDowell and Oduro, 1981).

Vimala et al. (2009), in their research work involving different processing methods, postulated that the highest retention of TCs (79–84%) and β-carotene (83–95%) was observed in the samples that were oven-dried. Boiling the samples gave retention of 71–84% for TCs and 74–84% for β-carotene. Vimala and his research team noticed that the colour of the flesh of the cassava changed to dark yellow, and orange, which they

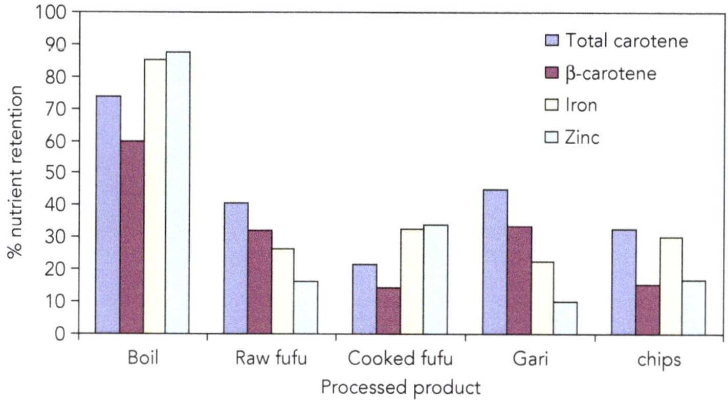

Figure 10 Retention of nutrients in cassava products. Source: HarvestPlus Report (2015).

attributed to gelatinization of starches. Frying the yellow flesh cassava into chips gave a range of 68–75% for TC retention and 45–75% for β-carotene retention. The sun-dried chips had the lowest retention ranges of 37–43% for TCs and 22–51% for β-carotene. These observations were in agreement with those made also by Nascimento et al. (2008) and Oliviera et al. (2008). Vimala concluded that differences in the retention of carotenoids and β-carotenes in yellow flesh cassava after processing could be due to the differences in the enzymatic oxidation during processing. They therefore recommended boiling, which is one of the most important modes for consuming yellow flesh cassava, as a good processing method with a fairly high retention range.

After consumption of the processed yellow flesh cassava with good retention of carotenoids, the quantity of the precursors for vitamin A that is actually available for use by the body, referred to as bioavailability, is also vital. The mode of processing can affect the bioavailability, and including a little fat to the meal could increase the bioavailability of the carotenoids present (La Frano et al. 2012). Preparations involving chopping and heating the yellow flesh cassava could effectively disrupt the plant cell walls and break down the carotenoid-protein complexes, resulting in an increase in the bioavailability of the carotenoids. The effect of variety and drying methods on the physico-chemical properties of high-quality flour from yellow cassava roots has been well investigated by Maziya-Dixon et al. (2005). Their findings revealed that the relatively high peak viscosity exhibited by most of the clones is indicative that the flour may be suitable for products requiring high gel strength and elasticity.

6 Quantification of carotenoid content in pro-vitamin A cassava varieties and food products

Carotenoids are tetraterpenoids (C-40) which are usually made from eight isoprenoid (C-5) units conjoined in an extensive conjugated double-bond system. Carotenoids are a light-absorbing chromophore and are responsible for the red, orange and yellow colouration in fruits and vegetables (Fig. 11) (De Moura et al., 2015). The light absorption properties

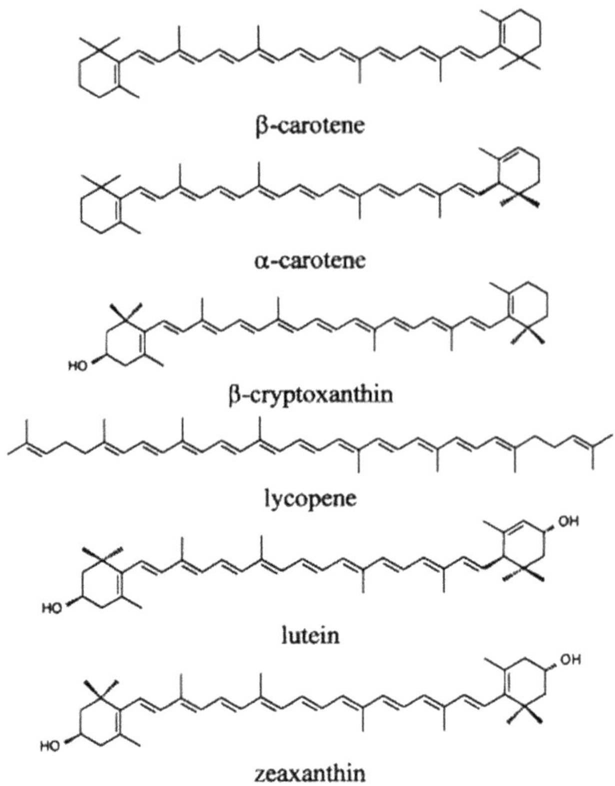

Figure 11 Structures of some principal carotenoids in foods compared with zeaxanthin. Source: Rodriguez-Amaya and Kimura (2004).

are fundamental to quantification methods. The ability of the conjugated double bonds in carotenoids to form chromophores allows them to be detected using a wide range of UV/visible absorption spectrum techniques. The majority of carotenoids absorb maximally at three wavelengths, which results in a three-peak spectrum. An increase in the number of conjugated double bonds leads to the shifting of the λ_{max} (maximum wavelength) to longer wavelengths. This means that unsaturated acyclic carotenoids like lycopene will absorb maximally at the longest wavelengths. Cyclization and cis-isomerism of a chromophore's double bond cause a decrease in the absorbance of light. Carotenoids in solution are therefore quantified according to the Beer-Lambert Law, which asserts that the absorbance of UV/visible light is directly proportional to the carotenoid concentration (Rodriguez-Amaya and Kimura, 2004).

There are several methods used for determining the carotenoid content in foods. However, the simplest of these include spectrophotometric and chromatographic methods and a more recent and quicker spectrophotometric method called the iCheck CAROTENE, designed by BioAnalyt, GmbH, Germany. Spectrophotometric methods quantify the TCs as β-carotene equivalents, and the HPLC methods quantify the TCs as well as the individual carotenoids in the sample. The iCheck system extracts and quantifies the TCs in one step (Islam and Schweigert, 2015).

Both spectrophotometric and HPLC methods have been widely used to determine the TCs and β-carotene contents in cassava. The spectrophotometer at an absorbance of 450nm has been used by many researchers to measure the TCC in cassava roots, and the HPLC method to quantify the carotenes present (Carvalho et al., 2012; Ceballos et al. 2013; Maziya-Dixon and Dixon, 2015). Gupta et al. (2015) and Saini et al. (2015) suggested that spectrophotometric methods can be used to determine the carotenoid content of green leafy vegetables including cassava leaves, although only HPLC methods can tell the difference between a *cis*- or *trans*-isomer, as well as the different carotenoids present.

The iCheck CAROTENE device has gained popularity among some researchers to determine the TCC of cassava and its products (Maroya et al. 2012). Spectrophotometric and HPLC methods, although standard for determining the TCs in cassava and its products, are time-consuming, expensive and require a complex laboratory setup, and adequately trained personnel to operate. On the other hand, the iCheck device is easy to use, is highly portable, and gives an estimation and quantification of the TCs quickly, with results that are similar to the standard spectrophotometric and HPLC methods. It can be used in the field, with no electricity or refrigeration, and by personnel who require very little training compared with the conventional methods (Ceballos and Parkes, 2014; Islam and Schweigert, 2015). The iCheck device and the conventional spectrophotometric method both operate at an absorbance of 450 nm and have good correlation of their results with an average r^2 of 0.77 for replicated samples. Results from HPLC are also strongly correlated with the standard spectrophotometric method and have an average r^2 of above 0.90. However results from HPLC and the iCheck show a positive correlation of $r^2 = 0.56$ which is somewhat low (Maroya et al. 2012).

Results from spectrophotometric readings tend to overestimate the carotenoid concentrations as compared to HPLC readings. HPLC readings show a lower carotenoid concentration than the standard spectrophotometric methods due to saponification losses as a result of extensive alkali treatments (Islam and Schweigert, 2015). It is important to note that the different methods used for quantification of carotene have diverse advantages based on the duration of extraction process, time and cost as referenced by different authors (Table 4).

Near-infrared spectroscopy (NIRS) is a method that makes use of the near-infrared region of the electromagnetic spectrum (from about 700 to 2500 nm). By measuring light scattered off of and through a sample, NIRS can be used to quickly determine a material's properties without altering the sample. Capable of examining irregular surfaces with the same ease as a carefully prepared sample, NIRS is non-destructive, and requires little or no sample preparation. It can also be used to analyse multiple constituents in a single scan.

NIRS predictions were highly satisfactory for dry matter content (Root dry matter content, $R^2 = 0.96$), TCC, ($R^2 = 0.92$) and total beta carotene ($R^2 = 0.93$). NIRS could also distinguish roots with high or low cyanogenic potential ($R^2 = 0.86$). Hunter colour parameters could also be used for predictions, but with lower accuracy than NIRS. NIRS or chromameter improve selection protocols, allowing faster gains from breeding. Results also demonstrate that TBC and DMC can be improved simultaneously (required for the adoption of bio-fortified cassava (T. Sánchez et al, 2014). This has been very successfully used by our CIAT partners.

Table 4 Review of methods used in quantification of carotene in crops

S.no	Quantification method	Distinctiveness	References
1	Use of flatbed scanner or JPEG images and ImageJ software	• This method bypasses the extraction process and performs the analysis on solid samples • It is also less time-consuming and avoids the use of expensive organic solvents and generation of organic waste, thus minimizing negative impact on the environment	Chen et al. (2012)
2	UHPLC-UV method	• Rapid separation and quantification of major carotenoids commonly found in fruits and vegetables • Cost effective as a result of the lesser amount of mobile solvent used • Increased sample throughput	Maurer et al. (2014)
3	HPLC and spectrophotometry	• Total carotene content was determined by use of visible absorption spectrophotometry • Centrifugation was used to separate solid and liquid phases (not filteration) during extraction of carotenoids • β-carotene was quantified using HPLC (High-Pressure Liquid Chromatography)	Ortiz et al. (2010)
4	HPLC and chromatography	• Quantification of the β-carotene was carried out using a Water Alliance HPLC System (Waters Chromatography, Milford, MA) • Samples were eluted through YMC carotenoids C30 column and detected with a photodiode array detector (PDA)	Vignest et al. (2012)
5	Near-infrared spectroscopy (NIRS)	• Can analyse a large number of samples per day • Fast method of predicting carotenoids in cassava • Ability to predict content of carotenoids and other pigments in different crops • It offers better R^2 (compared to HPLC standard) than Minolta chromameter • Ability to predict other variables like DMC, HCN simultaneously	Sánchez et al.(2014)
6	Minolta Chromameter (colour intensity and Hunter colour meter readings)	• Portable and can be used under field condition • Less expensive when compared to NIRS • Ability to predict better TBC than TCC value (though NIRS has better R^2 value) • Ability to predict TBC value linear to HPLC and Photospectrometer but TCC value tends to be overestimated • DMC has effect on the efficiency of prediction of Chromameter as R^2 are higher in samples with low DMC than for those with high DMC, especially when TCC/TBC is below 12µg/g • When pre-selection of cassava germplasm for high TCC/TBC value is based on the use of chromameter, special attention should be paid to DMC of the samples selected	Sánchez et al.(2014)

(*Continued*)

S.no	Quantification method	Distinctiveness	References
7	iCheck™ Carotene device	• Samples are prepared, and 0.4 mL is taken from the slurry sample and injected into the reagent vial (iEx Carotene) included in the test kit • Measurement can be taken at about five minutes after extraction of carotenoids in the samples • Concentration of TCC in cassava root is determined by multiplying the value obtained from the device by the dilution factor (total sample volume in water/ sample weight) • Portable and can be used under field conditions	Kulakow et al.(2015)
8	Portable fluorometer (iCheck FLUORO)	• Can analyse pro-vitamin A contents simply, reliably and safely • There is a linear relationship between the value of pro-vitamin A content in food measured with fluorometer and HPLC • The minimum of pro-vitamin A detected (LOD) and quantified (LOQ) by HPLC is lower in comparison to the portable fluorometer, making it slightly less reliable than the HPLC • No systematic and observable bias between the data obtained with the fluorometer and the HPLC, and adequacy was good. Good linearity between the results obtained with the fluorometer and the results obtained with HPLC with R^2> 0.91, making it as sensitive as HPLC	Laillou et al. (2014)

Colour intensity is closely related to carotenoids in cassava roots (Sánchez et al., 2006). While visual selection is useful for separating out white from yellow roots, it cannot efficiently distinguish the subtle differences between deep-yellow roots. As breeding for higher carotenoid levels in cassava advanced, the precision of measurement needed to increase in order to select among genotypes with small differences. Spectrophotometer and HPLC accurately quantifies differences in carotenoid content between deep-yellow roots in terms of carotenoid concentration. But this technique is not suitable for evaluating large populations of such roots, because it can handle only a limited number of samples per day. The roots need to be evaluated while still fresh because of their short shelf-life and also because carotenoids are lost during drying and/or storage when roots are processed to preserve them (Chávez et al., 2007; Çinar, 2004). NIRS, which has found many applications in research involving plant tissues, offers a solution to this problem. It is a non-destructive and, more importantly, a fast method for predicting carotenoid concentration in cassava roots.

7 Conclusion and future trends

The breeding, nutritional and delivery research and development programmes for high pro-vitamin A cassava have demonstrated high potential for impact and benefits for both growers and consumers. Growers and seed producers have additional income from a new value-added product, and consumers gain nutritional and health benefits. These benefits are starting to accrue within just a few years of the variety introduction. A multi-stakeholder platform engages both public and private partners for a fully functional value chain approach. This ensures that access to more nutritious foods can continue to increase.

HarvestPlus breeding activities from 2008 until the present time include strategic crosses using parents with background from the international cassava gene bank of CIAT, combined with elite African germplasm to improve TC level and other key traits like root dry matter content, root mealiness, high yield and disease resistance. Elite materials have been advanced for testing and some have been put into crossing blocks to further improve the bio-fortification breeding pipeline. There is an ongoing set of nationally coordinated research programme testing trials for performance, led by the National Root Crops Research Institute Nigeria, in testing 14 promising bio-fortified cassava variety candidates for future release. Information about the association between genes for carotenoid content and dry matter content located on the same region of Chromosome 1 promises faster progress towards impact from combining high carotenoid content with other key traits in cassava.

Strategic awareness delivery programmes of yellow flesh cassava have led to the increase of nutrition consciousness, hence the rapid adoption of bio-fortified food products to complement national and international interventions to reduce VAD, such as supplementation, food fortification and education. This has made it possible to mainstream bio-fortification in major agriculture and health programmes for countries involved in the breeding of yellow flesh cassava.

With the high level of awareness created on the health and nutrition benefits of yellow flesh cassava, farmers and consumers in countries where cassava breeding programmes exist have also increasingly become nutrition conscious, evidenced by the acceptance and gradual introduction of bio-fortified food products into their dietary systems. For instance, in Nigeria, yellow pro-vitamin A *fufu* has become the preferred product for *fufu* consumers in Akwa Ibom while yellow *gari* has recorded a fast-growing preference for consumers in traditionally white *gari* consumption areas. Nonetheless, there is continuing need for awareness-raising on the health benefits of consumption of yellow cassava as a source of micronutrients in diets. In addition, new varieties with even further increased nutritional value and more attractive yellow colour need to be developed.

Cassava breeding programmes need to continue to produce high-yielding and competitive pro-vitamin A cassava varieties to address existing and potential challenges in the value chain. Ideally, varieties with >15 μg/g of TCC and >30% dry matter content will revolutionize the pro-vitamin A cassava sector. Breeding for root mealiness to improve cooking quality of freshly harvested yellow cassava roots will reduce nutrient losses during processing and make more nutrients available to consumers. Expanding the fresh market sector (boiled roots) through research focused on developing cassava varieties with a high level of mealiness will boost the fresh market for cassava as well. Research focused on prolonging the shelf-life of harvested roots by reducing post-harvest deterioration (PPD) will strongly complement the value of bio-fortified cassava for producers, processors and consumers.

Increased participation of the private sector in seed production, processing and marketing should be strengthened for long-term sustainability. This requires additional studies on profitability analysis, cost benefit ratio, tested business models for the different enterprises along the value chain and in depth market analysis for guidance. There is evidence, based on the delivery efforts during the past years, that increasing product diversification will broaden market opportunities and sustain private sector interest in the bio-fortified food sector. Cassava processed food products have been reported to have low glycaemic index (Eli-Cophie et al., 2017). Efforts should be intensified in the use of pro-vitamin A cassava in weaning foods as well as validating *gari* and *fufu* as low glycaemic index foods which decrease the risk of cardiovascular disease, type 2 diabetes and metabolic syndrome. Different food-to-food fortification options including pro-vitamin A cassava, cowpea, soybean and groundnut should be developed to improve the protein levels in foods that are targeted to infants and young children. The use of 'High Quality Flour' from bio-fortified cassava in the production of pasta and bread should be profiled nutritionally, and if justified, they can be advanced towards commercialization.

There is a need to continue to bridge demand and supply gaps using both conventional and innovative marketing technological tools with a growing interest in online or e-market. There should be emphasis on increasing the number of marketing outlets to improve product distribution and access to consumers. Medium-scale food processors have started to distribute more nutritious pro-vitamin A cassava foods in supermarkets and restaurants (especially *gari* and *fufu*) in the big cities; thus, there is a pressing need to improve on packaging and branding of the products. Bio-fortification initiatives for cassava should invest more time in market intelligence and planning, strengthening linkages among market operators and building capacities to increase efficiencies.

8 Appendix: Procedure for carotenoid determination using iCheck™ methodology

Laboratory: sample preparation	Procedure
	Step1. Wash and peel roots
	Step 2. Dry clean the washed roots with hard tissue
	Step 3. Cut each of the five roots longitudinally into four equal parts and take the two opposite parts and chop them into cubes with a rust-free knife
	Step 4. Mix the chopped pile together and divide into four equal parts. Select two opposite parts of the pile, mix together. The other opposite pile should be kept for dry matter estimate
Preparation of homogenous slurry	**Step 1.** 5g of the chopped roots is weighed and macerated using mortar and pestle
	Step 2. Add 25ml of distilled water gradually as the homogenization proceeds
	Step 3. Transfer the resultant slurry into calibrated falcon tube and shake the solution
Measurement	**Step 1.** Using a needle and syringe, measure and inject 0.4ml of the slurry into the reagent vial
	Step 2. Shake the vial vigorously for 10 seconds
	Step 3. Allow the vial to stand and settle for 5 minutes
	Step 4. Place the vial in the iCheck™ device and take your reading
	Step 5. Document the reading shown on the iCheck display screen
	Step 6. Calculate your dilution factor and the total carotene

9 Where to look for further information

For information on bio-fortified cassava the readers are referred to the HarvestPlus annual reports and a recent publication which is available in *AJFAND* Volume 17 No. 2 (2017) - Special Issue on Biofortification.

Cassava bio-fortification work is funded by HarvestPlus and the research component is handled by research institutions and national partners as listed below.

1 National Root Crops Research Institute (NRCRI), Umudike, Nigeria
2 Crop Research Institute (CSIR-CRI), Ghana
3 National Agricultural Research, Malawi
4 INERA-DRC, Congo
5 CNRA-Côte-d'Ivoire
6 Sierra Leone, Agricultural Research Institute (SLARI)
7 Mozambique Agricultural Research Institute
8 Rwanda Agricultural Research Institute (RAB)
9 Tanzania Agricultural Research Institute (ARI)
10 NARO-NaCCRI, Uganda
11 HarvestPlus-Kinshasa, DRC
12 International Center for Tropical Agriculture (CIAT), Colombia
13 International Institute of Tropical Agriculture (IITA), Nigeria, Zambia and Liberia

10 Acknowledgements

We thank HarvestPlus for its support; other cassava breeders C. Egesi, P. Kulakow, P. Ilona, A. G. O. Dixon and B. Maziya Dixon; and all our national partners.

11 References

Akinwale, M. G., Aladesanwa, R. D., Akinyele, B. O., Dixon, A. G. O. and Odiyi, A. C. (2010). Inheritance of β-carotene in cassava (Manihotesculentacrantz). *International Journal of Genetics and Molecular Biology*, 2, 198–201.

Apea-Bah, F. B, Oduro, I., Ellis, W. O. and Safo-Kantanka, O. (2009). Principal components analysis and age at harvest effect on quality of gari from four elite cassava varieties in Ghana. *African Journal of Biotechnology*, 8(9), 1943–9.

Black, R. E., Allen, L. H., Bhutta, Z. A. and Maternal and child undernutrition study group (2008). Maternal and child undernutrition: Global and regional exposures and health consequences. *Lancet*, 371(9608), 243–60.

Bokanga, M. (2001). Cassava: Post-harvest operations, éd. International Institute of tropical Agriculture (IITA), Ibadan, Nigeria.

Brand-Miller, J., Foster Powell, K. and Colagiuri, S. (1996). *The Glycemic Index Factor: The Glycemic Index Solution. The Scientific answer to weight reduction, peak sports performance and blood sugar control.* Hodder & Stoughton, 10–16 South street, Rydalmere, NSW, 2116.

Carvalho, L. M. J., Oliviera, A. R. G., Godoy, R. L. O., Pacheco, S., Nutti, M. R., de Carvalho, J. L. V., Pereira, E. J. and Fukuda, W. G. (2012). Retention of total carotenoid and b-carotene in yellow

sweet cassava (*ManihotesculentaCrantz*) after domestic cooking. *Food & Nutrition Research*, 56, 15788. DOI: 10.3402/fnr.v56i0.15788.
Ceballos, H. and McClafferty, B. (2006). Biofortified Cassava. Breeding crops for better nutrition, HarvestPlus.
Ceballos, H., Morante, N., Sánchez, T., Ortiz, D., Aragón, I., Chávez, A. L., Pizarro, M., Calle, F. and Dufour, D. (2013). Rapid cycling recurrent selection for increased carotenoids content in cassava roots. *Crop Science*, 53, 2342–51.
Ceballos, H. and Parkes, E. (2014). Measuring Pro-vitaminA Content in Crops. Conference Brief. The 2nd Global Conference on Biofortification: Getting Nutritious Foods to People, Kigali, March 31 – April 2, 2014.
Chavez, A. L., Sanchez, T., Javamillo, G., Bedoya, J. M. I., Echeverry, J., Bolanos, E. A., Ceballos, H. and Iglesias, C. A. (2005). Variation of quality traits in cassava roots evaluated in landraces and improved clones. *Euphytica*, 143, 125–33.
Chavez, A. L., Sanchez, T., Ceballos.H., Rodriguez-Amaya, D. B., Nestel, P., Tohme, J. and Ishitani, M. (2007). Retention of carotenoids in cassava roots submitted to different processing methods. *Journal of the Science of Food and Agriculture*, 86, 634–9.
Chen, A., Chou, C., Jensen, B., Lerer, J., Pan, S. and Williams, J. (2012). Quantification of total and Pro-vitamin A Carotenoids from Solid Samples of Corn and Cassava using flatbed scanner or JPEG Images and ImageJ Software. Poster Presentation, South California Conferences for Undergraduate Research, 17 November 2012.
Darnton-Hill I., Webb P. and Harvey, P. W. (2005). Micronutrient deficiencies and gender: Social and economic costs. *American Journal of Clinical Nutrition*, 81(5), 1198S–1205S.
De Moura, F. Miloff, A. and Boy, E. (2015). Retention of Pro-vitamin A carotenoids in staple crops targeted for biofortification in Africa: Cassava, maize, and sweet potato. *Critical Reviews in Food Science and Nutrition*, 55, 1246–69.
Eli-Cophie, D., Agbenorhevi, J. K. and Annan, R. A. (2017), Glycemic index of some local staples in Ghana. *Food Science & Nutrition*, 5, 131–138. doi:10.1002/fsn3.372.
GAIN (2014). Market Opportunity Analysis for Nutrient – Rich cassava in Nigeria. Working draft.
Gichuki, S., Ada, M. and Mark, M. (2010). Consuming cassava as a staple food places children 2 to 5 years old at risk for inadequate protein intake, an observational study in Kenya and Nigeria. *Nutrition Journal*, 9, 9–15.
Gupta, P., Sreelakshmi, Y. and Sharma, R. (2015). A rapid and sensitive method for determination of carotenoids in plant tissues by high performance liquid chromatography. *Plant Methods*, 11(1), 5. http://dx.doi.org/10.1186/s13007-015-0051-0.
HarvestPlus (2014a). Biofortification Progress Briefs, August 2014, p. 12. Retrieved from www. HarvestPlus.org.
HarvestPlus Nigeria (2014b). *GIS Mapping of Vitamin A Cassava Dissemination and Multiplication Sites*. HarvestPlus Nigeria Operational Database.
Iglesias, C., Mayer, J., Chávez, L. and Calle, F. 1997. Genetic potential and stability of carotene content in cassava roots. *Euphytica*, 94, 367–73.
Ilona Paul (2017). Delivery of pro-vitamin A cassava in Nigeria-Lesson leant. Presentation at the Hplus Crop meeting at ISTRC-AB 13 Symposium 6–10 March 2017, Dar Salam, Tanzania.
Islam, K. M. S. and Schweigert, F. J. (2015). Comparison of three spectrophotometric methods for analysis of egg yolk carotenoids. *Food Chemistry*, 172, 233–7.
Jenkins, D. J. Jenkins, A. I., Wolever, T. M., Vuksan, V., Venket, R. A., Thompson, I. U. and Jasse, R. G. (1994). Low Glycemic index carbohydrate and physiological effect of altered food frequency. *American Journal of Clinical Nutrition*, 59, 706–9.
Kawuki, R. S., Pariyo, A., Nuwamanya, E., Ssemakula, G., Tumwesigye, S., Bua, A., Baguma, Y., Omongo, C., Alicai, T. and Orone, J. (2011). A breeding scheme for local adoption of cassava (ManihotesculentaCrantz). *Journal of Plant Breeding and Crop Science*, 3(37), 120–30.
Kimura, M., Kobori, C. N., Rodriguez-Amaya, D. B. and Nestel, P. (2007). Screening and HPLC methods for carotenoids in sweet potato, cassava and maize for plant breeding trials. *Food Chemistry*, 100, 1734–46.

Kulakow, P. A., Parkes, E. Y., Friedrich Simone, K., Schweigert, F., Alamu, O., Maroya, N. and Maziya-Dixon, B. (2015). Linearity, reproducibility and comparison of iCheckTM CAROTENE with spectrophotometer and HPLC for evaluation of total carotenoids in cassava roots. *European Journal of Nutrition and Food Safety, Special issue*, 5(5), 1040–1.

La Frano, M. R., Woodhouse, L. R. Burnett, D. J. and Burri, B. J. (2012). Biofortified cassava increases β-carotene and vitamin A concentrations in the TAG-rich plasma layer of American women. *British Journal of Nutrition*, 110, 310–20. DOI: 10.1017/S0007114512005004.

Laillou, A., Renaud, C., Berger, J., Moench-Pfanner, R., Fontan, L. and Avallone, S. (2014). Assessment of a portable device to quantify vitamin A in fortified foods (flour, sugar, and milk) for quality Control. *Food and Nutrition Bulletin*, 35(4), 449–57.

Luis, S., Willett, W. C. and Stampfer, M. J., Hu, F. B, Franz M., Sampson L, Hennekens, C. H. and Manson, J. E. (2000). A Prospective study of dietary glycemic load, carbohydrate intake and risk of coronary heart disease in U.S. women. *American Journal of Clininical Nutrition*, 71(6), 445–61.

Maroya, N. G., Kulakow, P. A. Parkes, E. Y., Friedrich, S. K., Schweigert, F., Alamu, E. and Maziya-Dixon, B. (2012). Quantification of total carotene content of yellow root cassava genotypes using iCheckTM Carotene in comparison with standard spectrophotometer method. Conference paper, June 2012. DOI: 10.13140/2.1.4534.4007.

Maurer, M. M., Mein, J. R., Chaudhuri, S. K. and Constant, H. L. (2014). An improved UHPLC-UV method for separation and quantification of carotenoids in vegetable crops. *Food Chemistry*, 165, 475–82.

Maziya-Dixon, B. B. and Dixon, A. G. O. (2015). Carotenoids content of yellow-fleshed cassava genotypes grown in four agroecological zones in Nigeria and their Retinol Activity Equivalents (RAE). *Journal of Food, Agriculture & Environment*, 13(2), 63–9.

Maziya-Dixon, B., Adebawale A. A., Onabanjo, O. O. and Dixon, A. G. O. (2005). Effect of variety and drying methods on the physico-chemical properties of high quality cassava flour from yellow cassava roots. *African Crop Science Conference Proceedings*, 7, 635–41.

Morillo, C. Y. (2009). Inheritance of carotenoids content in cassava roots (*Manihot esculenta* crantz). Ph.D. Dissertation. National University of Colombia, Palmira Campus.

Nair, S. G. and Unnikrishnan, M. (2007). Recent trends in cassava breeding in India. In: Ortiz R. and Nassar, N. M. A. (Eds), *Proceedings of the First International Meeting on Cassava Breeding, Biotechnology and Ecology*. Universidade de Brazilia, Brazil, pp. 37–53.

Nascimento, P., Fernandes, N. and Kimura, M. (2008). Assessment of ß- carotene retention during processing of cassava. In abstracts of First Sientific Meeting of Global Cassava Partnership GCPI – meeting the challenges of new millennium. SPOI-11: A-10, 35.

Nassar, N. M. A. and Ortiz, R. (2007). Review – Cassava improvement: Challenges and impacts. *Journal of Agricultural Science*, 145, 163–71.

Njenga, P., Edema, R. and Kamau, J. (2014). Combining ability for beta-carotene and important quantitative traits in a cassava F_1 population. *Journal of Plant Breeding and Crop Science*, 6(2), 24–30

Nweke, F. I. (2001). The cassava transformation in Africa. Food and Agriculture Organization (FAO). Retrieved 14 May 2016.

Obasi, N. E. and Chukwuma, C. S. (2015). Quality Evaluation of Cassava Crackers Made From Yellow Root Cassava (*Manihotesculenta*). *IOSR Journal of Environmental Science, Toxicology and Food Technology*, 9(5), 93–101.

Okafor, E., Erukainure, O., Ozumba, A., Adewale, C. and Kayode, F. (2017). Cassava flour substitution modulates glycemic responses and glycemic index of wheat breads in apparently healthy volunteers. *Journal of Dietary Supplements*, 14(4), 446–52.

Okogbenin, E., Setter, T. L., Ferguson, M., Mutegi, R., Ceballos, H., Olasanmi, B. and Fregene, M. (2013). Phenotypic approaches to drought in cassava: review. *Frontiers in Physiology*, 93(4), 1–15.

Oliviera, A. R. G., Carvalho, L. M. J., Nutti, M. R. and Carvalho, J. L. V. (2008). Carotene in raw and home-cooked sweet yellow cassava: A preliminary study. In: Abstracts of first scientific meeting global of GCPI- meeting the challenges of new millennium. SPOI-07, p. 33.

Oliviera, L., Fukuda, W., Watanabe, E., Nutt, M., Carvalho, J. L. and Kimura, M. (2008). Total carotenoid concentration and retention in cassava products. In: Abstracts of First Scientific Meeting of Global Cassava Partnership GCPI – meeting the challenges of new millennium. SPOI-16: A-15, p. 39.

Oparinde, A., Banerji, A., Birol, E. and Ilona, P. (2016). Information and consumer willingness to pay for biofortified yellow cassava: Evidence from experimental auctions in Nigeria. *Agricultural Economics*, 41(2), 215–33.

Ortiz, D., Sanchez, T., Morante, N., Ceballos, H., Pachon, H., Duque, M. C., Chavez, A. L. and Escobar, A. F. (2010). Sampling strategies for proper quantification of carotenoid content in cassava breeding. *Journal of Plant Breeding and Crop Science*, 3(1), 14–23.

Rice, A. L., West, K. P. and Black, R. E. (2004). Vitamin A deficiency. In Ezzati, M., Lopez, A. D., Rodgers, A. and Murray, C. J. L. (Eds), *Comparative Quantification of Health Risks: Global and Regional Burden of Disease Attributable to Selected Major Risk Factors*. World Health Organization, Geneva, pp. 211–56.

Rodriguez-Amaya, D. B. and Kimura, M. (2004). *HarvestPlus Handbook for Carotenoid Analysis*. HarvestPlus Technical Monograph 2, p. 13.

Saini, R. K., Nile, S. H. and Park, S. W. (2015). Carotenoids from fruits and vegetables: Chemistry, analysis, occurrence, bioavailability and biological activities. *Food Research International*, 76, 735–50.

Salmeron, J. A. and Ascherio, A., Rimm, E. B. and et al. (1997). Dietary fibre glycemic load and risk of NIDDM in men. *Diabetes Care*, 20(4), 545–50.

Sánchez, T., Ceballos, H., Dufour, D., Ortiz, D., Morante, N., Calle, F., ZumFelde, T., Dominguez, M. and Davrieux, F. (2014). Prediction of carotenoids, cyanide and dry matter contents in fresh cassava root using NIRS and Hunter color techniques. *Food Chemistry*, 151, 444–51. https://doi.org/10.1016/j.foodchem.2013.11.081

Ssemakula, G., Dixon, A. G. O., Maziya-Dixon, B. and et al. (2007). Stability of total carotenoid concentration and fresh yield of selected yellow-fleshed cassava (ManihotesculentaCrantz). *Journal of Tropical Agriculture*, 45, 14–20.

Vignesh, M., Firoz, H., Nepolean, T., Supradip, S., Agrawal, P. K., Guleria, S. K., Prasanna, B. M. and Gupta, H. S. (2012). Genetic variability for kernel b-carotene and utilization of crtRB1 3'TE gene for biofortification in maize (Zea mays L.). *Indian Journal of Genetics and Plant Breeding*, 72(2), 189–94.

Vimala, B., Nambisan, B., Thushara, R. and Unnikrishnan, M. (2009). Variability of Carotenoids in yellow-fleshed cassava (ManihotesculentaCrantz) clones. *Gene Conserve*, 8(31), 1–6.

Vimala, B., Thushara, R. and Nambisan, B. (2009). Carotenoid retention in yellow-fleshed cassava during processing. *Proceedings from the 15th Triennial ISTRC Symposium Peru*, pp. 21–6.

Welsch, R., Arango, J., Bär, C., Salazar, B., Ál-Babili, S., Beltrán, J., Chavarriaga, P., Ceballos, H., Tohme, J. and Beyer, P. 2010. Pro-vitamin A - accumulation in cassava (Manihot esculenta) roots driven by a single nucleotide polymorphism in a phytoene synthase gene. *Plant Cell* 22, 3348–56.

Zundel, C., Chibikom, R., Scheidegger, U., Nagel, P. and Hanna, R. (2010). Developing cassava cultivars based on farmers' needs and on the agro-ecological conditions of north-western Cameroon. *African Journal of Root and Tuber Crops*, 8(2), 23–33.

Chapter 9

Breeding cassava to meet consumer preferences for product quality

Adebayo Abass, International Institute of Tropical Agriculture (IITA), Tanzania; Wasiu Awoyale, International Institute of Tropical Agriculture (IITA), Liberia and Kwara State University Malete, Nigeria; and Lateef Sanni and Taofik Shittu, Federal University of Agriculture, Nigeria

1 Introduction

Cassava (*Manihot esculenta* Crantz) is ranked as the sixth most important source of calories in the human diet worldwide (Alfredo et al., 2000). An estimated 800 million people depend on the starchy roots and their products as sources of dietary energy (Akparobi et al., 1998; Nweke et al., 2002). Cassava leaves provide high protein content and many micronutrients, such as the carotenoid precursors of vitamin A. Cassava's combined abilities to produce high yields under poor conditions and suitability for underground storage make it a classic 'food security crop' (Hahn, 1989). Africa dominates the global production of cassava with 53.7% of global output being produced in the continent (FAOSTAT, 2014).

New high-yielding Tropical Manioc Selection (TMS) varieties from IITA have transformed cassava from a low-yielding, famine-reserve crop to a high-yielding cash crop for both rural and urban consumers in Africa (Manyong et al., 2000; Nweke et al., 2002; Johnson et al., 2003; Phillips et al., 2004). Although mean yields of cassava at the national level are in the range of 7–11 metric tonnes per hectare on farmers' fields in Africa, recent breeding work by both national and international research centres has made yields up to 30–45 metric tons per hectare obtainable when improved agronomic practices are combined with improved varieties. The opportunity provided by this high yield in terms

http://dx.doi.org/10.19103/AS.2016.0014.20

of lower production cost per hectare has made cassava a potential crop for large-scale or commercial-scale production.

This chapter looks at the range of uses of cassava and what this means for target properties in breeding new varieties. It reviews the range of breeding preferences amongst farmers, and a range of nutritional, sensory, processing and product properties that could be used in setting future breeding priorities for cassava.

2 Uses of cassava

Cassava roots are typically prepared by boiling, steaming, roasting, frying or baking. Common processing steps include:

- removal of the peel (composed of periderm and cortex) because it is less palatable and nutritious than the root flesh, and to get rid of a high proportion of the cyanogenic glucosides, which are concentrated there (Cyanogenic glucosides, held in cell vacuoles, are enzymatically degraded after cell damage to release the very toxic hydrocyanic acid (HCN). Low and high cyanogenic potential (CNP) varieties are often referred to as 'sweet' and 'bitter', respectively.);
- mechanically pressing and drying for removal of water;
- fermentation of the roots to soften the tissue, eliminate cyanogens, and develop the sour taste and aroma in the fermented food product, preferred by many consumers;
- roasting or steaming the grated cassava (in its fermented or unfermented forms) to partially gelatinize and develop the desired texture and flavour in the resulting cooked dry or moist products;
- drying in the sun or in artificial dryers to improve storability.

Dried cassava is often milled into flour for bread. More recently cassava has been processed to make noodles, biscuits and other industrial products like glucose syrup and sweeteners. Cassava has also been used as a corn starch substitute in products such as salad dressing or adhesives, as well as a thickener in soups, baby foods, sauces, etc. Cassava is now also used as an ingredient in feeds for cattle, goats, pigs and poultry, by mixing with protein sources such as maize, fish meal, groundnut and mineral salts (Kavana et al., 2005).

Apart from fresh roots and leaves, there are five common groups of traditionally processed cassava food products marketed by the farmers and food processors in Africa (Nweke, 2003):

- dried chips
- steamed paste
- wet paste
- toasted granules
- steamed granules

There are many different names for these products:

- Chips/flour, is known as *kokonte* in Ghana; *Lafun* in Nigeria; *cossette* in DRC; *chinyanya*, *makopa*, *abutega*, *omundyenge*, *udaga*, *kondowale*, *bada*, *dimwehe*, *uhu*, *udunnule*, *mbute*, *makonga*, *obusanobwe* or *kivunde* in Tanzania; *obuite*, *udilimbe*, *habalogala*, *ajobo* or *malwa* in Uganda (Nweke, 1994).

- Steamed paste is called *chickwangue* in Zaire, *placali* in Cote d'Ivoire, and *myondo* or *boboo* in Cameroon and is made from a fine mash (Nweke, 1994).
- Unsteamed (wet) paste is called *akpu* or *fufu* in Nigeria, *agbalima* in Ghana and water *fufu* in Cameroon (Nweke, 1994).
- Toasted granules are widely known as *gari* in West Africa countries except Cote d' Ivoire where it is called *atoukpou* (Nweke, 1994).
- Steamed granules are called *attieke* in Cote d' Ivoire where its production is concentrated.

There have been a range of studies into factors affecting the sensory properties of these products, such as that of Sanni et al. (1999) on the composition and sensory qualities of *lafun*. There have also been extensive studies on factors affecting the quality of *fufu*, for example by Okpokiri et al. (1985), Akingbala et al. (1991), Sanni et al. (1999), Sanni and Akingbala (2000), Tomlins et al. (2007) and Sanni (2015).

Cassava processing and consumption patterns and preferences differ among regions, thereby influencing the type of cassava varieties adopted by farmers. There is a general relationship between climate and the type of cassava produced (sweet or bitter), with bitter cassava production tending to decrease from humid through sub-humid to non-humid zones while sweet cassava production follows a reverse trend. The proportion of cassava processing increases or decreases in parallel with the proportion of bitter varieties.

The Collaborative Study of Cassava in Africa (COSCA) revealed that in the humid regions, the most important cassava products are (Nweke et al., 1994):

- fresh root (3%),
- paste (17%),
- granules (24%),
- flour (56%).

The important products in the sub-humid regions are:

- fresh root (1%),
- paste (13%),
- granules (27%),
- flour (59%).

In non-humid regions, the products are:

- fresh root (4%),
- granules (12%),
- flour (84%).

As an example of regional variations, Kamau et al. (1998) found that the use of cassava in Kenya varies widely from one region to another and so do cassava varieties. In western Kenya, cassava is processed and dried before use; therefore even bitter varieties are conventionally grown for their high yields. In the coast region, cassava is mainly boiled, fried or roasted, while in central Kenya cassava is chewed raw/boiled alone or mixed with green maize and pigeon pea (vegetable) and/or with meat.

3 Farmer, processor and end-user preferences

This wide range of uses and processes creates significant challenges for establishing breeding objectives. One challenge is how to balance the requirements of farmers with those of processors and end users, especially in the case where there may be a trade-off in productivity for varieties with highest expression of processor and consumer-preferred traits (Egesi et al., 2012). In one study, for example, Kamau et al. (2001) showed that farmers preferred a local variety to two improved varieties developed, in the view of the breeders, for better appearance, taste and texture. Farmers tend to prefer varieties with characteristics such as resistance to pests and diseases as well as more established quality characteristics such as low cyanogenic potential (Khizzah et al., 2013). Nweke et al. (1994) also added that the most important farmers' preferences are characteristics such as disease and pest tolerance, early bulking, high root yield, weed suppression and good in-ground storability – all traits which will not be of importance to consumers and processors, unless related to product cost.

A study done in Ghana showed that in villages where *gari* is the main product, 60% of farmers preferred early bulking varieties, 20% preferred high-yielding varieties, 10% preferred good cooking varieties and another 10% preferred high leaf-yield varieties (Otoo et al., 1995). In a village where fresh roots were the main product, the authors found that 67% of farmers preferred early bulking varieties while 33% wanted varieties with good pounding qualities to prepare *fufu*. Among villages where dry cassava chips (*kokonte*) are the main product, 50% of farmers preferred early bulking varieties, and 50% wanted varieties with a low cyanide potential. In a village where fermented cassava dough (*agbelima*) is the main product, farmers preferred varieties which were resistant to diseases and pests. A 2015 study of a Nigerian town focused on processing cassava for *gari* found a preference for early maturing varieties with good shelf-life during storage (Onyenwoke and Simonyan, 2014).

In Malawi, cassava is mostly used in the fresh unprocessed form by chewing it or by roasting or boiling the dried cassava (makaka or dried cassava chips). Malawi farmers prefer characteristics such as low cyanogenic potential and high dry matter which produces good quality flour and nsima (dough) (Salipira et al., 1995). An assessment of seven elite clones in Malawi showed that TMS 91934, TMS 63397, TMS 60142, TMS 4(2)1425, MK 90/1160 and Mk 90/1151 met these requirements while MK91/478 was difficult to cook (Salipira et al., 1995).

Experience in Malawi showed that where landraces grown by different ethnic/cultural groups were specifically selected for a set of desirable traits, the groups would readily adopt new varieties if the required desirable traits were available (Mkumbira, et al., 2004). A study of attributes of varieties desired by farmers in Malawi showed that early maturity was the main attribute desired by farmers followed by low cyanogenic potential, high yield, and disease/pest resistance (Mkumbira et al., 2003). The preference for low cyanogenic potential is key in locations where cassava is consumed raw or with minimum processing.

4 Target traits: nutritional and sensory properties

4.1 Nutritional properties

As mentioned previously, a key priority for many farmers is low cyanogen content or cyanogenic potential (CNP) in the roots. As a result, breeders at the National Root Crop

Research Institute (NRCRI) in Nigeria and IITA have sought to develop varieties that can be boiled and eaten without the laborious processes involved in reducing the cyanogen content of the high CNP varieties to safer levels before consumption. Nonetheless, only two out of a list of 17 improved cassava varieties released in Nigeria by the end of the 1990s were classified as low-CNP varieties (TMS 4(2) 1425 and TMS 3001) while the rest are high CNP (NR series: 8082, 8083, 8310777, 8208, 8212, 401044, and TMS series; 3055, 50395, 30572, 91934, 90257, 84537, 82/00058, 82/00661, 81/00110) (NRCRI, 1997; Nnodu and Dixon, 2001). There was however a significant reduction in cyanogenic potential of new experimental clones developed by IITA in the late 90s (Mkumbira et al., 2004). Some studies have shown that dry matter, total yield and number of roots per plant are significantly negatively correlated to CNP, suggesting potential problems in further reduction of CNP (Mahungu et al., 1994; Khizzah et al., 2013). However, these correlations are not consistent with many studies done in Latin America, with a broader, more diverse genetic base.

From a nutritional point of view, cassava's chief advantage is as a source of carbohydrate. The roots of currently available varieties are relatively low in protein, although in Central Africa substantial amounts of protein are derived from cassava leaves (Nweke, 1994). However, cassava has other potential nutritional benefits. Vitamin A is an essential micronutrient for human growth and development, and for the functioning of the visual and immune systems (Combs, 1998; ACC/SCN, 2000). Maroya *et al.* (2012) estimated that in sub-Saharan Africa, three million children under the age of five years suffer total or partial blindness caused by vitamin A deficiency.

According to Sanchez et al. (2012), cassava is an ideal vehicle for delivering pro-vitamin A carotenoids (pVAC) to human populations affected by deficiency of this important micronutrient. Improving the vitamin A status of children reduces mortality rates from 23 to 30% (ACC/SCN, 1992; Beaton et al., 1993). Research has demonstrated that micronutrient-enrichment traits are available within the genomes of some major staple food crops including cassava that could allow for substantial increase in the levels of pVAC without negatively impacting crop yield (Welch, 2001; Maroya et al., 2012).

Breeding for pVAC cassava to increase nutritional status of the consumers did not begin by national research institutions in countries such as Nigeria until the 1990s (Nweke et al., 1999), though IITA had developed varieties with β-carotene concentration range of 1.28–9.19 μg/g fresh weight (IITA, 1987; Maroya et al., 2012). A study by Egesi et al. (2012) to assess pVAC varieties with high productivity and end-user preferences in advanced trials showed that the fresh root yield ranged from 15 to 47 t/ha fresh weight (dry root yield from 3.7 to 14 t/ha), DMC ranged from 13.5% to 40.4% and total carotene ranged from 0.96 to 8.13 μg/g. The variety TMS 05/1632 had the highest carotene content of 8.13 μg/g. The carotenes confer a yellow colour to the roots, with greater intensity of yellow being closely related to carotene content.

In order to retain a large amount of the β-carotene content of fresh cassava in the processed products, processing needs to be controlled. Exposure to light, heat and oxygen may lead to loss of β-carotenes or pro-vitamin A potency. Similarly, the occurrence of post-harvest physiological deterioration (PPD) and microbial spoilage of yellow cassava decreases the amount of β-carotenes in cassava roots. Studies have been conducted on retention of β-carotene in processed food items. Awoyale et al. (2014) reported that starches from yellow-fleshed cassava could be stored for longer periods at temperatures ≤42°C, as their monolayer moisture content falls within the acceptable range for storage (<10%). In addition, negative (but not significant) correlations exist between β-carotene and the monolayer moisture content of the yellow-fleshed cassava root starches.

4.2 Sensory properties

Cassava varieties with cyanogen contents of less than approximately 50 mg/kg fresh roots are classified as sweet; low-CNP, intermediate or moderate-CNP roots have 50–100 mg/kg, and bitter or toxic roots have more than 100 mg cyanogens per kg fresh weight. However, some cassava varieties may exhibit different degrees of bitterness as well as different levels of cyanogens from one season to the next (rainy and dry) within the same growing cycle.

Texture and consistency, followed by taste, colour or appearance, were the most critical quality characteristics desired by processors for processed cassava products (Nweke, 1994). Texture is an important consideration in processed cassava food products across sub-Saharan Africa because of the widespread style of eating. This style consists of rolling steamed or cooked pasty foods into balls usually with the fingers and then dipping into sauce or soup of choice. Consumers look for a sour taste (the strength of the sourness will depend on regional preferences) which is characteristic of proper fermentation (Nweke, 1994). Jones (1959) reported that genotype (variety) also has an effect on the flavour of final products and consumer acceptability. A wide range of flavour is possible and consumers are often very selective in their purchases. Consumers often associate the sensory qualities of boiled, roasted, fried or baked cassava roots with their colour, flavour, taste, sweetness, firmness, fibrosity, stickiness, mouth-feel, cohesiveness, etc. In addition, the food quality of cassava roots may be related to the physicochemical properties of the flour or starch such as peak viscosity.

The main quality attributes noted by consumers of boiled cassava root are mealiness and softness; both of them should be achieved in a minimum cooking time (Bakayoko et al., 2009). Safo-Kantanka and Acquistucci (1996) reported that mealiness of cooked cassava roots depends largely on the total starch content of the root. Some cassava varieties give soft, mealy and easy-to-mash boiled roots. In some parts of Africa, after boiling, the roots from these types of varieties are pounded into a smooth paste called *fufu*. Other varieties give roots which when boiled remain hard and are waxy, and cannot be pounded into *fufu*. IITA and several African countries have breeding programmes to develop the mealy-type cassava varieties. Wherever boiled cassava is consumed, mealiness is the main quality characteristic (FAO, 1999).

During the dry season, cassava usually drops its leaves. At the onset of rains, a dramatic shift in root quality takes place, probably due to a remobilization of carbohydrates towards new leaf formation. The mealy texture of the boiled cassava root is often lost, and roots can no longer be used for this purpose (FAO, 1999). A study done in Côte d'Ivoire by Bakayoko et al. (2009) observed that the melting index (inverse of paste viscosity) of boiled cassava roots was positively correlated with mealiness index. Hence, the softness attribute of boiled cassava and storage mealiness are related and one could thus be predicted from the other. Storage mealiness is the powdery, soft, dry or crumbly texture of a boiled cassava root during storage. Similarly, Padonou et al. (2005) showed that sweet cassava cultivars had a significantly lower firmness than bitter ones. Mealiness is due to cell breaking on chewing; if the cohesive forces between cells are higher, then the sensory mealiness is lower. This may be linked to the presence of residual cell walls in the boiled root slices, which can limit the swelling of starch trapped within the cell structure (Padonou et al., 2005).

5 Target traits: processing properties

There are a number of properties that affect processing of cassava into a range of food and other products, including starch and fibre content, PPD, ease of peeling and water content.

5.1 Starch properties

Cassava starch aids the thickening and solidifying process, gives a sticky consistency and preserves the balance of water in foods such as soup, candy, pudding, sausages, bread, ice cream, and noodles, and serves as a binder in pelletized medicines. Cassava starch is used for the manufacture of glucose, fructose, soft drinks and canned foods, among many other uses. According to Pérez et al. (2012), about 74–85% of dry root weight of cassava is starch. Starch yield can vary significantly according to variety (Tremkumar et al., 2001). Dry matter content (DMC) strongly influences the amount of extractable starch from cassava roots as well as its shelf-life. Processors usually pay differential prices for the fresh roots depending on their DMC. Variation in root cortex or peel thickness also affects the amount of extractable starch, since the peel has a lower starch content than the root flesh (Kawiki, 2009).

Pre- and post-harvest factors such as genetic factors, environmental conditions during the growth of the plant and the method of starch isolation have been found to have a profound effect on the properties of cassava starches. The physicochemical, functional and pasting properties as well as granule morphology of starches from 40 different experimental cassava varieties (36 cassava-mosaic-disease-resistant (CMD) clones) and currently released cassava varieties in Nigeria (TMS 30572, 4(2) 1425, TME 1 and 82/00058) from the experimental farm of IITA, Ibadan, Nigeria, were investigated (Onitilo *et al.*, 2007a). Onitilo *et al.* (2007b) also reported on the physicochemical and functional properties of sour starches in different cassava varieties. This and previous studies revealed that there were significant differences in the properties of the starches (such as water absorption capacity, swelling properties, solubility, gelation and peak viscosity) from the different CMD resistant varieties (Onitilo et al., 2004).

Peak viscosity indicates the water-binding capacity of the starch or mixture, and is often correlated with final product quality (Onitilo, 2003). Amylose is also important in the gelling properties of starch. Amylose content of cassava starch is in the range of 13.6–23.8% with an average of 16.6% (Rickard et al., 1991; Tanticharoen, 1992). A positive correlation also exists between amylose content and peak viscosity and pasting properties. Low solubility also indicates a high quality starch with high paste viscosity. Granule size also affects pasting properties (Charles et al., 2005). It has been shown that novel starch types are technically possible through the introduction of desired genes into cassava (Munyikwa, 1997). One of the objectives of the cassava-breeding project at CIAT is the identification of clones with special root quality characteristics such as starch content and quality (Ceballos et al., 2007).

5.2 Fibre content

Cassava roots with high fibre content are typically less suitable for human consumption (Mkumbira et al., 2004). Pectinic substances, which are a type of viscous dietary fibre, exist in the cell-wall of cassava roots in two forms – water soluble (salts of pectinic and pectic acids) and water insoluble (protopectin, which is a form of pectic substance that is combined with cellulose). The pectic substances contribute to the texture and hardness of cassava, which in turn are responsible for the *mouth feel* of cooked or processed foods from cassava roots (Infante et al., 2013). The fraction of water-soluble pectin is often lower than the insoluble pectin in fresh cassava roots. The total amount of pectin and the ratio of the total soluble pectin to the insoluble form may alter the texture and mouth feel of different cassava varieties. It is also suspected that this ratio may vary in the roots of a growing cassava plant from one season to the next within the same growing year and also with the age of the plant (Infante et al., 2013). This variation may affect the texture and eating quality of the roots of some cassava varieties from season to season, but further study in this area is needed, particularly for some cassava varieties grown in Africa, where a wide range of processing forms depend on root texture. Furthermore, Infante et al. (2013) recommended the consumption of cassava bread in the daily diet, since the dietary fibre and pectin content would contribute to the benefit of consumer health. In addition, Moorthy et al. (1996) found that the presence of fibres in cassava flour delays gelatinization and gives a lower peak viscosity by limiting the access of water into the granules. Therefore, it will be necessary to continue evaluation of a wide range of genetic diversity of cassava from diverse geographic regions in relation to pectin content.

5.3 Post-harvest physiological deterioration

Cassava is prone to rapid PPD that occurs within 24–48 hrs after harvesting. Deterioration begins with light-brown discolouration in the parenchymatous tissue (vascular streaking) and a blue-black pigmentation of the xylem. Physiological deterioration is often followed by microbial spoilage. PPD constitutes a major constraint faced in large-scale production and marketing of fresh cassava roots. It substantially reduces the eating quality, transportation range, and market value of cassava (Booth, 1976; Buschmann et al., 2000; Westby, 2002; Lyer et al., 2010). The consequence is a reduction in income for the growers, traders and processors in the cassava value chain, and added cost for consumers.

Due to the rapid perishability or PPD of cassava after harvest, the traditional method of storage of cassava in the ground until it is needed is very widespread, especially in many African countries where processing facilities are limited. As has been noted, farmers often prefer varieties with good in-ground storability (Makambila, 1981). Breeding to introduce genes that prevent lignification of roots in underground storage has the potential to reduce the loss of quality of cassava across multiple harvest periods.

Ilona and Dixon (2001) carried out a study to identify cassava varieties with delayed PPD by evaluating the relationship between vascular streaking and root DMC, mealiness, flesh colour, cyanogenic potential and moisture loss after harvest. They suggested that the negative relationship between PPD and root dry matter might require a compromise in selection of new varieties if a longer shelf-life for fresh roots is desired. This implied that the higher the root DMC, the lower the rate of deterioration. However, high root dry matter is normally favoured in cassava because of its positive relationship to starch or flour yield and eating quality. This conflicting relationship between root DMC and PPD remains a problem in cassava breeding. However, although there is a generally negative relationship between PPD and DM content, there does appear to be potential for selection of both high DM content and low PPD if a broad genetic base is accessed (H. Ceballos, personal comm.).

In Nigeria, some improved lines and landraces (TMS 86/1, TMS91/01730, TMS 91/02319, TMS 92/0053, TMS 92/00099 TMS 90/01554, TMS92/0396 CB5-10980411, TMS 91/00457, TMS 81/000000942, TMS82/00661, TMS 91/00457, TMS 1095-D and ATU, MS 6, 'Ege dudu' and 'Isu nikan kinya') have tendencies for low PPD. The most susceptible genotype was TMS 30001 (Ilona and Dixon, 2001). As there is a close relationship between PPD and mechanical damage, and between injury and root desiccation, storage methods based on controlling desiccation by storing roots in humid environments such as in plastic bags have shown good prospects. To complement the progress that breeders have made in improving cassava yield, PPD needs to be addressed to strengthen strong links between small-scale producers and new markets (Wheatley et al., 1995). Simple coating of fresh roots with melted wax (140-160°C) is commercially used in Latin America to reduce the rate of PPD and extend the shelf-life of fresh roots for export. The technique has recently been tested in Uganda to keep cassava root in its fresh form and quality for at least 14 days, making it possible to market it over longer distances.

In addition, resistance to PPD, along with ease of peeling, will influence processing waste and product yield. Mechanized harvesting and processing are constrained by the irregular shape and size of cassava roots and are influenced by variety, growing environment, and age. Consequently, it will be important for cassava breeders to work with food process engineers for the development of cassava roots that could be easily harvested and processed mechanically. Efficient mechanization of peeling and grating operations may be significantly constrained by multiple factors. Roots with peels that can be rolled or rubbed off will have less waste and require less labour/time of processing than peels that fix tightly to the flesh and would have to be manually removed with knives.

It has been observed by post-harvest processing experts that the cost of drying of cassava for making some types of products (such as flour or chips) could be reduced if genotypes with higher DM content (i.e. less moisture) are developed and made available (Nweke et al., 1999). The higher the moisture content of cassava roots, the lower the product yield; the higher the drying costs, the lower the quality of processed products. Therefore, many preferred traits relate to reduction in processing time and costs; improved safety, quality and marketability of the final products; and consequently increased profitability and income for the farmers and processors. Targeted breeding can contribute to reducing processing costs that arise from high labour needs, long time of processing and high energy inputs (Nweke et al., 1999).

6 Target properties: products

One of the most important potential uses of cassava for resource-scarce countries is as a wheat substitute in bread, cake, biscuits, noodles, etc. (Abass et al., 2001). Bread baked with 10% and 20% High Quality Cassava Flour (HQCF) is generally acceptable to consumers in terms of appearance, taste and texture.

Information on pasting profile of flours has been used to assess the functionality of starchy food ingredients in processes like baking (Idowu et al., 1996; Rojas et al., 1999; Adebowale et al., 2012) and extrusion cooking (Ruales et al., 1993). High peak viscosity and stability (or low breakdown viscosity) are associated with cassava starches which produce acceptable bread (Adeyemi et al., 1978; Adebowale et al., 2012).

An assessment carried out in Nigeria on the chemical composition and pasting properties of 43 cassava-mosaic-disease-resistant varieties of cassava (Sanni et al., 2008) showed that peak viscosity ranged between 289 RVU and 491 RVU. Variations due to interactions between clone and agro-environment (clone × location) were significant ($P < 0.05$) for chemical composition and pasting properties. Sugar, protein and ash content have been shown to inversely affect the pasting profile of cassava while higher starch and amylose content improve the pasting profile (Dixon et al., 2007; Shittu et al., 2007; Sanni et al., 2008). Erikson (2013) evaluated quality parameters – including moisture, pH and starch contents and acidity as well as functional properties in water such as pasting characteristics, swelling power and water-binding capacity of flours – from three different local varieties of cassava (*Afisiafi*, *Bankye hemmaa* and *Doku Duade*) compared with wheat flour. The results showed that cassava flour had a whiter colour and higher swelling power, higher solubility and higher water-binding capacity than the wheat flour conventionally used for bread baking. This was attributed to higher starch content in cassava flour and also looser association of starch molecules in cassava starch granules. These functional properties of the starch also affected the pasting profile since cassava flours exhibited an early gelatinization, high peak viscosity, large paste breakdown and low retrogradation tendency compared to wheat flour. An increased proportion of cassava flour substitution in wheat-cassava composite bread caused reduced loaf specific volume and increased loaf density and hardness due to a reduction of wheat proportion and thus of gluten-forming proteins in the flour blends. The weakening of the dough resulted in a reduced leavening ability and loaf size. However, a substitution level of up to 20% cassava flour in composite bread was acceptable to consumers in terms of appearance, taste and texture since it was not perceived as different from 100% wheat bread. Therefore, there is need for cassava breeders to work on the possibilities of incorporating glutinous proteins into the cassava root.

Research on the use of cassava roots for making fried chips suggests that the age of cassava roots affects the quality of fried cassava chips. Cassava varieties with high sugar content give chips that turn to unattractive brown during frying while varieties with high DMC become too hard. Roots of varieties harvested very early or too late do not produce fried chips of good quality (Tremkumar et al., 2001).

7 Conclusions

The use of laboratory measurements in profiling the quality of large breeding populations (e.g. root flesh colour, starch pasting properties, fibre, stability after cooking, sugar content and amylose and amylopectin contents) could be rapid, cost effective and accurate in pre-determining the suitability of a breeding line for different applications. For example, analysis of pasting properties can guide breeders to target different cultivars to different end users (household-level food security needs, industrial uses such as for making adhesives, baking bread and brewing). Pasting properties could be used to predict the behaviour of cassava flour/starch during and after cooking, providing a proxy for predicting the texture of the final cassava products. Laboratory methods can be used for rapid screening of a large population of germplasm at the early stages of the breeding cycle by correlating the physicochemical properties of breeding lines with the desired quality parameters of the targeted derivate products. This approach could save research funds, reduce the time it takes to go through several cycles of breeding and selecting suitable varieties, and may also ensure that novel varieties with specialized value-added properties are not discarded during breeding.

8 References

Abass, A. B., Onabolu, A. O. and Bokanga, M. 2001. Impact of the High quality cassava flour Technology in Nigeria. In Akoroda, M. O. and Ngeve, J. M. (Eds), Root Crops in the 21st Century. Proceedings of the 7th Triennial Symposium of the International Society for Tropical

Root Crops-Africa Branch (ISTRC-AB) Centre International des Conferences, Cotonou Benin. 11–17 October 1998.

Adebayo, K. and Sangosina, M. A. 2005. Processors' perception of the effectiveness of some cassava processing innovations in Ogun State, Nigeria. Paper presented at the 19th Farm Management Association of Nigeria (FAMAN) National Conference *Economic Reforms and the Management of Nigerian Agriculture* held between Tuesday 11 and Thursday 13 October 2005 at Delta State University, Asaba Campus.

Adebowale, A. A., Sanni, S. A. and Koleoso, Y. S. 2012. Chemical composition, functional and pasting properties of high quality cassava flour in Nigerian markets. In Okechukwu, R. U. and Ntawuruhunga, P. (Eds.), Tropical Roots and Tuber Crops and the Challenges of Globalization and Climate Changes. Proceedings of 11 ISTRC-AB symposium, Kinshasa, Democratic Republic of Congo, 4–8 October 2010.

Akingbala, J. O., Oguntimein, G. B. and Abass, A. B. 1991. Effect of processing methods on the quality and acceptability of fufu from low cyanide cassava. *J. Sci. Food Agr.*, 57: 151–4.

Akparobi, S. O., Togun, S. O. and Ekanayake, I. J. 1998. Assessment of cassava genotypes for resistance to cassava mosaic disease, cassava bacterial blight and cassava green mite at lowland and midaltitude site in Nigeria. *Afr. Crop Sci. J.*, 6: 385–96.

Alfredo, A. C., Alves, A. A. C. and Tim, L. S. 2000. Response of cassava to water deficit: Leaf area growth and abscisic acid. *Crop Sci.*, 40: 133–7.

Awoyale, W., Sanni, L. O., Shittu, T. A., Adegunwa, M. O., Gueye, B. and Kulakow, P. 2014. Varietal effect on the Moisture Adsorption Isotherm of Yellow-fleshed Cassava Root Starches. *Afr. J. Root Tuber Crops*, 11(1): 41–9.

Bakayoko, S., Tschannen, A., Nindjin, C., Dao, D., Girardin, O. and Assa, A. 2009. Impact of water stress on fresh tuber yield and dry matter content of cassava (*Manihot esculenta Crantz*) in Côte d'Ivoire. *Afr. J. Agr. Res.*, 4(1):021–7.

Ceballos, H., Sánchez, T., Morante, N., Fregene, M., Dufour, D., Smith, A. M., Denyer, K., Pérez, J. C., Calle, F. and Mestres, C. 2007. Discovery of an Amylose-free Starch Mutant in Cassava (*Manihot esculenta* Crantz). *J. Agric. Food Chem.*, 55(18): 7469–76. doi:10.1021/jf070633y.

Charles, A. L., Sriroth, K. and Huang, T. C. 2005. Proximate composition, mineral contents, hydrogen cyanide and phytic acid of 5 cassava genotypes. *Food Chem.* 92: 615–20.

Dixon, A. G. O., Akoroda, M. O., Okechukwu, R. U., Ogbe, F., Ilona, P., Sanni, L. O., Ezedinma, C., Lemchi, J., Ssemakula, G., Yomeni, M. O., Okoro, E. and Tarawali, G. 2007. Fast track participatory approach to release of elite cassava genotypes for various uses in Nigeria's cassava economy. *Euphytica.*, 160(1): 1–13. doi:10.1007/s10681-007-9487-2.

Egesi, C. N., Olasanmi, B., Ekwe, K. C., Ndukwe, I., Ilona, P., Okogbenin, E., Onwuka, S. and Ewa, F. 2012. Participatory breeding for-vitamin A enriched cassava in Nigeria. In Okechukwu, R. U. and Ntawuruhunga, P. (Eds), Tropical Roots and Tuber Crops and the Challenges of Globalization and Climate Changes. Proceedings of 11 ISTRC-AB symposiums, Kinshasa, Democratic Republic of Congo, 4–8 October 2010.

Erikson, E. 2013. Flour from three local varieties of Cassava (*Manihot Esculenta Crantz*): Physico-chemical properties, bread making quality and sensory evaluation. Unpublished thesis submitted to Department of Food Science, SLU. http://stud.epsilon.slu.se

FAO. 1977. Cassava process; FAO Plant Production and Protection series No. 3.

FAO. 1999. Cassava: Post-harvest Operations. Edited by Mejia, D. and Lewis, B., p. 36.

Hahn, S. K. 1989. An overview of African traditional cassava processing and utilization. *Outlook Agr.*, 18: 110–18.

Ilona, P. C. and Dixon. A. G. O. 2001. Genetic variability for vascular streaking in cassava. In Akoroda, M. O. and Ngeve, J. M. (Eds), Root Crops in the 21st Century. Proceedings of the 7th Triennial Symposium of the International Society for Tropical Root Crops- Africa Branch (ISTRC-AB) Centre International des Conferences, Cotonou Benin. 11–17 October 1998.

Infante, R. B., García, O. O. and Rivera, C. 2013. Characterization of dietary fibre and pectin of cassava bread obtained from different regions of Venezuela. *Rev. Chil. Nutr.*, 40(2): 169–73.

Kamau, J. W., Kinama, J. M., Nguluu, S. N., Muhammad, L., Whyte, J. B. A., Ragwa, S. M., Migwa, E. N. and Simiyu, P. M. 1998. Farmers' evaluation of cassava varieties in semi-arid areas of Kenya. In Akoroda, M. O. and Ngeve, J. M. (Eds), Root Crops in the 21st Century: Proceedings of the 7th Triennial Symposium of the ISTRC-AB, pp. 378–83.

Kamau, J. W., Kinama, J. M., Nguluu, S. N., Muhammad, L., Whyte, J. B. A., Ragwa, S. M., Migwa, E. N. and Simiyu, P. M. 2001. Farmer's evaluation of cassava varieties in the semi-arid area of Kenya. In Akoroda, M. O. and Ngeve, J. M. (Eds), Root Crops in the 21st Century. Proceedings of the 7th Triennial Symposium of the International Society for Tropical Root Crops-Africa Branch (ISTRC-AB) Centre International des Conferences, Cotonou Benin. 11–17 October 1998.

Kavana, P. Y., Mtunda, K., Abass, A. and Rweyendera, V. 2005. Promotion of cassava leaves silage utilization for smallholder dairy production in Eastern coast of Tanzania. *Livest. Res. Rural Dev.*, 17 (4): 43. http://www.cipav.org.co/lrrd/lrrd17/4/kava17043.htm.

Kawiki, R. S. 2009. Variation in cassava (Manihot esculenta Crantz) based on single nucleotide polymorphisms, simple sequence repeats and phenotypic traits. Ph.D. Dissertation, Department of Plant Sciences, Faculty of Natural and Agricultural Sciences, University of Free State. Bloemfontein, South Africa, pp. 56–89.

Khizzah, B. W., Whyte, J. B. A. and Van der Grift, R. 2013. Inter-relationships between yield components and cyanogenic potential and their influence on selection. In Akoroda, M. O. and Ekanayake, I. J. (Eds), Root Crops and Poverty Alleviation. Proceedings of the Sixth Triennial Symposium of the International Society for Tropical Root Crops–Africa Branch August 1998.

Maroya, N., Isaac, G., Asante, K. and Dixon, A. 2012. Harvest Plus Papers Genotype by environment interaction effect on beta-carotene of yellow root cassava (Manihot esculenta Crantz) genotypes in Ghana. In Okechukwu, R. U. and Ntawuruhunga, P. (Eds), Tropical Roots and Tuber Crops and the Challenges of Globalization and Climate Changes. Proceedings of 11 ISTRC-AB symposium, Kinshasa, Democratic Republic of Congo, 4–8 October 2010.

Maziya-Dixon, B., Dixon, A. G. O. and Adebowale, A. A. 2006. Targeting different end uses of cassava: Genotypic variations for cyanogenic potentials and pasting properties. *Int. J. Food Sci. Tech.* 42: 969–76.

Mkumbira, J., Chiwona-Karltun, L., Lagercrantz, U., Mahungu, N. M., Saka, J., Mhone, A., Bokanga, M., Brimer, L., Gullberg, U. and Rosling, H. 2003. Classification of cassava into 'bitter' and 'cool' in Malawi: From farmers' perception to characterisation by molecular markers. *Euphytica*, 132: 7–22.

Mkumbira, J., Okechukwu, R. and Dixon, A. G. O. 2004. Breeding for root quality traits in cassava. In Proceedings of 9th Symposium of the ISTRC-AB, Kenya.

Nnodu, E. C. and Dixon, A. G. O. 2001. Evaluation of newly improved low cyanide cassava varieties in different agroecozones of Nigeria. In Akoroda, M. O. and Ngeve, J. M. (Eds), Root Crops in the 21st Century. Proceedings of the 7th Triennial Symposium of the International Society for Tropical Root Crops-Africa Branch (ISTRC-AB) Centre International des Conferences, Cotonou Benin. 11–17 October 1998.

Nweke F. I. 1994. Processing potentials for cassava production growth in sub-Sahara Africa. COSCA Working Paper N° 11. COSCA, IITA: Ibadan, Nigeria.

Nweke, F. I. 2003. New Challenges in the Cassava Transformation in Nigeria and Ghana. Conference Paper No. 8. Paper presented at the InWEnt, IFPRI, NEPAD, CTA conference. 'Successes in African Agriculture' Pretoria December 1–3 2003.

Nweke, F. I., Spencer, D. S. C. and Lynam, J. K. 2002. *The Cassava Transformation: Africa's Best-Kept Secret*. International Institute of Tropical Agriculture: Ibadan, Nigeria.

Nweke, F. I. 1992. Cassava Production Prospects in Africa, Collaborative Study of Cassava in Africa. COSCA Working paper No.13. IITA: Ibadan.

Nweke, F. I., Dixon, A. G. O., Asiedu, R. and Folayan, S. A. 1994. Cassava varietal needs of farmers and the potential for production growth in Africa. COSCA Working Paper No. 10. Collaborative Study of Cassava in Africa (COSCA), IITA: Ibadan, Nigeria, p. 239.

Nweke, F. I., Ugwu, B. O., Dixon, A. G. O., Asadu, C. I. A. and Ajobo, O. 1999. Cassava Production in Nigeria: A Function of Farmer Access to Markets and to improved production and processing technologies. Collaborative Study of Cassava in Africa, Working Paper No. 20.

Okpokiri, A. O., Ijioma, B. C., Alozie, S. O. and Ejiofor, M. A. N. 1985. Production of improved fufu. *Niger. Food J.*, 3: 145–8.

Onitilo, M. O. 2003. Effects of cassava varieties and processing methods on tapioca quality. M. Sc. Thesis, University of Agriculture, Abeokuta, pp. 36, 38–9.

Onitilo, M. O., Sanni, L. O., Oyewole, O. B. and Maziya-Dixon, B. 2007b. Physicochemical and Functional properties of Sour Starches from different Cassava varieties. *Int. J. Food Prop.*, 10: 607–20.

Onitilo, M. O., Sanni, L., Daniel, I., Maziya-Dixon, B. and Dixon, A. 2007a. Physicochemical and functional properties of native starches from cassava varieties in Southwest Nigeria. *J. Food Agr. Environ.*, 5(3 and 4): 108–14.

Onitilo, M., Sanni, L. O., Maziya-Dixon, B. and Dixon, A. 2004. Varietal difference in the physicochemical, functional, pasting properties and granules size of starch from different CMD resistant varieties. In: Proceeding of 9th Symposium of the ISTRC-AB, Kenya, pp. 494–511.

Onyenwoke, C. A. and Simonyan, K. J. 2014. Cassava post-harvest processing and storage in Nigeria: A review. *Afr. J. Agr. Res.*, 9(53): 3853–63

Otoo, J. A., Al-Hassan, R., Kissiedu, A. F. K. and Alhassan, A. Y. 1995. Cassava processing technology as a determinant of commercialization and farmer's varietal requirements in Ghana. - Root Crops and Poverty Alleviation: Proceedings of the Sixth Triennials Symposium of the International Society for Tropical Root Crops – AB held in Lilongwe, Malawi, 22–28 October 1995. 113–15

Padonou, W., Mestres, C. and Nago, M. C. 2005. The quality of boiled cassava roots: Instrumental characterization and relationship with physicochemical properties and sensorial properties. *Food Chem.*, 89: 261–70.

Pérez, J. C., Lenis, J. I., Calle, F., Morante, N., Sánchez, T., Debouck, D. and Ceballos, H. 2012. Heritability of root peel thickness and its influence in extractable starch from cassava (Manihot esculenta Crantz) roots. In Okechukwu, R. U. and Ntawuruhunga, P. (Eds.), Tropical Roots and Tuber Crops and the Challenges of Globalization and Climate Changes. Proceedings of 11 ISTRC-AB symposium, Kinshasa, Democratic Republic of Congo, 4–8 October 2010.

Phillips, T., Taylor, D. S., Sanni, L. and Akoroda. M. 2004. *A Cassava Industrial Revolution in Nigeria. The Potential for a New Industrial Crop.* IFAD/FAO: Rome, p. 43.

Rickard, J. E., Asaoke, M. and Blanshard, J. M. V. 1991. The physicochemical properties of cassava starch. *Trop. Sci.*, 31, 189–207.

Safo-Kantanka, K. O. and Acquistucci, R. 1996. The Physicochemical properties of cassava starch in relation to the texture of the cooked root. *Ghana J. Agr. Sci.*, 28/29: 69–80.

Salipira, K. L., Kalua, D. and Mkumbira, J. 1995. Farmer participatory evaluation of post-harvest aspect of elite clones in Malawi. Proceedings of the sixth triennial symposium of the International Society for Tropical Root Crops(ISTRC), Lilongwe, Malawi 22–28 October 1995 ISTRC (International Society for Tropical Root Crops), IITA (International Institute of Tropical Agriculture) and Government of Malawi, p. 342.

Sanni, L. O., Adebowale, A. A., Maziya-Dixon, B. and Dixon, A. G. O. (2008). Chemical composition and pasting properties of CMD resistant cassava clones planted at different locations. *J. Food Agr. Environ.*, 6(2): 97–104.

Sanni, L. O. 2015. Drying for Wealth, Food Security and Nation Building. 49th Inaugural Lecture. Federal University of Agriculture, Abeokuta, Nigeria, p. 124. http://unaab.edu.ng/attachments/professor_sanni_inaugural.pdf.

Sanni, L. O. and Akingbala, J. O. 2000. Effects of drying methods on physicochemical and sensory qualities of *fufu*. *Drying Technol.*, 18(1and 2): 421–31.

Sanni, L. O., Kolawole, A. G., Akingbala, J. O. and Kuye, A. 1999. Effect of drying methods on moisture isotherms of *fufu* at three different temperatures. *Dry. Tech.*, 17(1and2): 285–97.

Sanni, L. O., Kosoko, S. B., Adebowale, A. A. and Adeoye, R. J. 2004. The influence of palm oil and chemical modification on the pasting and sensory properties of *fufu* flour. *Int. J. Food Prop.*, 7(2): 229–37.

Sanni, L. O., Oyewole, O. B. and Olowogbade, D. V. 1998. Effect of different drying methods on the qualities of *lafun* (fermented cassava product). *Trop. Sci.*, 38, 1–4.

Shittu, T. A., Raji, A. O. and Sanni, L. O. 2007. Bread from composite cassava-wheat flour: I. Effect of baking time. *Food Res Int*. 40: 280–90.

Tanticharoen, M. 1992. Cassava in Thailand. In de Groot, P. (Ed.), *Cassava as a Cash Crop for Food, Fuel, Fodder and Chemicals*. CSC (92) AGR-17, PR 290. Agriculture Program of Commonwealth Science Council: London, pp. 19–68.

Tomlins, K., Sanni, L., Oyewole, O. B., Dipeolu, A., Ayinde, I., Adebayo, K. and Westby, A. 2007. Consumer acceptability and sensory evaluation of a fermented cassava product (Nigerian fufu). *J. Sci. Food Agric.*, 87: 1949–56.

Part 3

Managing pests and diseases

Chapter 10

Diseases affecting cassava

James Legg, International Institute of Tropical Agriculture (IITA), Tanzania; and Elizabeth Alvarez, International Center for Tropical Agriculture (CIAT), Colombia

1 Introduction

Cassava is primarily a vegetatively propagated crop, and consequently, it is vulnerable to infection by pathogens that pass from one cropping cycle to the next through the planting material, which is typically stem cuttings. Diseases have been some of the most important constraints to cassava production since the early days of published crop disease reports. Notably, the cassava mosaic-causing viruses that remain a major constraint today were first reported as far back as the end of the nineteenth century (Warburg, 1894). One of the earliest reports of bacterial disease from Latin America is also more than a century old (Berthet and Bondar, 1915). All the major plant disease-causing agents – bacteria, fungi, viruses and phytoplasmas – cause economically important damage to cassava through large parts of the regions, countries and continents within which the crop is grown. However, there are important differences in patterns of distribution. Although this point will be discussed in detail later in this chapter, it is important to highlight here that these differences mean that there continue to be significant risks of the spread of pathogens with a limited current distribution to parts of the world in which they do not currently occur. Although many of the diseases of cassava are most readily spread through vegetative propagules, several have important interactions with insect vectors. A key feature of this interaction is that changes in abundance levels or patterns of distribution of vectors can have significant consequences for the transmission and geographical spread of the

http://dx.doi.org/10.19103/AS.2016.0014.10

pathogens that they vector. Since insects can be particularly sensitive to abiotic factors, such as temperature, this also means that anticipated climate change will have a significant impact on transmitted pathogens.

Cassava diseases affect all parts of the plant: leaves, stem, flowers and roots, although the most common symptoms are chlorosis and blights in leaves and soft or dry rot in the tuberous roots. Pathogens often occur together, but they rarely interact in a synergistic way. The most important exceptions to this are the cassava mosaic begomoviruses (CMBs), which may cause greater than anticipated disease (synergism) in mixed infections, but which may also prevent super-infection by related strains (cross protection), and cassava frogskin disease (CFSD), where the severest symptoms occur when phytoplasmas infect plants together with viruses. Losses caused by cassava diseases vary greatly depending on the region, local environmental conditions and cassava variety. Overall, however, is it likely that more than a third of global cassava production, equivalent to several billion US dollars, is lost annually. Although control efforts have become more concerted and geographically extensive through the early part of the twenty-first century, much remains to be done to improve and extend management efforts. The importance of such initiatives is heightened since cassava is predicted to be the staple crop with greatest resilience to the deleterious impacts of anthropogenic climate change (Jarvis et al., 2012).

2 Viral cassava diseases in Africa

There are more than 20 species of virus that have been reported to cause disease in cassava. Although the greatest diversity of viruses has been reported from South America – cassava's native continent – a similar number of species are recognized from Africa (Legg et al., 2015 a). Only two virus species have so far been identified as important pathogens of cassava in Asia. This section will concentrate on diseases in Africa, while the next section turns to those from Latin America and Asia.

There are several viruses that have been detected from cassava plants in Africa, but that have not so far been linked clearly with disease. These include: Cassava virus C (Calvert and Thresh, 2002), *Cassava green mottle virus* (Lennon et al., 1987), *Cassava Ivorian bacilliform virus* (Fargette et al., 1991) and Cassava Kumi viruses A and B (Calvert and Thresh, 2002). There are only two groups of disease-causing viruses in Africa. These are the CMBs, which cause cassava mosaic disease (CMD), and the cassava brown streak ipomoviruses (CBSIs), causing cassava brown streak disease (CBSD).

2.1 Cassava mosaic begomoviruses

Basic biology

CMBs are single-stranded, positive sense, circular DNA viruses. They have bipartite genomes (DNA-A and DNA-B), each of which is approximately 2800 bp long (Stanley and Gay, 1983). The genome is encapsidated in a geminate icosahedral protein coat, approximately 22 nm × 38 nm in size (Böttcher et al., 2004). There are six open-reading frames (ORFs) on DNA-A, two in the virion sense (AV1, AV2) and four in the complementary sense (AC1-4), and two ORFs on DNA-B, one each in the virion (BV1) and complementary sense (BC1). ORFs on DNA-A code for the coat protein (AV1) as well as factors associated with replication and interference in the plant's host defence. BC1 facilitates long-distance virus movement, whilst BV1 codes for a nuclear shuttle protein (Hanley-Bowdoin et al., 1999). CMBs replicate by reprogramming gene expression of infected cells and recruiting host DNA polymerases. Replication is achieved by using both rolling circle and recombination-mediated mechanisms (Jeske et al., 2001). Cassava responds to infection by CMBs with post-transcriptional gene silencing, although begomoviruses have, in turn, evolved mechanisms to counter this. Virus titre and consequent symptom severity is a function of the relative balance between these two processes. Where cassava plants are infected by more than one CMB species, the combined 'anti-silencing' responses of the viruses can result in a synergistic interaction, leading to greatly increased titres of both viruses and severe disease (Vanitharani et al., 2004; Pita et al., 2001; Fondong et al., 2000).

Symptoms and effects on yield

CMD typically produces very conspicuous symptoms on the leaves of infected plants (Storey and Nichols, 1938) (Fig. 1). These comprise a yellow or yellow-green chlorotic mosaic on the leaf lamina which is often associated with leaf shape distortion and/or cupping. In severe infections most of the leaf lamina is chlorotic and leaf size is greatly reduced. Where synergistic mixed species virus infections occur, severe 'candle-stick' symptoms can be produced, which are characterized by etiolation, down-turned petioles, tiny leaf size and the loss of some leaves. Symptoms of this type were widespread in parts of East and Central Africa affected by the severe CMD pandemic (Otim-Nape et al., 1997; Legg, 1999). Patterns of CMD symptom expression often allow inferences to be made about the source of infection – whether cutting or whitefly-borne. In cutting-derived infection, the lowermost first-formed leaves show symptoms, whilst for whitefly-borne infection upper leaves are symptomatic and lower leaves remain symptom-free (Sseruwagi et al., 2004). In resistant varieties symptoms may be very mild and appear as only mild chlorotic flecks on a small proportion of leaves (Thresh et al., 1994a). Completely symptomless infection is rare. Symptom recovery during crop growth and reversion (the production of virus-free cuttings from infected parent stems) are both important features of CMD resistance (Fargette et al., 1994; Gibson et al., 1994). CMD affects cassava tuberous root production by reducing photosynthetic activity. As a consequence, there is a direct correlation between the severity of foliar symptoms and yield loss (Thresh et al., 1994b). Yield losses vary greatly depending on the virus(es) infecting the plant, the susceptibility/resistance of the cassava variety and the environmental conditions, and losses from almost nothing up to almost complete crop loss have been recorded. In a study of the effects of individual virus species and mixtures on a CMD-susceptible cultivar in Uganda, Owor et al. (2004) demonstrated losses ranging from 12% (for single ACMV infection) to 82% (for mixed ACMV+EACMV-Uganda infection). The most recent continent-wide assessment of annual cassava production losses attributable to CMD estimated them at 35 million tonnes, equivalent to more than US$ 1 billion (Legg et al., 2006).

Epidemiology

Nine species of CMBs have been described from Africa. These are: *African cassava mosaic virus* (ACMV), *African cassava mosaic Burkina Faso virus* (ACMBFV), *East African cassava mosaic virus* (EACMV), *East African cassava mosaic Cameroon virus* (EACMCV), *East African cassava mosaic Zanzibar virus* (EACMZV), *East African cassava mosaic Kenya virus* (EACMKV), *East African cassava mosaic Malawi virus* (EACMMV), *South African cassava*

Figure 1 Symptoms of major cassava diseases. 1. Cassava mosaic disease; 2. Cassava brown streak disease; 3. Severe cassava mosaic associated with mixed infection of cassava mosaic begomovirus species; 4. Cassava bacterial blight; 5. Dry brown necrotic rot in tuberous roots caused by cassava brown streak disease; 6. Cassava witches' broom; 7. Cassava frogskin disease; 8. Cassava root rot caused by *Phytophthora palmivora*.

mosaic virus (SACMV) and *Cassava mosaic Madagascar virus* (CMMGV). CMD was first observed in the coastal region of Tanzania, and sequence analysis has suggested that the CMBs probably originated in East Africa (Ndunguru et al., 2005). ACMV occurs in virtually all cassava-growing regions of Africa, with the exception of coastal East Africa. Both EACMV and EACMCV occur widely, but EACMKV, EACMMV, EACMZV and SACMV have restricted distributions in East and Southern Africa, whilst CMMGV has only been reported from Madagascar (Legg et al., 2015 a). All of the CMBs are transmitted by the whitefly vector, *Bemisia tabaci*, in a persistent manner (Dubern, 1994). Since adult *B. tabaci* are thought to remain viruliferous for their entire lifespan, it is considered that long distance vector-borne infection is an important factor in the regional epidemiology of the CMBs (Fauquet et al., 1988; Legg, 2010). In common with all other viruses infecting cassava, the CMBs are propagated through infected planting material, which facilitates both long and short distance virus spread. At the field level, external sources of inoculum are important and result in the establishment of environmental CMD gradients within newly infected fields, as there are higher incidences of infection on windward field borders (Fargette et al., 1990). Although CMBs have also been recorded from more than ten wild and cultivated plants (Alabi et al., 2015), these are not thought to play a significant role in the epidemiology. At the regional level, 'first-contact' epidemics of CMD spread through all cassava-growing areas of Africa in the 1920s to 1930s. In more recent years, the severe disease resulting from synergistic interactions between ACMV and East African cassava mosaic virus-Uganda (EACMV-UG) caused devastation in Uganda in the 1980s to 1990s, before spreading as a pandemic through large parts of East and Central Africa over the following two decades. New spread of this severe CMD continues to occur in eastern DR Congo and the western part of Central Africa, with the latest report of EACMV-UG coming from Equatorial Guinea (Valam-Zango et al., 2015). The occurrence of EACMV-UG is not confined to parts of Africa affected by the severe CMD, and this strain occurs elsewhere in mixed infection with ACMV in the absence of rapid epidemic-like disease spread. Large increases in the abundance of *B. tabaci* have been the key determinant of the rapid spread observed in pandemic-affected parts of East and Central Africa. Although the reasons for this change in the vector are still uncertain, there is evidence suggesting that genetic change in *B. tabaci* populations has enhanced their adaptation to cassava (Legg et al., 2014a).

Control

Efforts to control CMD in Africa have focused primarily on cultural methods and the development and deployment of host plant resistance (Legg and Fauquet, 2004; Thresh and Cooter, 2005). One of the most widely used control practices is the production of clean planting material. Thermotherapy combined with virus indexing is increasingly used for the production of virus-tested tissue culture plantlets, and this process facilitates the transfer of germplasm between countries. Mass multiplication of tissue culture plantlets is also being used as the first step in the development of clean seed stocks within systems set up to deliver healthy planting material to farmers. For the farm-level management of CMD, both the selection of healthy stems for planting material and the removal of symptomatic plants (roguing) have been recommended for use by farmers. Although the benefits of using selected stems as planting material have been clearly demonstrated (Mallowa et al., 2011), there is currently no evidence to show that farm-level roguing is economically justified.

The first attempts to develop host plant resistance to CMD involved introgressing putative sources of resistance from wild relatives through backcrossing experiments (Nichols, 1947).

The most promising sources of resistance were derived from *Manihot glaziovii* (Jennings, 1957). The focus for CMD resistance breeding efforts shifted from East to West Africa in the 1960s, and the *M. glaziovii*-derived multigenic resistance, subsequently referred to as CMD1, formed the basis for a major resistance breeding programme established at IITA in Nigeria. Resistant varieties produced through this effort made up a large proportion of the 206 varieties that were released by the national research programmes of 20 African countries between 1970 and 1998 (Manyong et al., 2000). New single-gene sources of resistance identified from West African landraces include CMD2 (Akano et al., 2002) and CMD3 (Okogbenin et al., 2012). These offer potential for the application of marker-assisted approaches to combine different resistance sources and thereby enhance both the strength and durability of resistance (Okogbenin et al., 2013). Research to identify and develop transgenic-based resistance strategies have been conducted over several decades (Taylor et al., 2004; Patil et al., 2009). Some success has been achieved with strategies exploiting RNA interference, in which target cassava varieties have been transformed with constructs expressing hairpin loop structures coding for viral replication genes, such as Rep (AC1) (Chellappan et al., 2004; Zhang et al., 2005), REn (AC2) (Zhang et al., 2005) or TrAp (AC3) (Zhang et al., 2005). However, the moderate success of these approaches, coupled with the high levels of resistance to CMD already available through conventional approaches, has led to a reduced focus on transgenics for CMD control.

2.2 Cassava brown streak ipomoviruses

Basic biology

CBSIs (Family, Potyviridae; Genus, *Ipomovirus*) are single-stranded positive-sense RNA viruses with a single genome component of *ca* 9 kb (Mbanzibwa et al., 2009). Virions are flexuous rods, *ca* 700 nm in length and 15 nm wide, and occur primarily in phloem tissue in leaves, stems and roots. The RNA genome comprises short untranslated regions at both 5′ and 3′ ends, and is typical of potyvirus genomes in coding for a large polyprotein that is subsequently cleaved by viral proteases to produce functional proteins (Allison et al., 1986; Dougherty and Carrington, 1988). Protein products of the proteolytic cleavage include: P1, P3, 6K1, CI, CK2, Vpg, NIaPro, HAM1 and CP, whilst PIPO is produced as a P3 fusion product through a +2 frame shift within the P3 sequence (Mbanzibwa et al., 2009; Patil et al., 2015). Unusual features of the CBSIs when compared with other potyviruses are: the absence of HC-Pro (with its role in the suppression of gene silencing played by the serine protease – P1) and the presence of the pyrophosphatase – HAM1, which is thought to reduce the rate of mutagenesis (Mbanzibwa et al., 2009).

Symptoms and effects on yield

Cassava plants infected by CBSIs typically express symptoms in leaves, stems and roots, although symptoms may be present in one, more or none of these plant parts (Nichols, 1950). Symptom expression is known to be affected greatly by the cassava variety as well as the environmental conditions. Varieties are highly variable in the manner in which they manifest infection by CBSIs; some express symptoms in foliage but not in the roots, whilst others have the opposite response. Cool temperatures enhance symptom expression. Symptoms in leaves are most pronounced on lower leaves, and comprise feathery chlorosis in secondary and tertiary veins and/or chlorotic blotching (Fig. 1). Dark-brown necrotic lesion 'streaks' may appear on the green portion of stems and symptoms in tuberous roots take the form of a dry brown necrotic rot (Nichols, 1950) (Fig. 1). Root symptoms become increasingly severe as the plant matures, and may be exacerbated by secondary infection by fungal soft rots. In severely affected plants, stem shoot tips may become necrotic and dry, leading to 'die-back'. The yield of CBSD-diseased plants may be reduced by a combination of the reduction in plant growth coupled with the loss of roots due to rot (Hillocks et al., 2001; Ndyetabula et al., 2016). Farmers are often forced to harvest early, which may result in shortages of planting material for the subsequent season. In addition, there are significant extra labour costs where farmers choose to try to remove rotten portions of affected roots. Hillocks et al. (2001) estimated that yield losses were as great as 70% for the most susceptible cultivars. More recently, survey data from Tanzania were used to calculate a countrywide loss due to CBSD of 17%, equivalent to > US$ 51 million annually (Ndyetabula et al., 2016), which is congruent with a previous Africa-wide loss estimate of US$ 75 million (Manyong et al., 2012).

Epidemiology

Two species of CBSIs are currently recognized by the International Committee for the Taxonomy of Viruses (ICTV). These are *Cassava brown streak virus* (CBSV) and *Ugandan cassava brown streak virus* (UCBSV). Both species occur through most of the CBSD-affected parts of East and Central Africa, although UCBSV predominates in northern Uganda, Rwanda, Burundi and eastern DR Congo (Bigirimana et al., 2011; Mbanzibwa et al., 2011; Mulimbi et al., 2012). CBSD was first described from north-eastern Tanzania in the 1930s (Storey, 1936), and had a confined distribution in coastal East Africa and along the shores of Lake Malawi up until 2004. Since then, the disease has spread rapidly through previously unaffected parts of East and Central Africa, most notably in the Great Lakes region (Alicai et al., 2007; Legg et al., 2015b). The most recent outbreak has been recorded from eastern DR Congo (Mulimbi et al., 2012). CBSD-like symptoms have been observed in tuberous roots of cassava plants in western DR Congo, northern Angola and Madagascar, but none of these observations has been supported by positive identification using standard virus testing approaches. A limited amount of deep sequencing of suspected infected samples from these locations has so far failed to confirm the presence of CBSIs or any other virus-like nucleic acid. CBSIs are transmitted by the whitefly vector, *B. tabaci* (Maruthi et al., 2005) in a semi-persistent manner (Jeremiah, 2012). The propagation of virus through planting infected stem cuttings is an important source of disease spread, which is exacerbated by the fact that symptoms are cryptic, difficult to see in mature plants and therefore readily overlooked by growers looking to select planting material for their own use or distribution to others. The easy spread through planting material and relatively lower efficiency of long-distance vector spread are thought to be key factors behind the contrasts observed between the patterns of development of the severe CMD and CBSD pandemics (Legg et al., 2011). Both pandemics, however, are driven by the large increases in abundance of *B. tabaci* that have been recorded from parts of East and Central Africa since the 1980s.

Control

From the earliest observations of CBSD in Tanzania in the 1930s, there have been efforts to control the disease through the development of host plant resistance. As for CMD, early breeding efforts focused on introgressing resistance from wild relatives, and most success was achieved with *M. glaziovii* and *Manihot melonobasis* (Hillocks and Jennings, 2003). One of the most resistant hybrids – 46106/27 – continues to be grown in parts of coastal East Africa, where it is known as Kaleso in Kenya and Namikonga in Tanzania. More recent genetic studies have identified quantitative trait loci (QTLs) associated with CBSD resistance on two chromosomes, and molecular markers are being developed to speed up the selection of resistant clones. When selecting for CBSD resistance, cassava breeders have primarily focused on clones not expressing CBSD symptoms in tuberous roots. A corollary of this is that there has not been a strong focus on selecting plants that are apparently free from CBSD infection. As a consequence, most of the breeding lines selected for resistance to CBSD are commonly referred to as 'tolerant', since they are frequently infected by CBSIs, express foliar symptoms, but produce tuberous roots that are mostly free of necrosis. Kiroba is an example of a released variety of this type which is widely grown in the coastal regions of Tanzania. In view of the limited success in using conventional breeding approaches to develop resistance to CBSIs, major efforts were initiated in the mid-2000s to apply transgenic approaches. An RNAi-based approach using UCBSV CP-derived constructs was shown to be effective in controlling UCBSV and partially controlling CBSV in both controlled (Yadav et al., 2011) and field (Ogwok et al., 2012) situations. Experiments in confined field trials further demonstrated that this resistance was sustained into a second propagation cycle (Odipio et al., 2014). Recent evidence, however, indicates that the somatic embryogenesis procedure used to produce CBSD-resistant transgenic plants results in the loss of function of CMD2 resistance, rendering transformed plants stably susceptible to CMD (Beyene et al., 2016). This unusual phenomenon does not seem to affect plants with CMD1 or CMD3 resistance sources.

The relative difficulty of recognizing symptoms of CBSD, when compared with CMD, means that it is less readily controlled using phytosanitary measures, such as the selection of healthy stems for planting material. In addition, symptomless infection is frequent, and the tendency for symptoms to be more prominent on lower leaves means that during dry periods when lower leaves are shed it can be difficult to recognize infected plants. These features of the disease have led to the development and promotion of clean 'seed' systems that use regular monitoring and virus testing to assure the health of planting material of improved varieties. In view of the partial effectiveness of both host plant resistance and phytosanitation in managing CBSD, future management efforts will need to combine both approaches in an integrated control system which will also incorporate tactics to reduce populations of the whitefly vector.

3 Viral cassava diseases in Latin America and Asia

3.1 Viral cassava diseases in Latin America

Eleven virus species from seven genera have been recorded from cassava in Latin America (Calvert and Thresh, 2002; Carvajal-Yepes et al., 2014; Legg et al., 2015a). These include: *Cassava common mosaic virus* (CsCMV), *Cassava frogskin-associated virus* (CsFSaV), Cassava torrado-like virus (CsTLV), *Cassava vein mosaic virus* (CsVMV), *Cassava virus X* (CsVX), Cassava new alphaflexivirus (CsNAV), Cassava polero-like virus (CsPLV), *Cassava symptomless virus* (CsSLV), Cassava Caribbean mosaic virus (CsCaMV), *Cassava Colombian symptomless virus* (CsCSLV) and *Cassava American latent virus* (CsALV). Although this represents a substantially greater degree of diversity compared to viruses occurring in the 'old world', the Latin American viruses are much less damaging, and consequently little researched in comparison with their African counterparts. It is striking to note that none of the economically important viruses in Africa is present in Latin America, and of the frequently occurring viruses in the 'new world', only one – *CsCMV* – has been recorded from Africa (Aiton et al., 1988). The key factor driving this great contrast in the viromes of the main cassava-growing continents is the fact that the whitefly, *B. tabaci*, which transmits all of the disease-causing viruses of Africa and Asia, does not colonize cassava in the Americas (Carabali et al., 2005). *B. tabaci* is considered to be a cryptic species complex, and genotypes occurring in Latin America differ greatly from the cassava-colonizing genotypes in Africa and Asia (Dinsdale et al., 2010). In this short summary, we provide details for four of the most frequently occurring Latin American viruses (CsCMV, CsFSaV, CsTLV and CsVMV). The reviews of Calvert and Thresh (2002) and Legg et al. (2015a) provide information about several of the less widely reported species.

Biology, symptoms and effects on yield

The most important disease phenomenon associated with virus infection of cassava in the Americas is CFSD. This was first reported in southern Colombia in the early 1970s (Pineda et al., 1983) and has subsequently been observed in Brazil, Peru, Venezuela and Costa Rica (Calvert and Thresh, 2002; Carvajal-Yepes et al., 2014). Although the aetiology of CFSD has never been definitively proven, it is thought that both viruses and phytoplasmas are involved in pathogenesis, leading to symptom development. This is described in greater detail in the subsequent section on 'Cassava Phytoplasmas and Cassava Frogskin Disease'. Three virus species commonly occur in plants affected by CFSD. These include CsFSaV, CsTLV and CsNAV (Carvajal-Yepes et al., 2014). CsFSaV (Family Reoviridae; Genus *Oryzavirus*) virions are isometric, *ca* 70 nm in diameter, and have dsRNA genomes comprised of 10 components, which range in size from about 1000 to 4000 nt (Calvert et al., 2008). CsTLV (Family Secoviridae; Genus *Torradovirus*) is also commonly associated with CFSD (Carvajal et al., 2014). Virions are isometric, with a diameter of *ca* 30 nm. The genome is bipartite, comprising two single-stranded positive-sense RNA molecules – *ca* 7000 nt and 5000 nt in size (Carvajal-Yepes et al., 2014; van der Vlugt et al., 2015). CsTLV is commonly found in field-grown plants expressing chlorotic mosaic and leaf deformation symptoms in Colombia, and it has also been shown experimentally to induce foliar symptoms when in single infection (Carvajal-Yepes et al., 2014). CsCMV (Family Alphaflexiviridae; Genus *Potexvirus*) causes mild mosaic-like symptoms, which are more prominent during cool periods. Virions are elongated semi-flexuous rods, 15 nm × 495 nm in size (Kitajima et al., 1965). The CsCMV genome is a single-stranded positive-sense RNA molecule of *ca* 6400 nt (Calvert et al., 1996). CsCMV occurs most commonly in the southern cassava-growing regions of South America. Most recently, severe mosaic symptoms were reported from cassava fields in north-eastern Argentina. Samples collected from 2012 to 2014 were shown to be infected primarily by CsCMV, although CsFSaV was also detected (Di Feo et al., 2015). CsCMV has also been detected in cassava samples from Africa (Aiton et al., 1988) and Asia (Chen et al., 1981), but in neither case has this been associated with significant disease. Yield losses of up to 60% have been reported (Costa and Kitajima, 1972), but CsCMV typically causes only mild symptoms. CsVMV is the type species of the genus *Cavemovirus*, family Caulimoviridae. CsVMV infection is characterized by veinal chlorosis, leaf distortion and epinasty. Young plants may be stunted, but symptoms decline as the plant matures (Calvert et al., 1995). CsVMV virions are *ca* 50 nm in diameter and quasi-spherical. They have a single, circular, dsDNA genome of *ca* 8200 nt (de Kochko et al., 1998). CsVMV was found to be widespread in north-eastern Brazil, particularly in the semi-arid regions (Calvert et al., 1995), but the effects on yield are not considered to be significant (Santos et al., 1995).

Epidemiology

Incomplete knowledge about the viruses affecting cassava in the Americas has meant that little published information is available describing the epidemiology of the viruses and the disease that some of them cause. Torradoviruses have been confirmed to be transmitted by three whitefly species in Europe (*Trialeurodes vaporariorum*, *Trialeurodes abutilonea*, and *B. tabaci*) (van der Vlugt et al., 2015). None of these is present on cassava in Colombia, suggesting that one of the three known cassava-colonizing whiteflies is likely to be the vector of CsTLV, which if proven would make the genus *Torradovirus* the only genus of plant viruses to have vectors from four whitefly species. CsCMV is extremely stable and may be transmitted mechanically such as through the use of infected cutting tools. CsVMV seems to spread within fields, however, and although it is therefore assumed that there must be a vector, none has yet been identified. Although it has been suggested that cassava virus disease is becoming more prevalent in Latin America (Calvert and Thresh, 2002), there have been no reports of severe virus-associated damage occurring in regions previously unaffected, and consequently there have been no region-wide studies of cassava virus epidemiology in South or Central America. It seems apparent that the failure of *B. tabaci* genotypes to colonize cassava in the Americas is an important reason for the contrasting patterns of virus epidemiology between the old and new worlds. It is currently unclear why some *B. tabaci* genotypes are unable to colonize cassava, although it has been presumed that this may be linked to the cyanogenic glucosides produced by cassava. However, several genetically distinct *B. tabaci* populations in Africa and Asia are adapted to feeding on cassava as well as other whitefly genera in Latin America, and so there is a significant risk that a similar adaptation evolves amongst *B. tabaci* in Latin America, or cassava-colonizing *B. tabaci* nymphs from Asia or Africa are inadvertently carried to the Americas. Strict application of quarantine and regular monitoring of cassava throughout Latin America will be required to mitigate this threat.

Control

In view of the limited understanding of the epidemiology of viruses infecting cassava in the new world, control efforts have focused primarily on phytosanitary measures, such as the production of virus-tested plantlets through tissue culture, and field-based roguing and selection of healthy stems for planting material (Calvert and Thresh, 2002). CsCMV may be transmitted by using contaminated farm implements, and so it has been recommended that these are disinfested to reduce spread (Lozano and Nolt, 1989). Studies of whiteflies colonizing cassava in Latin America, notably *Aleurotrachelus socialis*, have identified sources of resistance in Ecuadorian cassava landraces, and there are on-going efforts to determine whether these might also be effective against *B. tabaci* in Africa. One of the most important control measures for viruses in South America is the strict application of virus elimination and indexing techniques for any germplasm in tissue culture that is to be exported to countries in either Africa or Asia, where almost all of the South American viruses are absent. The absence of any outbreaks of these viruses in Africa/Asia is evidence that to date these measures have proven to be highly effective.

3.2 Viral cassava diseases in Asia

CMD is the only significant virus disease affecting cassava in Asia. Up until 2015, CMD had only been reported from India and Sri Lanka. An isolated outbreak in Cambodia in late 2015, however, represents the first recorded spread into South-East Asia (Wang et al., 2015), which signifies a worrying development for a continent that was previously only lightly affected by these damaging pathogens. CMD was first described from India in the 1950s (Abraham, 1956). It has generally been a less prominent economic problem than CMD in Africa, although there are reports of increasing levels of incidence in recent years (Anitha et al., 2011).

Biology, symptoms and effects on yield

CMD in Asia is caused by two CMBs: *Indian cassava mosaic virus* (ICMV) (Malathi et al., 1985) and *Sri Lankan cassava mosaic virus* (SLCMV) (Saunders et al., 2002). Both viruses have similar biological properties to their African relatives, and the symptoms are also indistinguishable. It has been suggested that SLCMV is more virulent than ICMV, which may explain why it is now the dominant species, even in southern India (Anitha et al., 2011). Calvert and Thresh (2002) reviewed several yield studies carried out between 1974 and 1987 which reported losses attributable to a CMD of 9–42%. There have been no overall assessments of the impact of CMD on cassava in any of the three Asian countries affected.

Epidemiology

B. tabaci has been confirmed as the vector of both ICMV and SLCMV (Duraisamy et al., 2013). Although the whiteflies used in this study were not genotyped, the Asia II 5 species group of the *B. tabaci* cryptic species complex has been identified as the cassava-colonizing group in India (Ellango et al., 2015). Epidemiology experiments conducted in the 1980s in which initially virus-free cassava was exposed to nearby infection sources under field conditions revealed generally low levels of virus infection (Nair, 1985). The more recent survey data of Anitha et al. (2011), however, demonstrated both high levels of overall incidence (ranging from 45% to 95%) and significant proportions of whitefly-borne infection. These provide further evidence that the disease dynamics of CMD in South Asia have changed over the last 30 years.

Control

Measures applied for managing CMD in Asia are generally the same as those used against CMD in Africa, although in Asia there has been a greater focus on the production and propagation of virus-free stocks of planting material. This is a consequence of the much higher level of commercialization of cassava in Asia in comparison with Africa, which means that commercial companies, such as starch producers, are prepared to invest in tissue culture combined with meristem tip culture. Conventional resistance breeding programmes have made use of both wild cassava relatives as well as clones derived from

West African landraces carrying the CMD2 resistance gene (Sheela et al., 2012). Transgenic varieties are being developed with both RNAi (Ntui et al., 2015) and the *virE2* gene from *Agrobacterium tumefaciens* (Resmi et al., 2015). This work remains at an early stage of development, however, and it is likely to be several years before transformed cassava plants will be ready for field trials. The recent occurrence of CMD in South-East Asia has highlighted the importance of applying strict phytosanitary controls to movements of germplasm between regions and across borders. It seems likely that Cambodia, Vietnam, Laos and Thailand will all need to establish rigorous procedures for minimizing the risk of further CMD spread within South-East Asia.

4 Bacterial blight, phytoplasmas and frogskin disease

4.1 Cassava bacterial blight

Biology, symptoms and effects on yield

Cassava bacterial blight (CBB) is caused by *Xanthomonas axonopodis* pv. *manihotis* (Xam). Xam is a rod-shaped, gram-negative bacteria from the family Xanthomonadaceae. In South America, Colombian Xam populations show a high degree of diversity and have been grouped into seven clusters. A correlation was observed between the clusters and their geographical origin (Restrepo and Verdier, 1997). Diversity of Xam is generally greatest in Latin America, but genetic variability is also strongly correlated with geographical location in the other parts of the world where Xam is an important pathogen (Bart et al., 2012).

CBB is a disease of global economic importance. Following the first report in Brazil in 1912, this disease has become a serious constraint to the production of cassava in Africa, South America, South-east Asia and southern China. For example, yield losses of between 12% and 92% have been reported from Thailand in the 1970s (Mansfield et al., 2012; Lozano, 1986). CBB infection results in the development of necrotic water-soaked spots on leaves (Fig. 1) which increase in frequency and size over time, eventually producing dark brown to black angular necrotic leaf spots and larger areas of necrotic blight (Lozano and Booth, 1974; Hillocks and Wydra, 2002). Severe infection of younger plants may give rise to die-back of shoot tips, whilst leaf necrosis may result in the drying out and shedding of the lower leaves of mature plants.

Epidemiology

Xam bacteria are most readily spread through the use of cuttings obtained from an infected parent crop. Transmission may also occur, however, by rain-splash or through the use of contaminated tools. *Xanthomonas* bacteria are known to survive for several weeks on field trash and may also persist on weeds. Damage caused to cassava stems either by hail stones or by the insect *Pseudotheraptus devastans* may facilitate infection by Xam (Maraite and Meyer, 1975). Since infected stems or stem cuttings are often asymptomatic, their movement has allowed the pathogen to disperse over large distances, including the inadvertent continental introductions to Africa and Asia (Verdier et al., 1997). To prevent introducing the pathogen into disease-free regions, a polymerase chain reaction (PCR) diagnostic method has been developed to detect Xam in cassava-infected stems (Verdier et al., 2001).

Control

The main control approach for managing CBB has been the development of cassava varieties with strain-specific and multigenic resistance (Jorge et al., 2000). Strain-specific resistance loci may explain the high genotype by environment interaction observed when selecting cassava genotypes for resistance to CBB. It was therefore proposed that selection for resistance should be done by inoculating cassava genotypes with different Xam pathotypes (Wydra el at., 2004).

Over a period of some 25 years, the CIAT cassava breeding program routinely screened the genebank and hybrid populations in the eastern savanna region of Colombia, where CBB is consistently present at very high levels. This strategy allowed the identification of high levels of resistance and the combination of good resistance with high yield and high dry matter. The resistance identified was stable across other regions of Colombia, and these sources were broadly used in Asia breeding programmes.

Biological control has also been proposed as a potential technique for managing CBB. Studies have shown that *Pseudomonas putida* strains significantly reduce disease symptoms on cassava (Hernandez et al., 1986). In spite of this preliminary success, however, this approach has not been widely applied, probably in view of the relatively high cost. Phytosanitary measures are an important tactic for managing a systemically infecting pathogen such as Xam. At farm level, farmers can be encouraged to select healthy stems for replanting, and within multiplication and propagation schemes, certification protocols can be applied to ensure that the planting material produced conforms to the required standards. General crop hygiene and good agronomic practice can help to limit the infection and impact of CBB. Useful measures of this type include the removal and burning of crop residues after harvesting, and crop rotation to avoid recycling cassava on the same plot.

4.2 Cassava phytoplasmas and cassava frogskin disease

There are various groups and subgroups of phytoplasma that have been reported to be associated with cassava diseases in Latin America (Alvarez et al., 2009; Cardoso et al., 2016) and Asia (Alvarez et al. 2013, Hoat et al., 2015; Graziosi et al., 2016). Also, '*Candidatus* Phytoplasma aurantifolia' (16SrII group) has been reported from cassava in Uganda (Arocha et al., 2009). The aetiology of CFSD has yet to be definitively characterized, but both phytoplasmas and viruses appear to be associated with this condition.

Biology, symptoms and effects on yield

Phytoplasmas are mycoplasma-like organisms that are often mistaken for plant pathogenic viruses. They are specialized bacteria that are obligate parasites of plants and that infect phloem tissue. They are often difficult to diagnose and detect. The most important cassava diseases associated with phytoplasma infection are CFSD and Cassava Witches' Broom (CWB). Discovered in the 1960s, phytoplasmas are pathogens that are becoming increasingly important in Asia, Latin America and the Caribbean.

CFSD has been reported in Colombia (Alvarez et al., 2009), Costa Rica (Pardo et al., 2014), Paraguay (Cardozo et al., 2016), Brazil (Santos de Oliveira et al., 2013), Mexico, Panama, Peru and Venezuela. The disease has not been reported in Asia or Africa. The most important symptom of CFSD is a corky corrugation of the tuberous roots of infected plants which results in variable levels of reduced yield (Calvert and Thresh, 2002) (Fig. 1). The peel also presents lip-like slits that, when they join, create a net-like or honeycomb pattern (Alvarez et al., 2009). The disease causes deep lesions in roots, eventually reducing their diameter. In many cassava cultivars, symptoms are observed only when the plants are harvested. In addition, the expression of CFSD symptoms is influenced by temperature and host genotype. Through its effect on root production, CFSD can cause severe economic loss. In Colombia and Costa Rica yield losses of more than 90% have been reported.

CFSD is associated with a phytoplasma group 16SrIII subgroup L (Alvarez et al., 2009) and interacts synergistically with infection by several virus species. CFSD-affected plants are commonly infected with CsFSaV (Carvajal-Yepes et al., 2014). While still under study, the most severe infections seem to occur where both CsFSaV and phytoplasma are present, which has led researchers to the conclusion that there is a synergistic interaction between these two pathogens (de Souza et al., 2014). In vitro isolation of a phytoplasma associated with CFSD was successfully accomplished using an artificial culture medium (Betancourth et al., 2014). Pathogenicity tests were conducted by stem injection using phytoplasma cells grown in vitro with a liquid medium. Typical root symptoms were observed six months after inoculation.

In South-East Asia, CWB disease is associated with a phytoplasma. Aster yellows phytoplasmas have been detected and restriction fragment length polymorphism analyses of nested PCR amplified fragments from Vietnamese and Cambodian CWB phytoplasmas indicated the presence of differentiable strains all related to the 16SrI group. Sequence analyses of partial 16S rDNA fragments showed that Vietnamese and Cambodian CWB phytoplasmas have 99% similarity with '*Candidatus* Phytoplasma asteris' (Alvarez et al., 2013). Phytoplasma-associated symptoms like small yellow leaves, short internodes, sprout proliferation, stem vascular necrosis and/or stunting were observed in fields in cassava-producing regions of Vietnam (Alvarez et al., 2013) (Fig. 1). CWB reduces starch content and yield, and in Vietnam, the disease is widespread and causes yield losses of up to 80% (Alvarez et al. 2013). Similarly, many cassava farms in Cambodia were affected in 2012, with losses of up to 50% being reported (Alvarez et al., 2014).

Epidemiology

CFSD is disseminated through the movement of infected cuttings. Since stems of diseased plants tend to be thicker than those of healthy ones (more photosynthates remain in the leaves and stems of the diseased plants rather than being translocated to the roots), farmers may preferentially select diseased plants to provide seed for the next planting season. It has also recently been demonstrated that the leafhopper *Scaphytopius marginelineatus* (Cicadellidae) is able to acquire the CFSD phytoplasma from infected plants and successfully transmit it to healthy plants (Alvarez and Betancourth, 2016). Further studies are required, however, to determine how significant this vector transmission is as a factor in the field and regional level epidemiology of phytoplasmas associated with CFSD. Since several of the viruses putatively associated with CFSD have only recently been characterized (CsTLV, CsPLV and CsNAV) (Carvajal-Yepes et al., 2014), there are currently no confirmed reports of vector transmission. Three whitefly species – *Bemisia tuberculata*, *A. socialis* and *Trialeurodes variabilis* – occur frequently in parts of Colombia where the viruses associated with CFSD occur, and it was suggested that one or more of these could be the vector of one of the causal agent(s) of CFSD (Angel et al.,

1987). However, none of the three has been proven to be the vector of any of the viruses associated with CFSD.

CWB has no known vector, but spread is known to be mediated by the use of planting material obtained from infected stems.

Control

An effective method of controlling phytoplasma diseases is through preventing infection by assuring the health and quality of planting material. CFSD can be effectively controlled by rigorous selection, following harvesting, although grafting and field screening have been used to identify varieties that are less affected. Improved quarantine procedures are required to prevent long-range disease spread. A thermotherapy treatment has been developed and implemented in Colombia for the mass-propagation of cassava planting material, and the same approach was subsequently validated in Brazil, Costa Rica and Paraguay (Alvarez, 2015). In germplasm evaluation studies, cassava genotypes HMC-1 and CM 1495-7 had low incidences of CFSD (5%) with mild symptoms (score 2), whilst MCOL 1505 and MNGA 11 were severely affected with incidences of about 60%. Use of a resistance inducer (Kendal®) produced a 20% increase in yield and reduced disease incidence to about 10% (Alvarez et al., 2014). Accurate and rapid diagnosis is a key component of disease management strategies, and a loop-mediated isothermal amplification (LAMP) assay has been developed to allow specific detection of the phytoplasma causing CWB from field-collected samples (Vu et al., 2016).

5 Cassava fungal diseases: foliar

5.1 Cassava super-elongation disease

Biology, symptoms and effects on yield

Super-elongation disease (SED) in cassava is caused by the fungus *Sphaceloma manihoticola* (Teleomorph: *Elsinoe brasiliensis*). The conidia of *S. manihoticola* isolates from cassava are small, thin-walled, ellipsoid, with one or two guttules. Conidiophores are phialides, hyaline to slightly pigmented and 0-to-1 septate. In Colombia a homogeneous and reproducible band of about 545 bp was obtained with PCR which, when digested by restriction enzymes, showed an equal pattern of bands for all isolates. The isolates thus belonged to one species. Random-amplified polymorphic DNA analysis revealed intraspecific genetic diversity, demonstrating a wide range of genetic diversity in the pathogen population. In Brazil a cluster analysis separated cassava and milkweed isolates (that also infect cassava) into two distinct groups corresponding to two species: *S. manihoticola* and *Sphaceloma krugii*. Therefore, more than one species of *Sphaceloma* infects cassava naturally (Zeigler and Lozano, 1983). RAPD and RFLP analyses provided genetic evidence of four well-characterized regional groups among isolates of *S. manihoticola*.

SED is one of the major diseases affecting cassava in Latin America and the Caribbean. It was observed for the first time in Brazil in 1950 (Bitancourt and Jenkins, 1950) in the species *M. glaziovii*. In Colombia, outbreaks were reported for the first time in 1972 (CIAT, 1973). SED has caused losses of more than 80% in susceptible cassava cultivars in Colombia, Brazil, Venezuela and Central America (Krausz, 1976). Disease symptoms consist mostly of an exaggerated lengthening of internodes in young stalks and petioles. This is caused by the hormone gibberellin A4 due to the expression of the *Sm450-2* gene. Other symptoms include deformation or involution of young leaves and the presence of cankers in stalks, petioles and leaf blades (Alvarez and Molina, 2000; Alvarez et al., 2003a).

Epidemiology

The causal organism is easily spread by wind, water and the use of infected planting material (Alvarez and Molina, 2000). Easy dispersion and the possible existence of physiological races or biotypes of the fungus are two aspects that hinder management of the disease (Valle et al., 1996). A majority of plant species in the genus *Manihot* are likely to be hosts for the fungus (Ziegler et al., 1980).

Control

The most effective means of preventing infection and the spread of SED is by planting disease-free cuttings. Strict quarantine and surveillance measures are required to prevent its introduction to Asia or Africa. Significant cultivar × isolate interactions have indicated pathogenic specialization. Resistance has been shown to be a quantitative trait controlled largely by additive genetic factors. The use of resistant parent genotypes in breeding programmes combined with phenotypic field selection under high disease pressure should enhance control (Kawano et al., 1983). At CIAT, breeding lines are typically screened in sites in the eastern savanna region (Llanos Orientales), where the disease is endemic and where high, uniform levels of infection allow good differentiation of resistance levels.

5.2 Cassava anthracnose disease

Biology, symptoms and effects on yield

Cassava anthracnose disease (CAD) is caused by the fungus *Colletotrichum gloeosporioides* f.sp. *manihotis* and is an economically important disease of cassava in Africa (Hahn and Keyser, 1985), South America (Lozano et al., 1981) and Asia (Chandrasekharan-Nair et al., 1979). Conidia of the fungus are cylindrical or ellipsoidal with obtuse ends. Anthracnose infection causes shoot tip die-back, cankers on stems, branches and fruits, and leaf spots. Damage to stems reduces the viability of cuttings used for establishing a new field. Small circular sunken spots are present on infected leaves, and developing lesions become pinkish as fruit bodies of the fungus are produced (Hahn et al. 1989). In susceptible cassava cultivars defoliation may occur. Stem symptoms are characterized by oval pale-brown depressions. Deeper cankers affect plant transportation and infected planting materials may have 50–75% less viability.

Epidemiology

Most CAD spread occurs through the use of infected stems for planting material. However, an insect vector, *P. devastans*, also plays an important role in disease spread (Fokunang et al., 2000). Rainfall is important in the development and dispersal of the pathogen. Rain-splash spreads spores to nearby cassava stems. The fungus may also spread by wind, as spores are carried from stem cankers.

Control

CAD can be controlled by crop rotation or by destroying infected plant debris. Fallow periods reduce inoculum build-up (Lozano et al., 1981). Clean planting material is required to produce CAD-free cuttings for propagation (Ikotun and Hahn, 1991). A biological control approach has been tested, involving the use of rhamnolip (a glycolipid that works as a biosurfactant) produced by *Pseudomonas aeruginosa* (Goswami et al., 2015). Cassava resistance to CAD has been associated with phenol metabolic changes (Friend, 1981) and peroxidase activity around the dark brown lesions formed. Lignification in cassava at the beginning of storage root development plays a role in cassava resistance (Fokunang et al., 1999).

5.3 Cassava brown leaf spot

Biology, symptoms and effects on yield

Cassava brown leaf spot (CBLS), caused by *Cercosporidium henningsii*, is the most economically important of several fungal pathogens that produce leaf spotting on cassava. Two others which occur widely are white leaf spot (*Cercospora caribaea*) and diffuse leaf spot (*Cercospora vicosae*). CBLS is caused by a fungus that produces two types of spores: macro-conidia and micro-conidia. The conidia of the fungus are filiform, 3–10 septate and 30–85 × 5–7 µm in size (Ayesu-Offei and Antwi-Boasiako, 1996). CBLS is found wherever cassava is grown (Jameson, 1970; Lozano and Booth, 1974). Symptoms comprise dry brown necrotic lesions, which are roughly circular in shape and 5–12 mm in size. Although symptoms are generally mild and most prominent on the lower leaves of maturing plants, severe incidences of the disease have been noted in several African countries, particularly in high rainfall areas. Severe infection may result in defoliation. The reduction in photosynthetic activity produced by damage to leaves, coupled occasionally with leaf loss, can reduce root yield (Terry and Oyekan, 1976) and yield losses of up to 20% have been reported (Hillocks and Wydra, 2002).

Epidemiology

Primary infections occur when wind or rain carry conidia that have dropped from wounds of infected tissues towards leaves of a new planting. If the humidity is sufficiently high, the conidia will germinate, producing branched germinal tubes that frequently anastomose. When lesions are mature, stromata appear from which conidiophores emerge. Secondary cycles of the disease are repeated throughout the rainy season, when wind or rain carries conidia to new susceptible tissues of the plant. The fungus survives the dry season in old lesions, frequently those of fallen leaves. It renews activity with the advent of the rainy season and the growth of new leaves of the host.

Control

Disease management includes burning or burying infected material, as well as crop rotation. Use of disease-free planting materials is recommended to control the spread of the disease. Wider spacing between cassava plants reduces moisture build-up in the crop stand which makes conditions less favourable for the development of the disease. Reducing excess humidity at planting time is recommended. Fungicides based on copper oxide and copper oxychloride, suspended in mineral oil, also provide good chemical control.

6 Cassava fungal diseases: root rots

Hillocks and Wydra (2002) listed publications for 16 root-rot causing pathogens of cassava. Root rot symptoms are encountered wherever cassava is grown, and particularly where cassava crops are over-mature and fields are poorly drained, or in farmland recently converted from humid forest (Ospina and Ceballos, 2002) (Fig. 1). Pathogens associated with root rot often occur in complex mixtures, and it can be difficult to determine the source of primary infection. Root rots initiated by primary infecting pathogens are frequently exacerbated by subsequent colonization by saprophytic species. In Colombia, up to 80% of cassava has been reported to be affected by root rots, whilst in southern India, production losses of 70% have occurred. Root rots are also a major constraint to cassava production in the forested areas of Central and West Africa. In this section we highlight several of the most important fungal pathogens causing root rot in cassava.

6.1 *Phytophthora* spp

Biology, symptoms and effects on yield

One of the pseudo-fungal diseases affecting cassava is root rot caused by *Phytophthora* species from the genus of plant-damaging Oomycetes. Seven species of *Phytophthora* have been reported to attack cassava. *Phytophthora palmivora* is characterized by oval and ellipsoid sporangia with a short pedicel. In *Phytophthora tropicalis* the sporangia are papillate and caducous with pedicels. Sporangia are obovoid, limoniform or ellipsoid and have tapered bases. PCR amplification of the ITS region (internal transcribed spacer) followed by digestion with restriction enzymes can be used to separate the species. A collection of *Phytophthora* strains obtained from different regions in Colombia showed high morphological and pathogenic variation. *P. palmivora* and *P. tropicalis* were identified as the most important species causing root rot in Colombia. Detection of the genus *Phytophthora* was accomplished by a hybridization technique of the ITS region with a specific probe. In India, isolate identifications were confirmed using an ITS marker. Host–pathogen interactions were also studied under greenhouse conditions (Sánchez, 1998). *Phytophthora* rots affect the root system, causing disintegration of tissues and producing a foul odour (Lozano and Booth, 1974). *Phytophthora* species attack young and mature cassava plants, especially under favourable conditions such as poor drainage, continuous cultivation without rotation, use of infected cuttings and late harvesting.

Epidemiology

Phytophthora species are spread through the use of stem cuttings obtained from an infected crop. Transmission also occurs through infested soils, which is common in fields which are flooded. The pathogen survives in the soil. Disease incidence is highest during the rainy season, on humid soils and in newly cleared land (Hillocks and Wydra, 2002).

To prevent introducing the pathogen into disease-free regions, a real-time PCR diagnostic method has been developed to detect *P. palmivora* in infected tissues.

Control

Strategies to control *Phytophthora* root rot have focused on the use of a thermotherapy system in which cuttings are immersed in water at 49°C for 49 minutes. Results comparable to the use of a fungicide were obtained when thermotherapy was preceded by a pre-treatment that involved dipping cuttings in water at 49°C for 10 min. In a biological control strategy, *Trichoderma harzianum* and *Trichoderma viridae* reduced the percentage of root infection by 83% and increased production by 11%. A high correlation was observed between cassava tuberous root production and concentration of *Trichoderma* in the soil. Thermotherapy and biological control strategies have been validated through technology packages tested with farmers (Alvarez et al., 2003b).

Sánchez (1998) reported cassava varietal resistance to *Phytophthora* species. Studies conducted by Alvarez and Barragan (1998) described variation in the reaction of cassava germplasm to the pathogen and identified sources of resistance. QTLs associated with resistance to *Phytophthora* root and stem rots have been mapped (Loke et al., 2004).

6.2 *Fusarium* spp

Biology, symptoms and effects on yield

Fusarium species are an important component of the fungal complex that causes cassava root rot. *Fusarium* species isolated from rotted cassava in Africa include *Fusarium moniliforme* sensu lato, *Fusarium semitectum*, *Fusarium oxysporum* and *Fusarium solani* (Akinyele and Ikotun, 1989). The *F. solani* and *F. oxysporum* species complexes have also been reported as pathogens of cassava in Colombia (Lozano, 1992). The main symptom is a breakdown in tissue of the mature tuberous roots, usually associated with a foul odour and changes in colour, which may be useful in distinguishing the pathogens involved. For example, storage tissue of tuberous roots primarily infected by a *Fusarium* species are often pink or yellow. Yield losses because of root rot average 0.5–1 t/ha, but losses may occasionally be as high as >3 t/ha, an equivalent of 15–20% of the total tuberous root yield. Studies in Cameroon showed that 30% of rotted roots were infected by *Fusarium* species 12 months after planting and that they represented 25% of all the fungal isolates recovered (El-Sharkaway, 1993).

Epidemiology

There has been little work investigating the epidemiology of the *Fusarium* species associated with root rot in cassava; however, many of the characteristic mechanisms of plant fungal pathogen spread are also used by these pathogens, including carriage on infected planting material, carry-over from one season to the next through crop debris and persistence in humid soils.

Control

Fusarium root rot can be controlled by rotating with non-susceptible crops. Cuttings should be sourced from healthy plantations and/or immersed in fungicide and tools should be disinfected. Laboratory assays, involving the inoculation of stem cuttings, root slices and whole roots have been used to identify resistant genotypes (Onyeka et al., 2005).

6.3 *Rhizoctonia solani*

Biology, symptoms and effects on yield

Rhizoctonia solani occurs throughout the world, causing severe damage on a broad range of crops and ornamental plants (Dolores, 2002; Parmeter and Whitney, 1970; Yang and Kharbanda, 1996). Disease symptoms consist of a root rot. The fungal mycelium is in the roots or in the base of the stem and spreads through the stem, roots and rhizosphere. When young plants are infected, germination may be reduced as plants are killed, whilst plants that are more than four months of age may suffer from partial or total wilting (Ospina and Ceballos, 2002).

Epidemiology

R. solani can survive in the soil for many years in the form of sclerotia. Sclerotia of *Rhizoctonia* have thick outer layers to allow for survival, and they function as the overwintering structure for the pathogen. The sclerotia/mycelium overwinter in plant debris, soil or host plants. The main source of inoculum is sclerotia produced by the fungus on the leaves and damaged parts of stems. Sclerotia spread during soil preparation and irrigation (Ceresini, 2011).

Control

This fungus is difficult to control because of its ability to survive under adverse soil conditions as sclerotia for many years, its saprophytic activity and its wide host range (Ogoshi, 1996). Biological control using *Trichoderma* spp., *Bacillus* spp. and *Pseudomonas* spp. has been successful (Lewis and Lumsden, 2001; Yu et al., 2002; Adesina et al., 2009). Several lytic enzymes, as well as many active compounds, have been isolated and described, which can inhibit the development of *R. solani* (Raaijmakers et al., 2002; Yu et al., 2002; Jung et al., 2003). A compound produced by *Paenibacillus* sp. IIRAC-30 is active against *R. solani* (Canova et al., 2010). The isolate *Bacillus pumilus* MAIIIM4a showed a strong inhibitory activity against *R. solani*. Pumilacidin was identified as the antifungal compound produced by *B. pumilus* (Melo et al., 2009).

7 Summary

From the end of the nineteenth century, diseases have been recognized as one of the primary constraints to cassava production. Viral, bacterial, phytoplasma and fungal pathogens cause important yield losses wherever cassava is grown. Although some of these pathogens and the diseases that they cause have attracted significant research attention, many are only partially characterized and understood, and many research gaps still need to be addressed. The limited progress in research into cassava diseases and in the management of these diseases is at least partly due to the nature of cultivation of the crop. Cassava is still predominantly produced for subsistence or local processing and sales rather than large-scale commercial purposes, particularly in Africa, which is the continent with the greatest number of farmers and overall production. The majority of cassava growers do not have sufficient resources to invest in effective disease management.

In this chapter we have presented information on a diverse set of cassava diseases and their causal agents. Several of these are of key importance today and will likely remain so through the first quarter of this century. In Africa, the dual pandemics of CMD and CBSD continue to spread through large swathes of prime cassava-producing country in East and Central Africa, and both now threaten West Africa. The significance of this threat has increasingly been recognized by the research, plant protection and extension communities within and outside the continent, and this heightened concern has led to the development of a planned 'War on Cassava Viruses' (Legg et al., 2014b). This and other related strategies from cassava stakeholder institutions is paving the way for an increased commitment to developing effective, affordable and durable solutions to this problem. In Latin America, great progress has been made in recent years in untangling the complex set of pathogen interactions that give rise to the devastating CFSD. Moreover, experience gained in researching phytoplasmas associated with this condition is having a powerful spin-off effect on new research being conducted on a related pathogen that is causing the CWB epidemic in South-east Asia. One of the most globally significant events for cassava diseases in recent times has been the occurrence for the first time of CMD caused by SLCMV in South-east Asia (Cambodia; Wang et al., 2015). This highlights the importance of one of the most important themes in the management of cassava diseases, which is the prevention of pathogen spread between countries and continents through the vigorous application of phytosanitary protocols. Cassava offers great promise as a 'climate change proof' crop of the future, which could guarantee the food security of hundreds of millions of people living in the tropics for the foreseeable future. If this potential is to be realized, it will be essential that human and technical capacity to enforce phytosanitary controls is greatly strengthened and that the latest ICT approaches are applied to ensure that all cassava stakeholders are fully informed of the risks of carrying cassava propagules between territories and continents.

8 Future trends in research

More cassava research is now being conducted than at any time in history, and research on cassava diseases has been very prominent in recent years. The trend towards an increasing volume of research on this crop is likely to continue, not least since cassava is expected to adapt better to climate change than any other major staple. Some of the key themes of future disease research on cassava are likely to be:

- **Official release of transformed virus-resistant varieties**. Varieties that have been genetically transformed for resistance to viruses will be advanced from confined field trials into the farming environment. It is expected that the first virus-resistant transgenic varieties of cassava will be officially released within the next 10–15 years.
- **Resistance pyramiding**. Both biotechnological and conventional approaches will be used to combine specific traits (resistance to viruses, CBB and whiteflies; delayed post-harvest physiological deterioration, biofortification, drought tolerance etc.) to produce farmer-preferred 'super varieties'.

- **Gene editing**. Various approaches are being explored for the targeted editing of the cassava genome (Chavarriaga-Aguirre et al., 2016). In particular, host proteins that bind TAL effectors produced by CBB are strong potential targets for gene-editing approaches that could produce non-transgenic resistant plants (Boch et al., 2014).
- **Digital disease diagnostics**. The ever-increasing resolution of digital imaging technology and its increasing availability through smartphones offer opportunities to develop point-of-care diagnostic methods in which users will take images of disease symptoms with a phone and immediately receive a diagnosis. These techniques are already being pioneered for cassava virus diseases in Africa (IITA, 2016).
- **Geostatistics for surveillance and disease risk forecasting**. The combination of digital imaging data with GIS software is providing new opportunities for producing region-wide maps of disease epidemics and developing forecasts for future patterns of disease change using geostatistics and modelling.
- **Biocontrol products/packages**. The role of biological agents in suppressing cassava disease infection and development remains poorly understood. As further research on these complex interactions is completed, and optimal treatments determined, there will be opportunities for the commercial development of biocontrol packages.
- **Seed systems promoting cassava disease management**. New commercial developments in cassava production are leading to a growing role for the formal seed sector. Certification guidelines for cassava are being evaluated in several countries in Africa, a process which is being enhanced by technological developments in cassava disease diagnostics. In the near future, these factors will coalesce to advance the application of phytosanitary measures within cassava 'seed' systems.
- **Global approaches to the phytosanitary management of cassava health**. Improved linkages between cassava stakeholders throughout the world, facilitated by rapid progress in ICT systems, will encourage wider recognition of the risks of cross-border movements of cassava germplasm, and will result in the increasingly stringent application of quarantine and phytosanitation to prevent the inadvertent spread of cassava pathogens.

9 Where to look for further information

The following materials will provide the reader with additional and more detailed content on cassava diseases:

Review articles

Calvert, L. A. and Thresh, J. M. (2002), 'The viruses and virus diseases of cassava', in *Cassava: Biology, Production and Utilization*, Bellotti, A. C., Hillocks, R. J. and Thresh, J. M. (Eds). CABI, Wallingford, UK. 332, pp. 237–60.

Hillocks, R. J. and Wydra, K. (2002), 'Bacterial, fungal and nematode diseases', in *Cassava: Biology, Production and Utilization*, Bellotti, A. C., Hillocks, R. J. and Thresh, J. M. (Eds), CABI, Wallingford, UK, 332, pp. 261–79.

Legg, J. P. (2012), 'Cassava Diseases: Ecology and Control', in *Encylopedia of Pest Management*. Taylor and Francis, London, UK. DOI: 10.1081/E-EPM-120041170.

Lozano, J. C. and Booth, R. H. (1974), 'Diseases of cassava (*Manihot esculenta* Crantz)', *PANS* 20, 30–54.
Msikita, W., James, B., Nnodu, E., Legg, J., Wydra, K. and Ogbe, F. (2000), 'Disease control in cassava farms. IPM Field Guide for Extension agents', IITA, Ibadan, Nigeria, 27pp.

Societies involved in cassava disease study

International Society of Tropical Root Crops (ISTRC). http://www.istrc.org/.
American Phytopathological Society (APS). http://www.apsnet.org/Pages/default.aspx
International Society for Plant Pathology (ISPP). http://www.isppweb.org/index.asp
Global Cassava Partnership for the 21st Century (GCP-21). http://www.gcp21.org/

Information sources about plant disease

CABI. http://www.cabi.org/isc/
Plantwise Knowledge Bank. http://www.plantwise.org/KnowledgeBank/Home.aspx

Image repositories

Plant Village. https://www.plantvillage.org/en/topics/cassava-manioc/diseases_and_pests_description_uses_propagation
CIAT Flickr. https://www.flickr.com/photos/ciat/sets/
IITA. Flickr. https://www.flickr.com/photos/45796762@N03/

International institutions involved in cassava disease research

Centro Internacional de Agricultura Tropical (CIAT). www.ciat.org
International Institute of Tropical Agriculture (IITA). www.iita.org
Roots, tubers and bananas programme of the CGIAR. www.rtb.cgiar.org

10 References

Abraham, A. (1956), 'Tapioca cultivation in India', Farm Bulletin No.17, Indian Council of Agricultural Research, New Delhi, India. 20pp.
Adesina, M. F., Grosh, R., Lembke, A., Vatchev, T. D. and Smalla, K. (2009), 'In vitro antagonists of *Rhizoctonia solani* tested on lettuce: rhizosphere competence, biocontrol efficiency and rhizosphere microbial community response', *FEMS Microbiol. Ecol.* 69, 62–74. doi:10.1111/j.1574-6941.2009.00685.x
Aiton, M. M., Lennon, A. M., Roberts, I. M. and Harrison, B. D. (1988), 'Two new cassava viruses from Africa', in *5th International Congress of Plant Pathology,* Kyoto, Japan, 1988, p. 43.
Akano, A. O., Dixon, A. G. O., Mba, C., Barrera, E. and Fregene, M. (2002), 'Genetic mapping of a dominant gene conferring resistance to cassava mosaic disease', *Theor. Appl. Genet.* 105, 521–35. doi:10.1007/s00122-002-0891-7
Akinyele, B. and Ikotun, T. (1989), 'Microorganisms associated with cassava tuber rot, IITA Root and Tuber and Plantain Improvement Program, IITA 1988 Annual Report, IITA, Ibadan, Nigeria, pp. 16–18.
Alabi, O., Mulenga, R. and Legg, J. P. (2015), 'Cassava mosaic disease'. In *Virus Diseases of Tropical and Sub-Tropical Crops*, Tennant, P. and Firmin, G. (Eds), CABI, Wallingford, UK, pp. 56–72.

Alicai, T., Omongo, C. A., Maruthi, M. N., Hillocks, R. J., Baguma, Y., Kawuki, R., Bua, A., Otim-Nape, G. W. and Colvin, J. (2007), 'Re-emergence of cassava brown streak disease in Uganda', *Plant Dis.* 91, 24–9.

Allison, R., Johnston, R. E. and Dougherty, W. G. (1986), 'The nucleotide sequence of the coding region of tobacco etch virus genomic RNA: Evidence for the synthesis of a single polyprotein', *Virology* 154(1), 9–20.

Alvarez, E. and Barragán, M. I. (1998), 'Evaluación de la tolerancia a la pudrición radical causada por *Phytophthora* spp. en variedades de yuca *Manihot esculenta* Crantz, bajo condiciones de invernadero', *Ascolfi Informa* 24(3), 11–12.

Alvarez, E. and Betancourth, C. A. (2016), 'Transmission of a 16SrIII-L phytoplasma by leafhoppers (*Scaphytopius marginelineatus*) to cassava (*Manihot esculenta* Crantz) in Colombia', http://www.apsnet.org/meetings/Documents/2016_meeting_abstracts/aps2016_47.htm. Accessed 9 October 2016.

Alvarez, E. and Molina, M. L. (2000), 'Characterizing the Sphaceloma fungus, causal agent of superelongation disease in cassava', *Plant Dis.* 84, 423–8.

Alvarez, E., Llano, G. A. and Mejía, J. F. (2012), 'Cassava diseases in Latin America, Africa and Asia', in *The Cassava Handbook: A Reference Manual based on the Asian Regional Cassava Training Course, Bangkok, Thailand*, Howeler, R. H. (Ed.), CIAT, CALI, Colombia, pp. 258–304.

Alvarez, E., Loke, J. B. and Llano, G. (2003b), 'Development of ecological practices to manage Phytophthora root rot of cassava (*Manihot esculenta*)', http://ciat-library.ciat.cgiar.org/articulos_ciat/phytophthora_root_rot_cassava_alvarez.pdf. Accessed 9th October, 2016.

Alvarez, E., Mejía, J. F., Llano, G. A. and Loke, J. B. (2007), 'Detection and characterization of a phytoplasma associated with frogskin disease in cassava', *Bull. Insectology* 60(2), 273–4.

Alvarez, E., Mejía, J. F., Llano, G., Loke, J., Calari, A., Duduk, B. and Bertaccini, A. (2009), 'Detection and molecular characterization of Phytoplasma associated with frogskin disease in cassava', *Plant Dis.* 93, 1139–45.

Alvarez, E., Mejía, J. F. and Valle, T. L. (2003a), 'Molecular and pathogenicity characterization of *Sphaceloma manihoticola* isolates from south-central Brazil', *Plant Dis.* 87, 1322–8.

Alvarez, E., Pardo, J. M., Mejía, J. F., Bertaccini A., Thanh, H. D and Hoat, T. X. (2013), 'Detection and identification of a 16SrI group phytoplasma associated with witches' broom disease of cassava in Vietnam', *Phytopathogenic Mollicutes* 3(2), 77–81.

Alvarez, E., Pardo, J. M., Truke and M. J. (2014), 'Detection and identification of '*Candidatus* Phytoplasma asteris'-related phytoplasma associated with a witches' broom disease of cassava in Cambodia', *Phytopathol.* 104 (Suppl. 3), S3.7.

Álvarez, E., Gómez, Y., Zacher, M. B. and Campo, R. 2014. Innovaciones tecnológicas en el manejo integrado del Cuero de Sapo de la yuca (*Manihot esculenta* Crantz): Estrategias para reducir el impacto de la enfermedad por efectos del cambio climático en Colombia, Costa Rica y Paraguay. IX Taller de Seguimiento Técnico de Proyectos Fontagro y 59 Reunión Anual del PCCMCA 2014. Managua, Nicaragua, abril 28–29 2014.

Álvarez, E., Pardo, J. M., Mejía, J. F., de Oliveira, S., Alves, S., Zacher, M., Cardozo, L. and Gómez, Y. 2015. *Manejo del'Cuero de sapo'enfermedad limitante de la yuca.* https://cgspace.cgiar.org/bitstream/handle/10568/70086/CIAT_MANEJO_DEL_CUERO_DE_SAPO_ENFERMEDAD_LIMITANTE_DE_LA_YUCA.pdf.

Angel, S. J., Nolt, B. L. and Pineda, L. B. (1987), 'Estudios sobre transmission por moscas blancas (Homoptera: Aleyrodidae) de virus asociados con el cuero de sapo en yuca (*Manihot esculenta* Crantz)', *Acta Agronomica* 37, 27–39.

Anitha, J., Makeshkumar, T. and Edison, S. (2011), 'Survey of cassava mosaic disease in Kerala. *J. Root Crops* 37, 41–7.

Arocha, Y., Echodu, R., Talengera, D., Muhangi, J., Rockefeller, E., Asher, O., Nakacwa, R., Serugga, R., Gumisiriza, G., Tripathi, J., Kabuye, D., Otipa, M., Vutseme, K., Lukanda, M. and Boa, E. (2009), 'Occurrence of '*Candidatus* Phytoplasma aurantifolia' (16SrII group) in cassava and four other species in Uganda', *Plant Pathol.* 58(2), 390.

Ayesu-Offei, E. N. and Antwi-Boasiako, C. (1996), 'Production of microconidia by *Cercospora henningsii* Allesch, cause of brown leaf spot of Cassava (*Manihot esculenta* Crantz) and tree cassava (*Manihot glaziovii* Muell.-Arg.)', *Ann. Bot.* 78(5), 653–7.

Bart, R., Cohn, M., Kassen, A., Mccallum, E. J., Shybut, M., Petriello, A., Krasileva, K., Dahlbeck, D., Medina, C., Alicai, A., Kumar, L., Moreira, L. M., Neto, J. R., Verdier, V., Santana, M. A., Kositcharoenkul, N., Vanderschuren, H., Gruissem, W., Bernal, A. and Staskawicz, J. (2012), 'High-throughput genomic sequencing of cassava bacterial blight strains identifies conserved effectors to target for durable resistance', *Proc. Natl. Acad. Sci.* 109(32), 13130.

Berthet, A. and Bondar, G. (1915), 'Molestia bacteriana da mandioca', *Boletin de Agricultura, Sao Paolo* 16, 513–24.

Betancourth, C. A., Pardo, J. M., Truke, M. J., Muñoz, J. E. and Alvarez, E. (2014), 'In vitro isolation of a phytoplasma associated with frogskin disease of cassava (*Manihot esculenta* Crantz)', *Phytopathology* 105 (Suppl. 1), S1.6.

Beyene, G., Chauhan, R. D., Wagaba, H., Moll, T., Alicai, T., Miano, D., Carrington, J. C. and Taylor, N. J. (2016), 'Loss of CMD2-mediated resistance to cassava mosaic disease in plants regenerated through somatic embryogenesis', *Mol. Plant Pathol.* 17(7), 1095–110.

Bigirimana, S., Barumbanze, P., Ndayihanzamaso, P., Shirima, R. and Legg, J. P. (2011), 'First report of cassava brown streak disease and associated *Ugandan cassava brown streak virus* in Burundi', *New Dis. Rep.* 24, 26.

Bitancourt, A. and Jenkins, A. E. (1950). Estudos sôbre as Miriangiales, II', *Arq. Inst. Biol. Sao Paulo* 20, 116.

Boch, J., Bonas, U. and Lahaye, T. (2014), 'TAL effectors – pathogen strategies and plant resistance engineering', *New Phytol.* 204(4), 823–32.

Böttcher, B., Unseld, S., Ceulemans, H., Russell, R. B. and Jeske, H. (2004), 'Geminate structures of African cassava mosaic virus', *J. Virol.* 78(13), 6758–65.

Calvert, L. A., Cuervo, M., Lozano, I., Villareal, N. and Arroyave, J. (2008), 'Identification of three strains of a virus associated with cassava plants affected by frogskin disease', *J. Phytopathol.* 156, 647–53.

Calvert, L. A., Cuervo, M. I., Ospina, M. D., Fauquet, C. M. and Ramirez, B-C. (1996), 'Characterization of cassava common mosaic virus and a defective RNA species. *J. Gen. Virol.* 77, 525–30.

Calvert, L. A., Ospina, M. D. and Shepherd, R. J. (1995), 'Characterization of cassava vein mosaic virus: A distinct plant pararetrovirus', *J. Gen. Virol.* 76, 1271–8.

Canova, S. P., Petta, T., Reyes, L. F., Zucchi, T. D., Moraes, L. A. and Melo, I. S. (2010), 'Characterization of lipopeptides from *Paenibacillus* sp. (IIRAC30) suppressing *Rhizoctonia solani*', *World J. Microbiol. Biotechnol.* 26(12), 2241–7.

Carabali, A., Bellotti, A. C., Montoya-Lerma, J. and Cuellar, M. E. (2005), 'Adaptation of *Bemisia tabaci* biotype B (Gennadius) to cassava, *Manihot esculenta* (Crantz)', *Crop Prot.* 24, 643–9.

Cardozo-Téllez, L., Pardo, J. M., Zacher, M., Torres, A. and Alvarez, E. (2016), 'First report of a 16SrIII phytoplasma associated with frogskin disease in cassava (*Manihot esculenta* Crantz) in Paraguay', *Plant Dis.* 100(7), 1492.

Carvajal-Yepes, M., Olaya, C., Lozano, I., Cuervo, M., Castaño, M. and Cuellar, W. J. (2014), 'Unraveling complex viral infections in cassava (*Manihot esculenta* Crantz) from Colombia. *Virus Res.* 186, 76–86.

Ceresini, P. (2011), '*Rhizoctonia Solani*', http://www.cals.ncsu.edu/course/pp728/Rhizoctonia/Rhizoctonia.html (Accessed 9 October 2016).

Chavarriaga-Aguirre, P., Brand, A., Medina, A., Prias, M., Escobar, R., Martinez, J., Diaz, P., Lopez, C., Roca, W. M., Tohme, J. (2016), 'The potential of using biotechnology to improve cassava: a review', *Vitro Cell. Dev. Biol. Plant* 45, 1–18.

Chen, C. T., Ko, N. C. and Chen, M. J. (1981), Report of the Taiwan Sugar Research Institute No. 93, 21–27.

CIAT. (1973), 'Sistemas de producción de yuca', Informe Annual, CIAT, Cali, Colombia, pp. 99–100.

Chadrasekharan-Nair, M., Menon, M. R., Suharban, N. and Verma, A. S. (1979), 'Anthracnose of cassava: a new record in India', *Curr. Sci.* 48, 443.

Chellappan, P., Masona, M., Vanitharani, R., Taylor, N. J. and Fauquet, C. M. (2004), 'Broad spectrum resistance to ssDNA viruses associated with transgene-induced gene silencing in cassava', *Plant Mol. Biol.* 56, 601–11.

Costa, A. S. and Kitajima, E. W. (1972), 'Cassava common mosaic virus', C. M. I./A. A. B. In Descriptions of Plant Viruses, No. 90.

de Souza, A. N., da Silva, F. N., Bedendo, I. P. and Carvalho, C. M. (2014), 'A phytoplasma belonging to a 16SrIII-A sub-group and dsRNA virus associated with cassava frogskin disease in Brazil', *Plant Dis.* 98(6), 771–9.

de Kochko, A., Verdaguer, B., Taylor, N., Carcamo, R., Beachy, R. N. and Fauquet, C. (1998), '*Cassava vein mosaic virus* (CsVMV), type species for a new genus of plant double stranded DNA viruses?', *Arch. Virol.* 143, 945–62.

Di Feo, L., Zanini, A., Rodríguez Pardina, P., Cuervo, M., Carvajal-Yepes, M. and Cuellar, W. J. (2015), 'First report of cassava common mosaic virus and cassava frogskin-associated virus infecting cassava in Argentina. *Plant Dis.* 99(5), 733.

Dinsdale, A., Cook, L., Riginos, C., Buckley, Y. M. and De Barro, P. (2010), 'Refined global analysis of *Bemisia tabaci* (Hemiptera: Sternorrhyncha: Aleyrodoidea: Aleyrodidae) mitochondrial cytochrome oxidase I to identify species level genetic boundaries', *Ann. Entomol. Soc. Am.* 103, 196–208.

Dolores, G. H. (2002), 'Estado actual de la taxonomía de *Rhizoctonia solani* Kuhn', *Rev. Mex. Fitopatol.* 200, 200–5.

Dougherty, W. G. and Carrington, J. C. (1988), 'Expression and function of potyviral gene products', *Ann. Rev. Phytopathol.* 26(1), 123–43.

Dubern, J. (1994), 'Transmission of African cassava mosaic geminivirus by the whitefly (*Bemisia tabaci*). *Trop. Sci.* 34, 82–91.

Duraisamy, R., Natesan, S., Muthurajan, R., Gandhi, K., Lakshmanan, P., Karuppusamy, N. and Chokkappan, M. (2013), 'Molecular studies on the transmission of *Indian cassava mosaic virus* (ICMV) and *Sri Lankan cassava mosaic virus* (SLCMV) in cassava by *Bemisia tabaci* and cloning of ICMV and SLCMV replicase gene from cassava', *Mol. Biotechnol.* 53(2), 150–8.

El-Sharkaway, M. A. (1993), 'Drought tolerant cassava for Africa, Asia and Latin America', *Bioscience* 43, 441–51.

Ellango, R., Singh, S. T., Rana, V. S., Priya, N. G., Raina, H., Chaubey, R., Naveen, N. C., Mahmood, R., Ramamurthy, V. V., Asokan, R. and Rajagopal, R. (2015), 'Distribution of *Bemisia tabaci* genetic groups in India'. *Environ. Entomol.* 44(4), 1258–64.

Fargette, D., Fauquet, C., Grenier, E. and Thresh, J. M. (1990), 'The spread of African cassava mosaic virus into and within cassava fields', *J. Phytopathol.* 130, 289–302.

Fargette, D., Roberts, I. M. and Harrison, B. D. (1991), 'Particle purification and properties of cassava Ivorian bacilliform virus', *Ann. Appl. Biol.* 119, 303–12.

Fargette, D., Thresh, J. M. and Otim-Nape, G. W. (1994), 'The epidemiology of African cassava mosaic geminivirus: reversion and the concept of equilibrium', *Trop. Sci.* 34, 123–33.

Fauquet, C., Fargette, D. and Thouvenel, J.-C. (1988), 'Some aspects of the epidemiology of African cassava mosaic virus in Ivory Coast'. *Trop. Pest Man.* 34, 92–6.

Fokunang, C. N., Akem, C. N., Ikotun, T., Dixon, A. G. O. and Tembe, E. A. (1999), 'Seed survival and transmission of cassava anthracnose disease and seed treatment effect on seedling growth', *Pak. J. Biol. Sci.* 2, 849–54.

Fokunang, C. N., Dixon, A. G. O., Akem, C. N. and Ikotun, T. (2000), 'Cultural, morphological and pathogenic variability in *Colletotrichum gloeosporioides* f.sp. *Manihotis* isolates from cassava (*Manihot esculenta*) in Nigeria', *Pak. J. Biol. Sci.* 3, 542–6.

Fondong, V., Pita, J. S., Rey, M. E. C., de Kochko, A., Beachy, R. N., and Fauquet, C. M. (2000), 'Evidence of synergism between African cassava mosaic virus and the new double recombinant geminivirus infecting cassava in Cameroon', *J. Gen. Virol.* 81, 287–97.

Friend, J. (1981), 'Plant phenolics, lignification and disease resistance', *Prog. Phytochem.* 7, 197–261.

Gibson, R. W. (1994), 'Long-term absence of symptoms in heat-treated African cassava mosaic geminivirus-infected resistant cassava plants', *Trop. Sci.* 34, 154–8.

Goswami, D., Borah, S. N., Lahkar, J., Handique, P. J. and Deka, S. (2015), 'Antifungal properties of rhamnolipid produced by *Pseudomonas aeruginosa* DS9 against *Colletotrichum falcatum*', *J. Basic Microbial.* 55(11), 1265–74.

Graziosi, I., Minato, N., Alvarez, E., Ngo, D. T., Hoat, T. X., Aye, T. M. and Wyckhuys, K. A. (2016), 'Emerging pests and diseases of South-east Asian cassava: a comprehensive evaluation of geographic priorities, management options and research needs', *Pest Man. Sci.* 72(6), 1071–89.

Hahn, S. K. and Keyser, J. (1985), 'Cassava: a basic food of Africa', *Outlook Agric.* 14, 95–100.

Hahn, S. K., Isoba, J. C. G. and Ikotun, T. (1989), 'Resistance breeding in root and tuber crops at the International Institute of Tropical Agriculture (IITA), Ibadan, Nigeria', *Crop Prot.* 8, 147–68.

Hanley-Bowdoin, L., Settlage, S. B., Orozco, B. M., Nagar, S. and Robertson, D. (1999), 'Geminiviruses: models for plant DNA replication, transcription, and cell cycle regulation', *CRC Crit. Rev. Plant Sci.* 18, 71–106.

Hernandez, J. M., Laberry, R. and Lozano, J. C. (1986), 'Observations on the effect of inoculating cassava (*Manihot esculenta*) plantlets with fluorescent pseudomonads', *Phytopathol.* 117, 17–25.

Hillocks, R. J. and Jennings, D. L. (2003), 'Cassava brown streak disease: a review of present knowledge and research needs', *Int. J. Pest Man.* 49, 225–34.

Hillocks, R. J., Raya, M., Mtunda, K. and Kiozia, H. (2001), 'Effects of brown streak virus disease on yield and quality of cassava in Tanzania', *J. Phytopathol.* 149, 1–6.

Hoat, T. X., Quan, M. V., Lan Anh, D. T., Cuong, N. N., Vuong, P. T., Alvarez, E., Nguyen, D. T., Thuy T. D., Wyckhuys, K., Paltrinieri, S., Pardo, J. M., Mejía, J. F., Thanh, N. D., Dickinson, M., Duong, C. A., Kumasaringhe, N. C. and Bertaccini, A. (2015), 'Phytoplasma diseases on major crops in Vietnam', *Phytopathogenic Mollicutes* 5(1 suppl.), S69–S70.

IITA (2016), 'Cassava disease surveillance network', http://cassavadiseasenet.org/ (Accessed10 October 2016).

Ikotun, T. and Hahn, S. K. (1991), 'Screening cassava cultivars for resistance to anthracnose disease', *Proc. 9th Symp. ISTRC*, 20–26 October 1991, Accra, Ghana, 178–83.

Jameson, J. D. (1970), *Agriculture in Uganda*, 2nd edn, Oxford University Press, Oxford, UK, 116–276.

Jarvis, A., Ramirez-Villegas, J., Campo, B. V. H. and Navarro-Racines, C. (2012), 'Is cassava the answer to African climate change adaptation?', *Trop. Plant Biol.* 5(1), 9–29.

Jennings, D. (1957), 'Further studies in breeding cassava for virus resistance', *E. Afr. Agric. J.* 22, 213–19.

Jeremiah, S. (2012), 'The role of whitefly (Bemisia tabaci) in the spread and transmission of cassava brown streak disease in the field', PhD thesis, University of Dar es Salaam, Tanzania, 229pp.

Jeske, H., Lutgemeier, M. and Preiss, W. (2001), 'DNA forms indicate rolling circle and recombination-dependent replication of Abutilon mosaic virus', *EMBO J.* 20, 6158–67.

Jorge, V., Fregene, M. A., Duque, M. C., Bonierbale, M. W., Tohme, J. and Verdier, V. (2000), 'Genetic mapping of resistance to bacterial blight disease in cassava (*Manihot esculenta* Crantz)', *Theor. Appl. Genet.* 101(5–6), 865–72.

Jung, W. J., An, K. N., Jin, Y. L., Park, R. D, Lim, K. T., Kim, K. Y. and Kim, T. H. (2003), 'Biological control of damping-off caused by *Rhizoctonia solani* using chitinase-producing *Paenibacillus illinoisensis* KJA-424', *Soil Biol. Biochem.* 35, 1261–4.

Kawano, K., Umemura, Y. and Kano, Y. (1983), 'Field assessment and inheritance of cassava resistance to superelongation disease', *Crop Sci.* 23(2), 201–5.

Kitajima, E. W., Wetter, C., Oliveira, A. R., Silva, D. M. and Costa, A. S. (1965), 'Morfologia do virus do mosaico comun da mandioca', *Bragantia* 24, 247–60.

Krausz, J. P. (1976), 'The superelongation disease of cassava', PhD thesis, Cornell University, Ithaca, New York, USA.

Legg, J. P. (1999), 'Emergence, spread and strategies for controlling the pandemic of cassava mosaic virus disease in east and central Africa', *Crop Prot.* 18(10), 627–37.

Legg, J. P. (2010), 'Epidemiology of a whitefly-transmitted cassava mosaic geminivirus pandemic in Africa', In Bemisia: *Bionomics and Management of a Global Pest*, Stansly, P. A. and Naranjo, S. E. (Eds), Springer, Dordreecht-Heidelberg-London-New York, pp. 233–57.

Legg, J. P. and Fauquet, C. M. (2004), 'Cassava mosaic geminiviruses in Africa', *Plant Mol. Biol.* 56, 585–99.

Legg, J. P., Jeremiah, S. C., Obiero, H. M., Maruthi, M. N., Ndyetabula, I., Okao-Okuja, G., Bouwmeester, H., Bigirimana, S., Tata-Hangy, W., Gashaka, G., Mkamilo, G., Alicai, T. and Lava Kumar, P. (2011), 'Comparing the regional epidemiology of the cassava mosaic and cassava brown streak pandemics in Africa', *Virus Res.* 159, 161–70.

Legg, J. P., Lava Kumar, P. and Kanju, E. E. (2015), 'Cassava brown streak virus disease', In *Virus Diseases of Tropical and Sub-Tropical Crops*, Tennant, P. and Fermin, G. (Eds), CABI, Wallingford, UK, pp. 42–55.

Legg, J. P., Lava Kumar, P., Makeshkumar, T., Ferguson, M., Kanju, E., Ntawuruhunga, P., Tripathi, L. and Cuellar, W. (2015a), 'Cassava virus diseases: biology, epidemiology and management', *Adv. Virus Res.* 91, 85–142.

Legg, J. P., Owor, B., Sseruwagi, P. and Ndunguru, J. (2006), 'Cassava mosaic virus disease in East and Central Africa: Epidemiology and management of a regional pandemic', *Adv. Virus Res.* 67, 355–418.

Legg, J. P., Somado, E. A., Barker, I., Beach, L., Ceballos, H., Cuellar, W., Elkhoury, W., Gerling, D., Helsen, J., Hershey, C., Jarvis, A., Kulakow, P., Kumar, L., Lorenzen, J., Lynam, J., McMahon, M., Maruthi, G., Miano, D., Mtunda, K., Ntawuruhunga, P., Okogbenin, E., Pezo, P., Terry, E., Thiele, G., Thresh, M., Wadsworth, J., Walsh, S., Winter, S., Tohme, J. and Fauquet, C. (2014b), 'A global alliance declaring war on cassava viruses in Africa', *Food Secur.* 6, 231–48.

Legg, J. P., Sseruwagi, P., Boniface, S., Okao-Okuja, G., Shirima, R., Bigirimana, S., Gashaka, G., Herrmann, H. -W., Jeremiah, S. C., Obiero, H. M., Ndyetabula, I., Tata-Hangy, W., Masembe, C. and Brown, J. K. (2014a). Spatio-temporal patterns of genetic change amongst populations of cassava Bemisia tabaci whiteflies driving virus pandemics in East and Central Africa', *Virus Res.*, 186, 61–75.

Lennon, A. M., Aiton, M. M. and Harrison, B. D. (1987), 'Purification and properties of cassava green mottle, a previously undescribed virus from the Solomon Islands', *Ann. Appl. Biol.* 110, 545–55.

Lewis, J. A. and Lumsden, R. D. (2001), 'Biocontrol of damping-off of greenhouse-grown crops caused by *Rhizoctonia solani* with a formulation of Trichoderma spp', *Crop Prot.* 20, 49–56.

Loke, J. B., Alvarez, E., Vallejo, F. A., Marín, J., Fregene, M., Rivera, S. and Llano, G. A. (2004), 'Análisis de QTLs de la resistencia a pudrición de raíz causada por *Phytophthora tropicalis* en una población segregante de yuca (*Manihot esculenta* Crantz)', *Acta Agronómica* 53(3/4), 35–41.

Lozano, J. C. (1986), 'Cassava bacterial blight: a manageable disease', *Plant Dis.* 70, 1089–93.

Lozano, J. C. (1992), 'Overview of integrated cassava diseases', *Fitopathol. Brasil* 17, 18–22.

Lozano, J. C. and Nolt, B. L. (1989), 'Pest and pathogens of cassava', in *Plant Protection and Quarantine: Selected Pests and Pathogens of Quarantine Significance, Vol* 2, Kahn, R. P. (Ed.), CRC Press, Boca Raton, FL, USA, pp. 174–5.

Lozano, J. C., Bellotti, A. C., Reyes, J. A., Howland, R. H., Leihner, D. and Dol, J. (1981), *Field Problems in Cassava*, CIAT, Colombia, CA, pp. 16–25.

Malathi, V. G., Nair, N. G. and Shantha, P. (1985), 'Cassava mosaic disease', *Technical Bulletin Series 5*, Central Tuber Crops Research Institute, Trivandrum, India.

Mallowa, S. O., Isutsa, D. K., Kamau, A. W. and Legg, J. P. (2011), 'Effectiveness of phytosanitation in cassava mosaic disease management in a post-epidemic area of western Kenya', *J. Agric. Biol. Sci.* 6(7), 8–15.

Mansfield, J., Genin, S., Magori, S., Citovsky, V., Sriariyanum, M., Ronald, P. et al. (2012), 'Top 10 plant pathogenic bacteria in molecular plant pathology', *Mol. Plant Pathol.* 13, 614–29.

Manyong, V. M., Dixon, A. G. O., Makinde, K. O., Bokanga, M. and Whyte, J. (2000), *The Contribution of IITA-improved Germplasm to Food Security in Sub-Saharan Africa: An Impact Study*, IITA, Ibadan, Nigeria.

Manyong, V. M., Maeda, C., Kanju, E. and Legg, J. P. (2012), 'Economic damage of cassava brown streak disease in sub-Saharan Africa', In *Tropical Root and Tuber Crops and the Challenges of Globalization and Climate Change*, Okechukwu, R. U. and Ntawuruhunga, P. (Eds), *11th*

ISTRC-AB Symposium, 4–8 October 2010, Kinshasa, Democratic Republic of Congo. IITA, Ibadan, Nigeria, pp. 61–8.

Maraite, H. and Meyer, J. A. (1975), '*Xanthomonas manihotis* (Arthaud-Berthet) Starr, causal agent of bacterial wilt, blight and leaf spots of cassava in Zaire', *PANS Pest Articles & News Summaries* 21(1), 27–37.

Maruthi, M. N., Hillocks, R. J., Mtunda, K., Raya, M. D., Muhanna, M., Kiozia, H., Rekha, A. R., Colvin, J. and Thresh, J. M. (2005), 'Transmission of *Cassava brown streak virus* by *Bemisia tabaci* (Gennadius)', *J. Phytopathol.* 153, 307–12.

Mbanzibwa, D. R., Tian, Y., Mukasa, S. B. and Valkonen, J. P. (2009), 'Cassava brown streak virus (Potyviridae) encodes a putative Maf/HAM1 pyrophosphatase implicated in reduction of mutations and a P1 proteinase that suppresses RNA silencing but contains no HC-Pro', *J. Virol.* 83, 6934–40.

Mbanzibwa, D. R., Tian, Y. P., Tugume, A. K., Patil, B. L., Yadav, J. S., Bagewadi, B., Abarshi, M. M., Alicai, T., Changadeya, W., Mkumbira, J., Muli, M. B., Mukasa, S. B., Tairo, F., Baguma, Y., Kyamanywa, S., Kullaya, A., Maruthi, M. N., Fauquet, C. M. and Valkonen, J. P. T. (2011), 'Evolution of cassava brown streak disease-associated viruses', *J. Gen. Virol.* 92, 974–87.

Melo, F. M. P. D., Fiore, M. F., Moraes, L. A. B. D., Silva-Stenico, M. E., Scramin, S., Teixeira, M. D. A. and Melo, I. S. D. (2009), 'Antifungal compound produced by the cassava endophyte *Bacillus pumilus* MAIIIM4A', *Scientia Agricola* 66(5), 583–92.

Mulimbi, W., Phemba, X., Assumani, B., Kasereka, P., Muyisa, S., Ugentho, H., Reeder, R., Legg, J. P., Laurenson, L., Weekes, R. and Thom, F. E. F. (2012), 'First report of *Ugandan cassava brown streak virus* on cassava in Democratic Republic of Congo', *New Dis. Rep.* 26, 11.

Nair, N. G. (1985), 'Nature and extent of spread of cassava mosaic disease', In Proceedings of the National Symposium, CTCRI, Trivandrum, India, 1985, pp. 175–8.

Ndunguru, J., Legg, J. P., Aveling, T. A. S., Thompson, G. and Fauquet, C. M. (2005), 'Molecular biodiversity of cassava begomoviruses in Tanzania: evolution of cassava geminiviruses in Africa and evidence for East Africa being a center of diversity of cassava geminiviruses. *Virol. J.* 2, 21.

Ndyetabula, I. L., Merumba, S. M., Jeremiah, S. C., Kasele, S., Mkamilo, G. S., Kagimbo, F. M. and Legg, J. P. (2016), 'Analysis of interactions between cassava brown streak disease symptom types facilitates the determination of varietal responses and yield losses', *Plant Dis.* 100(7), 1388–96.

Nichols, R. F. W. (1947), 'Breeding cassava for virus resistance', *E. Afr. Agric. J.* 12, 184–94.

Nichols, R. F. J. (1950), 'The brown streak disease of cassava: distribution, climatic effects and diagnostic symptoms', *E. Afr. Agric. J.* 15, 154–60.

Ntui, V. O., Kong, K., Khan, R. S., Igawa, T., Janavi, G. J., Rabindran, R., Nakamura, I. and Mii, M. (2015), 'Resistance to *Sri Lankan cassava mosaic virus* (SLCMV) in genetically engineered cassava cv. KU50 through RNA silencing', *PLoS ONE* 10(4), 1–23.

Odipio, J., Ogwok, E., Taylor, N. J., Halsey, M., Bua, A., Fauquet, C. and Alicai, T. (2014), 'RNAi-derived field resistance to cassava brown streak disease persists across the vegetative cropping cycle', GM Crops and Food, 5(1), http://dx.doi.org/10.4161/gmcr.26408.

Ogoshi, A. (1996), 'The genus *Rhizoctonia*', In Rhizoctonia *Species: Taxonomy, Molecular Biology, Ecology*, Sneh, B., Jabli-Hare, S., Neate, S., Dijst, G., (Eds), Springer, Dordrecht, the Netherlands, pp. 1–9.

Ogwok, E., Odipio, J., Halsey, M., Gaitan-Solis, E., Bua, A., Taylor, N., Fauquet, C. M. and Alicai, T. (2012), 'Transgenic RNA interference (RNAi)-derived field resistance to cassava brown streak disease', *Mol. Plant Pathol.* 13, 1019–31.

Okogbenin, E., Egesi, C. N., Olasanmi, B., Ogundapo, O., Kahya, S., Hurtado, P., Marin, J., Akinbo, O., Mba, C., Gomez, H., de Vicente, S., Baiyeri, S., Uguru, M., Ewa, F. and Fregene, M. (2012), 'Molecular marker analysis and validation of resistance to cassava mosaic disease in elite cassava genotypes in Nigeria', *Crop Sci.* 52(6), 2576–86.

Okogbenin, E., Moreno, I., Tomkins, J., Fauquet, C. M., Mkamilo, G. and Fregene, M. (2013), 'Marker-Assisted Breeding for Cassava Mosaic Disease Resistance', In *Translational Genomics for Crop*

Breeding: Biotic Stress, Volume 1, Varshney, R. K and Tuberosa, R. (Eds), John Wiley & Sons Ltd, Chichester, UK.

Onyeka, T. J., Dixon, A. G. O., Ekpo, E. J. A. (2005), 'Assessment of laboratory methods for evaluating cassava genotypes for resistance to root rot disease', *Mycopathologia* 159, 461–7.

Ospina, B. Y. and Ceballos, H. (2002). 'La yuca en el tercer Milenio: Sistemas modernos de producción, producción, procesamiento, utilización y comercialización', Centro internacional de Agricultura Tropical; Consorcio Latinoamericano y del Caribe de Apoyo a la Investigación y Desarrollo de la Yuca, Proyecto IP-3 Mejoramiento de Yuca, Colombia, CA, 586pp.

Otim-Nape, G. W., Bua, A., Thresh, J. M., Baguma, Y., Ogwal, S., Semakula, G. N., Acola, G., Byabakama, B. and Martin, A. (1997), *Cassava Mosaic Virus Disease in Uganda: The Current Pandemic* and *Approaches to Control*, Natural Resources Institute, Chatham, UK, 65pp.

Owor, B., Legg, J. P., Okao-Okuja, G., Obonyo, R. and Ogenga-Latigo, M. W. (2004), 'The effect of cassava mosaic geminiviruses on symptom severity, growth and root yield of a cassava mosaic virus disease-susceptible cultivar in Uganda'. *Ann. Appl. Biol.* 145(3), 331–7.

Pardo, J. M., Truke, M. J., Cardozo, L., Varela, I. and Alvarez, E. (2014), 'A real-time PCR assay to detect and quantify 16SrIII-L and 16SrI phytoplasmas associated with cassava frogskin disease in Costa Rica and Paraguay', *Phytopathol.* 105 (Suppl. 1), S1.3.

Parmeter, J. and Whitney, H. (1970), 'Taxonomy and nomenclature of the perfect stage', In Rhizoctonia solani *Biology and Pathology*. Parmeter, J. (Ed.), University of California Press, Berkeley, CA, USA, pp. 7–19.

Patil, B. L. and Fauquet, C. M. (2009), 'Cassava mosaic geminiviruses: actual knowledge and perspectives', *Mol. Plant Pathol.* 10, 685–701.

Patil, L. B., Legg, J. P., Kanju, E. and Fauquet, C. M. (2015), 'Cassava brown streak disease: a threat to food security in Africa', *J. Gen. Virol.* 96, 956–68.

Pineda, B., Jayasinghe, U. and Lozano, J. C. (1983), 'La enfermedad "cuero de sapo" en yuca (*Manihot esculenta* Crantz)', *ASIAVA (Colombia)* 4, 10–12.

Pita, J. S., Fondong, V. N., Sangare, A., Otim-Nape, G. W., Ogwal, S. and Fauquet, C. M. (2001), 'Recombination, pseudorecombination and synergism of geminiviruses are determinant keys to the epidemic of severe cassava mosaic disease in Uganda', *J. Gen. Virol.* 82, 655–65.

Raaijmakers, J. M., Vlami, M. and de Souza, J. T. (2002), 'Antibiotic production by bacterial biocontrol agents', *Antoine van Leeuwenhoek* 81, 537–47.

Resmi, T., Hohn, T., Hohn, B. and Veluthambi, K. (2015), 'The *Agrobacterium tumefaciens* Ti plasmid virulence gene *virE2* reduces *Sri Lankan cassava mosaic virus* infection in transgenic *Nicotiana benthamiana* plants', *Viruses* 7, 2641–53.

Restrepo, S. and Verdier, V. (1997), 'Geographical differentiation of the population of *Xanthomonas axonopodis* pv. *manihotis* in Colombia', *Appl. Environ. Microbiol.* 63(11), 4427–34.

Sánchez, J. (1998), 'Caracterización de *Phytophthora* spp. agente causal de pudrición en raíz de yuca (*Manihot esculenta* Crantz) utilizando pruebas de patogenicidad y técnicas moleculares', PhD thesis, Universidad Nacional de Colombia, Bogota, Colombia, 198pp.

Santos, A. P., Gonçalves, J. A., Queiroz, G. M. and Lima, R. N. (1995), 'Efeito do vírus do mosaico das nervuras sobre os componentes produtivos da mandioca, cv pretinha, em duas regiões fisiográficas do Ceará', *Fitopatologia Brasileira* 20, 506–8.

Santos de Oliveira, S. A., Medeiros Abreu, E. F., Silva Araujo, M. T., Oliveira, E. J., Chumbinho de Andrade, E., Pardo Garcia, J. M. and Alvarez, E. (2013), 'First report of a 16SrIII-L phytoplasma associated with frogskin disease in cassava (*Manihot esculenta* Crantz) in Brazil', *Plant Dis.* 98(1), 153.

Saunders, K., Salim, N., Mali, V. R., Malathi, V. G., Briddon, S. R., Markham, P. G. and Stanley, J. (2002), 'Characterisation of *Sri Lankan cassava mosaic virus* and *Indian cassava mosaic virus*: Evidence for acquisition of a DNA B component by a monopartite begomovirus', *Virology* 293, 63–74.

Sheela, M. N., Abraham, K. and Moorthy, S. N. (2012). 'Development of elite inbreds of cassava and their use in heterosis breeding'. *Annual Report 2011–12*. Thiruvananthapuram, India: CTCRI, p. 25.

Sseruwagi, P., Sserubombwe, W. S., Legg, J. P., Ndunguru, J. and Thresh, J. M. (2004), 'Methods of surveying the incidence and severity of cassava mosaic disease and whitefly vector populations on cassava in Africa: a review', *Virus Res.* 100, 129–42.

Stanley, J. and Gay, M. R. (1983), 'Nucleotide sequence of cassava latent virus DNA', *Nature* 301, 260–2.

Storey, H. H. (1936), 'Virus diseases on East African plants. VI. A progress report on the studies of the diseases of cassava', *E. Afr. Agric. J.* 2, 34–9.

Storey, H. H. and Nichols, R. F. W. (1938), 'Studies on the mosaic of cassava'. *Ann. Appl. Biol.* 25, 790–806.

Taylor, N., Chavarriaga, P., Raemakers, K., Siritunga, D. and Zhang, P. (2004), 'Development and application of transgenic technologies in cassava', *Plant Mol. Biol.* 56(4), 671–88.

Terry, E. R. and Oyekan, J. O. (1976), 'Cassava diseases of Africa reviewed', *SPAN* 19, 116–18.

Thresh, J. M. and Cooter, R. J. (2005), 'Strategies for controlling cassava mosaic virus disease in Africa', *Plant Pathol.* 54, 587–614.

Thresh, J. M., Fargette, D. and Otim-Nape, G. W. (1994a), 'Effects of African cassava mosaic geminivirus on the yield of cassava', *Trop. Sci.* 34, 26–42.

Thresh, J. M., Otim-Nape, G. W. and Jennings, D. L. (1994b), 'Exploiting resistance to *African cassava mosaic virus*', *Aspects Appl. Biol.* 39, 51–60.

Valam-Zango, A., Zinga, I., Hoareau, M., Tocko-Marabena, B. K., Mvila, A. C., Semballa, S. and Lett, J. M. (2015), 'First report of cassava mosaic geminiviruses and the Uganda strain of East African cassava mosaic virus (EACMV-UG) associated with cassava mosaic disease in Equatorial Guinea', *New Dis. Rep.* 32, 25.

Valle, T. L., Kanthack, R. A. D., Lorenzi, J. O., Ribeiro, I. J. A. and Monteiro, D. A. (1996), Superalongamento possível ameaça a cultura da mandioca na região Centro-sul do Brasil', *Congresso Brasileiro de Mandioca, IX. Resúmenes, no. 89, CERAT/UNESP-SBM*, São Pedro, SP, Brazil.

van der Vlugt, R. A. A., Verbeek, M., Dullemans, A. M., Wintermantel, W. M., Cuellar, W. J., Fox, A. and Thompson, J. R. (2015), 'Torradoviruses', *Ann. Rev. Phytopathol.* 53(1), 485–512.

Vanitharani, R., Chellapan, P., Pita, J. S. and Fauquet, C. M. (2004), 'Differential roles of AC2 and AC4 of cassava geminiviruses in mediating synergism and suppression of posttranscriptional gene silencing', *J. Virol.* 78(17), 9487–98.

Verdier, V., Ojeda, S. and Mosquera, G. (2001), 'Methods for detecting the cassava bacterial blight pathogen: a practical approach for managing the disease', *Euphytica* 120(1), 103–7.

Verdier, V., Restrepo, S., Boher, B., Nicole, M., Geiger, J. P., Alvarez, E. and Bonierbale, M. (1997), 'Cassava bacterial blight: recent achievements in understanding the disease', *Afr. J. Root Tuber Crops* 2, 64–8.

Vu, N. T., Pardo, J. M., Alvarez, E., Le, H. H., Wyckhuys, K., Nguyen, K. L. and Le, D. T. (2016), 'Establishment of a loop-mediated isothermal amplification (LAMP) assay for the detection of phytoplasma-associated cassava witches' broom disease', *Appl. Biol. Chem.* 59(2), 151–6.

Wang, H. L., Cui, X. Y., Wang, X. W. and Liu, S. S. (2015), 'First report of *Sri Lankan cassava mosaic virus* infecting cassava in Cambodia', *Plant. Dis.* 100(5), 1029.

Warburg, O. (1894), 'Die kulturpflanzen usambaras', *Mitteilungen aus den Deutschen Schutzgebieten* 7, 131.

Wydra, K., Zinsou, V., Jorge, V. and Verdier, V. (2004), 'Identification of pathotypes of *Xanthomonas axonopodis* pv. *manihotis* in Africa and detection of quantitative trait loci and markers for resistance to bactorial blight of cassava', *Phytopathol.* 94(10), 1084–93.

Yadav, J. S., Ogwok, E., Wagaba, H., Patil, B. L., Baqewadi, B., Alicai, T., Gaitan-Solis, E., Taylor, N. J. and Fauquet, C. M. (2011), 'RNAi-mediated resistance to cassava brown streak Uganda virus in transgenic cassava', *Mol. Plant Pathol.* 12, 677–87.

Yang, J. and Kharbanda, P. D. (1996), 'Characterization, virulence, and genetic variation of *Rhizoctonia solani* AG 9 in Alberta', *Plant Dis.* 80, 513–18.

Yu, G. Y., Sinclair, G. L., Hartman, B. L. and Bertagnolli, B. L. (2002), 'Production of iturin A by *Bacillus amyloliquefaciens* suppressing *Rhizoctonia solani*', *Soil Biol. Biochem.* 34, 955–63.

Zhang, P., Vanderschuren, H., Fütterer, J. and Gruissem, W. (2005), 'Resistance to cassava mosaic disease in transgenic cassava expressing antisense RNAs targeting virus replication genes', *Plant Biotechnol. J.* 3, 385–98.

Ziegler, R. S. and Lozano, J. C. (1983), 'The relationship of *Sphaceloma* and *Elsinoe* species pathogenic on cassava and other members of the Euphorbiaceae in Central and South America', *Phytopathology* 73, 293–300.

Ziegler, R. S., Powell, L. E. and Thurston, D. (1980), 'Gibberellin A4 production by *Sphaceloma manihoticola*, causal agent of cassava superelongation disease', *Phytopathology* 70, 589–93.

Chapter 11

Integrated management of arthropod pests of cassava: the case of Southeast Asia

Ignazio Graziosi and Kris A.G. Wyckhuys, International Center for Tropical Agriculture (CIAT), Vietnam

1 Introduction

Cassava, *Manihot esculenta* Crantz (Malpighiales: Euphorbiaceae), is a perennial woody plant native to South America and is cultivated throughout the tropics for its starchy roots (Olsen and Schaal 1999). Approximately 57% of the world's cassava is grown in Africa, where it provides food security for more than 200 million people (Prudencio and Al-Hassan 1994; Manyong 2000; FAOSTAT 2015). In Southeast Asia, cassava is of increasing importance as a food, cash and bio-energy crop. Since the turn of the twenty-first century, cassava productivity in Southeast Asia has nearly doubled, and the crop has fast become one of the key agricultural commodities of the region. Local cassava provides industrial starches, dry chips for animal feed, biofuel and a range of products for human consumption, for both domestic use and export (Hahn 1989; Maziya-Dixon et al. 2007).

Indonesia and Thailand are the leading Asian producers. Cassava is a staple crop in Indonesia, while Thailand largely grows the crop for the billion-dollar starch industry (WITS 2005; Ceballos 2012). Cassava has a well-earned reputation as a hardy 'survivor' crop, able to sustain yields under relatively low soil fertility and proving exceptionally resilient to climate change. Resilience is of particular value to the region's resource-poor farmers, who often cannot afford the inputs to ameliorate biotic and abiotic stresses (Devendra and Thomas 2002; Ceballos et al. 2011). In fact, cassava continues to be grown primarily by smallholder farmers, who often operate on marginal lands and cultivate plots that are regularly smaller than 1 ha (Howeler 2006; Valentin et al. 2008). Hence, stabilization and augmentation of cassava yields can have important spillover benefits on local and regional food security (e.g. in Indonesia, southern Philippines), household and livelihood security,

http://dx.doi.org/10.19103/AS.2016.0014.13

prosperity of rural economies and the overall human welfare (Piya et al. 2011; van Donge et al. 2012; De Koninck and Rousseau 2014).

Over the past decades, better agronomy and improved varieties had driven a steady growth of Southeast Asian cassava yields, which reached approximately 21 Tonne/Ha by the early 2000s. New high-yielding and high starch varieties had the earliest impact, but improved agronomy followed, especially in Thailand and Vietnam (Henry and Gottret 1996; Waddington et al. 2010; FAOSTAT 2015). This trend, however, is under threat, following the accidental introduction of several deleterious non-native arthropods and (insect-vectored) pathogenic microorganisms, and increased outbreaks of a number of long-time invaders and indigenous species (Bellotti et al. 2012a; Alvarez et al. 2013; Graziosi et al. 2016). In particular, multiple species of non-native mealybugs (Hemiptera: Pseudococcidae) have made their appearance in Southeast Asia, and are spreading aggressively, aided by natural and anthropogenic means of dispersal and facilitated by prolonged drought periods in some parts of the region (Muniappan et al. 2009; Parsa et al. 2012). A number of these pests and insect-vectored diseases have caused substantial yield losses and have triggered the widespread adoption of systemic insecticides and acaricides. The prophylactic use of chemical insecticides could aggravate the economic burden for smallholders, and might have long-lasting detrimental impacts on the environmental integrity of local agro-ecosystems and human health (Kogan 1998; Wilson and Tisdell 2001; Goulson 2013; Van der Sluijs 2015). Fortunately, as past work in Africa and Latin America has shown, cassava is a crop that is tailor-made for biological control. Hence, increased efforts should be made to avoid (irrational) insecticide use and safeguard the valuable principles of integrated pest management (IPM), to further mitigate the impact of emerging biotic threats and secure the sustainability and resilience of local cropping systems (Edwards 1989; Birch et al. 2011).

In recent years, global interest in resource-conserving crop and pest management strategies has grown dramatically. In developing countries in particular, there is an urgent need to implement cultivation practices that can limit costs for small-scale farmers, reduce chemical inputs, promote agro-ecosystem health and mitigate the effects of climatic change, including climate-triggered pests or diseases (Pretty et al. 2003, 2006). Low-input or chemical-free pest management nearly seems tailor-made to a semi-permanent crop such as cassava, as this crop is grown throughout the developing-world tropics by resource-poor smallholders, on small plots or degraded soils (Bouaguimbeck 2011; Radhakrishnan et al. 2015). Non-chemical pest management should also be promoted in more intensive, commercially oriented cassava production systems, as are increasingly common in Thailand, western Indonesia or southern Vietnam. To effectively respond to these trends, address invasive pest complexes and build broader systems resilience, there is an urgent need for applied research on environmentally sound tactics and cropping systems adaptations. In this chapter, we give an updated account of the arthropod pest complex that is affecting Southeast Asian cassava, describe current pest management options, and highlight future opportunities for development and scale-up of IPM and biological control technologies.

2 Cassava pests in Southeast Asia

The arthropod pest complex of Southeast Asian cassava crops is primarily constituted by non-native piercing-sucking herbivores, comprising both new and old-time invaders.

Two newly arrived exotics – the cassava mealybug *Phenacoccus manihoti* Matile-Ferrero (Hemiptera: Pseudococcidae) and the green mite *Mononychellus mcgregori* (Flechtmann and Baker) (Acari: Tetranychidae) – are herbivores that have co-evolved with their host plant in its native South America. Other species are generalist herbivores that have gradually adapted to cassava, and only in recent years are increasingly reported at heightened population levels (Bellotti et al. 2012b). Among these are several non-native mealybugs and well-known pests of many crops like the papaya mealybug, various red spider mites (Acari: Tetranychidae), and two cosmopolitan whitefly species (Hemiptera: Aleyrodidae). Some non-natives, like the globally important striped mealybug *Ferrisia virgata* (Cockerell) or *Pseudococcus jackbeardsleyi* Gimpel and Miller, have made it into Southeast Asia over the past century, but only recently have been suspected to cause yield losses (Bayubay and Corpuz-Raros 2006; Bellotti 2012a).

Among the endemic taxa are species within the red mite complex and herbivores often associated with cassava crops, such as white grubs or termites. Mealybugs, mites and whiteflies are the main taxonomic groups relevant to Southeast Asian cassava. We here present an overview of the composition and bioecology for these groups, and elucidate their occurrence and local infestation pressure (Fig. 2) (see Graziosi et al., 2016).

2.1 Mealybugs

Mealybugs (Hemiptera: Pseudococcidae) are soft-bodied sap-feeding insects, some of which possess a marked sexual dimorphism, with an ability to produce protective wax secretions (Gullan and Kosztarab 1997; Hardy et al. 2008). Mealybugs are among the world's most invasive groups of herbivorous insects (Miller and Miller 2002; Wang et al. 2010; Beltrà et al 2013). Their cryptic habit and minute size make them particularly difficult to detect upon plant inspection, thus facilitating their accidental introduction with plant stocks (Ben-Dov 1994). Furthermore, the wide host range and virus-vectoring capability of a number of species significantly enhance their respective invasiveness and pest status (Minafra and Hadidi 1998; Sether et al. 1998; Wetten et al. 2015). In cassava, mealybugs are typical dry-season pests, with quick population build-up triggered by lowered precipitation and increased temperatures (Bellotti et al. 1994; Calatayud et al. 1994). Distinctive symptoms of mealybug damage include leaf yellowing and curling, defoliation and bud death (Ayanru 1987; Bellotti 2000), with detrimental effects on plant growth and root yield (Nwanze 1982; Zeddies et al. 2001).

Morphological features are used to organize the different mealybug taxa, with a distinction of short- and long-tailed species (Sanders 1909). Short-tailed mealybugs reduce waxy filaments that are equal to, or shorter than 75% of, the body length, and produce ovisacs, which are wax-covered egg masses deposited on host plants. Long-tailed mealybugs are characterized by the conspicuous abdominal filaments (i.e. so-called 'tails'), the production of live nymphs and the absence of ovisacs. Among the 24 species of mealybugs known to feed on cassava worldwide, seven taxa are reported from Southeast Asia (Parsa et al. 2012). Five non-native short-tailed mealybugs are identified in the region (Table 1).

One of the most worrying recent invaders in Southeast Asia is the cassava mealybug *P. manihoti*, the neotropical herbivore that invaded Africa during the 1970s and devastated local cassava crops (Nwanze 1982; Zeddies et al. 2001). Compared with other (generalist) mealybugs occurring on Southeast Asian cassava, *P. manihoti* is an oligophagous species, affecting as little as nine host plant genera (Ben-Dov 1994). Parthenogenetic females lay hundreds of eggs in ovisacs, and the newly hatched nymphs disperse both within the plant

Table 1 Main arthropod taxa reported on Southeast Asian cassava

	Pest	Common name	Origin	Record SE Asia	Max yield loss
MEALYBUGS	**Short-tailed mealybugs**				
	Phenacoccus manihoti	Cassava mealybug	Neotr.	2008	84%
	Paracoccus marginatus	Papaya mealyb.	Near.	2008	40%
	P. solenopsis	Cotton mealyb.	Near.	2006	n. a.
	P. madeirensis	Madeira mealyb.	Neotr.	2004	n. a.
	Maconellicoccus hirsutus	Hibiscus mealyb.	Neotr.	1942	n. a.
	Long-tailed mealybug				
	Pseudococcus jackbeardsleyi	Jack Beardsley m.	Neotr.	1996	n. a.
	Ferrisia virgata	Striped mealyb.	Neotr.	1942	n. a.
MITES	**Green mite complex**				
	Mononychellus mcgregori	Green mite	Neotr.	2009	60%
	Red mite complex				
	Neotetranychus lek	Red mite	n. a.	2013	53%[1]
	Oligonychus thelytokous	Red mite	Afrotr.	1998	
	Tetranychus marianae	Red mite	Near.	1975	
	T. yusti	Red mite	Palear.	1975	
	T. urticae	Red mite	Afrotr.	1975	
	Eutetranychus africanus	African red mite	Neotr.	1975	
	O. biharensis	Red mite	Oriental	1975	
	T. cinnabarinus	Red mite	Palear.	1969	
	T. neocaledonicus	Red mite	Austral.	1962	
	T. truncatus	Red mite	Palear.	1962	
	T. kanzawai	Red mite	Palear.	1962	
	Eirtetrunychus orientalis	Red mite	Palear.	1962	
WHITE-FLIES	**Whiteflies**				
	Aleurodicus dispersus	Spiraling whitefly	Near.	1987	n. a.
	Bemisia tabaci	Silverleaf whitefly	Palear.	1933	n. a.
SECONDARY PESTS	**Scale insects**				
	Aonidomytilus albus	Cassava scale	Near.	1935	n. a.
	Parasaissetia nigra	Black scale	n. a.	n. a,	n. a.
	Saissetia miranda	Black scale	Near.	n. a.	n. a.
	Secondary pests				
	Isoptera: Rhinotermitidae	Termites	Oriental	n. a.	n. a.
	Coleoptera: Scarabaeidae	White grubs	Oriental	n. a.	n. a.

Modified from Graziosi et al. (2016). Pest Management Science.

[1] referred to whole red mite complex.

Figure 1 Arthropods affecting cassava crops in Southeast Asia. (a) Cassava mealybug *P. manihoti*: healthy nymphs (1), nymph parasitized by parasitic hymenopteran (= mummy) (2), mealybug cadaver after emergence of adult parasitoid (3). (b) Short-tailed mealybug *Pa. marginatus*. (c) Long-tailed mealybug *Ps. jackbeardsleyi* (photo: T. Chuong). (d) Appearance of cassava top infested by *P. manihoti*. (e) *Anagyrus lopezi*, biological control agent of the cassava mealybug (photo: G. Goergen, International Institute of Tropical Agriculture, IITA). (f) Lacewing (Neuroptera: Chrysopidae): larva predating mealybugs (1), and egg (2). (g) High population of red mites (Acari: Tetranychidae) on cassava-growing points. (h) Leaf damage caused by red mites (i) Nymphs of spiralling whitefly *A. dispersus* feeding on cassava leaf (j) Cassava scale *Ao. albus* on a cassava stem (k) Cassava stake infested by Caribbean black scale *S. neglecta*. (l) Damage on roots caused by termites (Isoptera: Rhinotermitidae).

and to neighbouring plants, settling on growing tips and going through four instars prior to adult moulting (Fig. 1a; Nwanze et al. 1979; Herrera et al. 1989). Due to the release of toxins during mealybug feeding, young leaves are deformed, and the infested plants can be identified by the clumped appearance of the growing point, that is, so-called bunchy-top (Fig. 1d). Following its first detection in eastern Thailand in 2008, this insect rapidly dispersed into Laos, Cambodia, Vietnam, Indonesia and Malaysia (Muniappan et al. 2009; Winotai et al. 2010; Parsa et al. 2012; Sartiami et al. 2015). Geographic spread and local proliferation were likely facilitated by unregulated trade of cassava planting material, favourable climatic conditions and absence of effective natural enemies. The fast-spreading insect heavily impacted cassava fields throughout the region, and yield drops up to 50% were reported from Thailand (Winotai et al. 2010; Bellotti 2012a).

The remaining 4 short-tailed mealybug species are well-known pests of other crops, and non-native generalist herbivores that adapted to *M. esculenta*. The papaya mealybug *Paracoccus marginatus* Williams and Granara de Willink (Fig. 1b) is a global pest of papaya and numerous other crops, which invaded Asia in 2008, and has since spread to Indonesia, India, Cambodia, Thailand and the Philippines (Miller and Miller 2002; Muniappan et al. 2008, 2011). In India, yield loss due to *Pa. marginatus* reached 40% (Myrick et al. 2014). The cotton mealybug *P. solenopsis* Tinsley and the Madeira mealybug *P. madeirensis* Green are also recent invaders proliferating in climatically suitable areas of Southeast Asia, while the hibiscus mealybug *Maconellicoccus hirsutus* Green has been present for a much a longer period of time (Takahashi 1942). For most of the above species, little is known about their impact on local cassava crops and their occurrence (Lu et al. 2004; Muniappan et al. 2009; Nagrare et al. 2009; Wang et al. 2010). Among the two long-tailed mealybugs, *F. virgata* is the oldest invader (Takahashi 1942). As a highly polyphagous and cosmopolitan herbivore, this species commonly reaches outbreak levels in coffee, sweet potato, citrus and cotton (da Silva-Torres et al. 2013). On cassava, *F. virgata* populations are thought to have increased substantially over the past decade, occasionally bringing about losses in root yield (Bellotti et al. 2012a). A major agricultural and horticultural pest, *Pseudococcus jackbeardsleyi* Gimpel and Miller (Fig. 1c) (Shylesha 2013), is also inhabiting Southeast Asian cassava fields, and its presence and yield impact on local crops is yet to be evaluated (Williams 2004; Mani et al. 2013).

In dry-season cassava plots, mealybugs are the most common group of arthropods and the pest complex with the highest incidence, occurring in 70% of fields (Fig. 2a) and infesting >25% of plants on average (Fig. 2b, 2c). In 2014, *P. manihoti* occurred in most of the region's key cassava-growing areas (Fig. 3). Overall, *P. manihoti* population levels were the highest, followed by (other) short-tailed and long-tailed taxa (Fig. 2f), thus highlighting the invasiveness of the former insect.

2.2 Mites

The global cassava mite complex includes approximately 50 taxa of red and green spider mites (Acari: Tetranychidae) (Byrne et al. 1983). In Asian cassava crops, one green mite species and at least 12 different red mite species have been reported (Table 1) (Byrne et al. 1983). Mites can rapidly attain outbreak levels during the dry season, owing to high reproductive rates and comparatively slow colonization and build-up of natural enemy populations, especially in larger monocultures (Bellotti 2008). In Indonesia, red mites cause yield reductions up to 50%, while they are also noted as major pests in eastern Thailand and central Vietnam (Byrne et al. 1983). Green mites are oligophagous herbivores and major

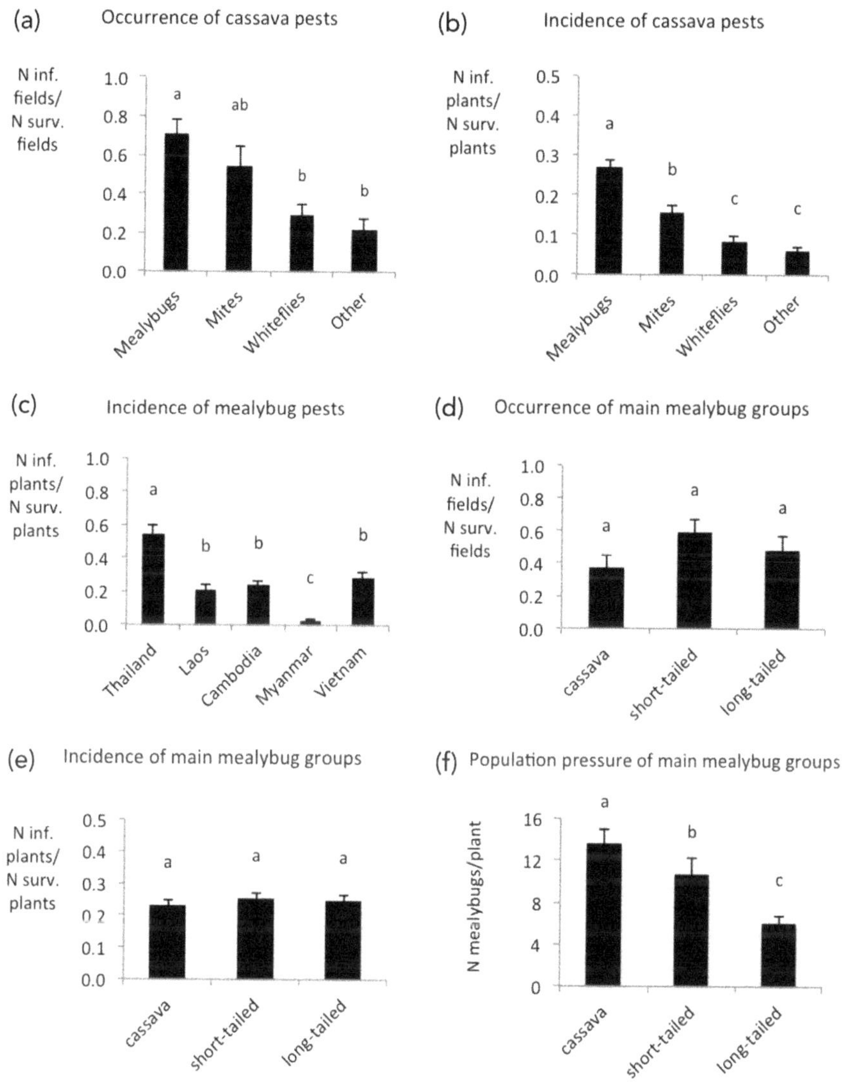

Figure 2 (a) Occurrence (proportion of affected fields per province) and (b) incidence (proportion of affected plants per affected field) of arthropod pests affecting cassava crops in Southeast Asia (Other arthropods include cassava scale, Caribbean black scale and termites). (c) Incidence of mealybug pests in different Southeast Asian countries. (d) Occurrence, (e) incidence, and (f) population pressure of main mealybug species or species groups. Means followed by the same letter do not differ (ANOVA, $\alpha = 0.05$) (429 fields in 26 provinces across 5 countries were monitored).

pests of cassava in Africa (Yaninek and Hanna 2003), while red mites are cosmopolitan and ubiquitous herbivores, with an exceptional ability to adapt to new hosts (Agrawal 2000). Most green mites preferentially infest young leaves and growing tips, and red mites usually

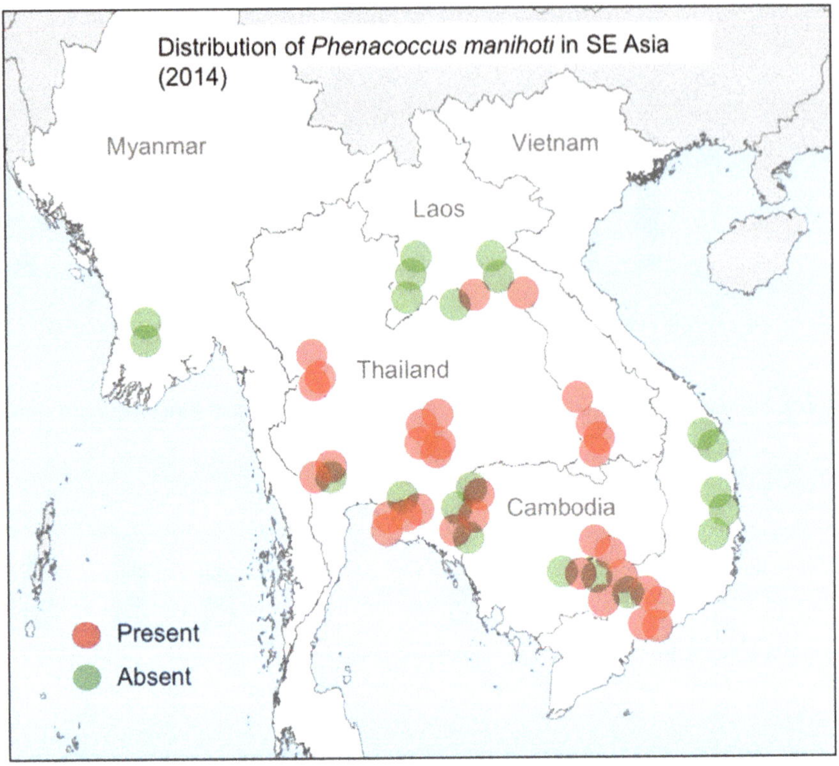

Figure 3 Distribution of *P. manihoti* in Southeast Asia cassava-growing areas (survey 2014).

feed on the underside of leaves located on the lower canopy, causing leaf yellowing and defoliation (Fig. 1g). When occurring at high population levels, red mites also affect the upper parts of the plant (Byrne et al. 1983; Yaninek et al. 1989) (Fig. 1h). At present, there is very little information on the composition and relative species abundance of the cassava mite complex in Asia, and novel species continue to be described (Flechtmann 2013). Some of the most common species are *Tetranychus urticae* Koch, *T. truncatus* Ehara, *T. cinnabarinus* Boisd., *T. kanzwai* Kishida and the African red mite *Eutetranychus africanus* (Tucker) (Table 1A) (Oei-Dharma 1969; Baker 1975a,b; Flechtmann 1981; Byrne et al. 1983; Hinomoto 2007; Sakunwarin et al. 2003, 2004; Bellotti et al. 2012b,c; Naing et al. 2014).

In addition to red mites, the neotropical green mite *M. mcgregori* has been identified in Vietnam and is thought to occur in China and Cambodia, but accurate information is needed regarding its status and occurrence in these countries (Bellotti et al. 2012a; Parsa et al. 2015). Furthermore, there is certain concern about an eventual presence of the cassava green mite *Mononychellus tanajoa*, a crucial pest of cassava capable of yield losses over 80% in Africa (Lu et al. 2014; Parsa et al. 2015).

During 2014 region-wide monitoring, mites were reported from over half of dry-season cassava fields (Fig. 2a), and on 16% of plants (Fig. 2b). Initial data show that the cassava mite complex was primarily composed of red mite species, with absence of *M. mcgregori* outbreaks.

2.3 Whiteflies

Whiteflies (Hemiptera: Aleyrodidae) are global pests, damaging a wide range of agricultural and horticultural crops in open-field and greenhouse settings (Gerling 1990; Bellotti and Arias 2001). Plant damage is caused by direct feeding, virus transmission and proliferation of sooty moulds on leaves due to honeydew secretions (Jones 2003). The most prominent whitefly-vectored viruses are the cassava mosaic virus (CMV), with variations in Africa, India and Sri Lanka; and cassava brown streak virus in Africa. While the Southeast Asia region had historically proven free of whitefly-vectored cassava viral diseases, this trend dramatically changed in late 2015. In December 2015, Wang et al. (2016) reported presence of the Sri Lanka strain of CMV from isolated cassava fields in eastern Cambodia. As this virus is vectored by whitefly and can also be spread through anthropogenic means (i.e. by infected planting material), there is substantial concern that cross-border trade flows and deficient quarantine systems will facilitate further disease spread throughout Southeast Asia. Even though a highly-diverse group of whitefly species impacts cassava in the neotropics, only two whitefly species are found across Southeast Asia, that is, *Aleurodicus dispersus* and *Bemisia tabaci* (Bellotti 2002; Bellotti et al. 2012a), where they infrequently reach high infestation levels and cause crop damage and defoliation (Bellotti and Arias; 2001, Bellotti 2008).

The spiralling whitefly *A. dispersus* Russel (Hemiptera: Aleyrodidae) (Fig. 1l) is a neotropical species impacting cassava and various other crops worldwide (Neuenschwander, 2009). In 1987, *A. dispersus* made its initial appearance in the Philippines, and has since been found throughout the region (Kajita et al. 1991; Mani and Krishnamoorthy 2002; Muniappan et al. 2009; Bellotti et al. 2012a). While *A. dispersus* attains its highest population densities during the dry season, nymphal development is also facilitated by high humidity (D'Almeida et al. 1998).

The silverleaf whitefly *B. tabaci* Gennadius is a cosmopolitan invader capable of developing on over 600 host plants, and is thought to have invaded Southeast Asia during the 1930s (Oliveira et al. 2001). The insect is a rainy-season pest; their numbers critically increase at the end of each rainy period, but background populations persist all year round (Bellotti and van Schoonhoven 1978, Robertson 1987). *B. tabaci* is a complex of morphologically identical species with distinct geographical distributions and genetic traits (Brown et al. 1995, De Barro et al. 2011). Multiple *B. tabaci* genotypes are described as feeding on cassava, and different genetic populations occur in Vietnam, Thailand, Malaysia and Indonesia (Berry et al. 2004; De Barro 2005; Sseruwagi et al. 2006). While a lot of information is available in the global literature on both *A. dispersus* and *B. tabaci*, the exact taxonomic composition, bioecology and incidence of whiteflies in Southeast Asian cassava fields are yet to be determined.

In region-wide monitoring efforts during the 2014 dry season, *B. tabaci* and *A. dispersus* were observed in less than one third of cassava fields, infesting on average 8% of plants (Fig. 2a, b) (Graziosi et al. 2016). While no information is available on whitefly populations during the rainy season, we expect infestation pressure to be somewhat higher.

2.4 Secondary pests

A number of other arthropods are known to feed on Asian cassava, and localized yield losses may occur. The cassava or tapioca scale *Aonidomytilus albus* (Fig. 1j) infests plants and 'stakes' (i.e. planting material, as cassava is vegetatively propagated) in India and

Thailand (Subramaniam et al. 1977; Wongkobrat et al. 1987). Black or brown scales (Fig. 1k) *Parasaissetia nigra* Nietner and *Saissetia miranda* Cockerell and Parell have also been reported from different cassava-growing areas (Peña and Waddill 1982; Bellotti et al. 2012b; Bellotti 2012c). Furthermore, white grubs (Coleoptera: Scarabaeidae) and termites (Isoptera: Rhinotermitidae) occasionally infest the lower part of the stem and roots (Fig. 1l), and cause differing levels of yield loss (Ananda and Rasdiman 1979; Miura et al. 1990; Atu 1993; Ranga Rao et al. 2006; Bellotti et al. 2012c). Scales or termites are occasionally reported in local cassava plots, occur in one fifth of fields and infest only 6% of plants (Fig. 2a, b) (see Graziosi et al. 2016).

3 Guidelines for non-chemical pest management

With the recent arrival of several invasive pests in Southeast Asia, chemically based pest management has fast gained a foothold in intensive cassava cropping systems (Parsa et al. 2012; Le and Wyckhuys, unpublished). For instance aqueous solutions of systemic neonicotinoids Thiamethoxam, Imidacloprid and Dinotefuran are now extensively used in Thailand and actively promoted in neighbouring countries for stake disinfection/disinfestation and as prophylactics targeting mealybugs (Parsa et al. 2012; Wyckhuys, unpublished). There is mounting concern about the efficacy, economic sustainability and environmental impact of the preventative use of systemic insecticides, and its incompatibility with long-established principles of IPM (Goulson 2013). Also, the extraordinary ability of the predominant cassava pests (i.e. mealybugs, mites) to rapidly build up their populations, and their propensity to develop insecticide resistance, can constrain the long-term sustainability of chemically based pest control in field crops (Herron et al. 1998; Franco et al. 2009; Koh et al. 2009; Van Leeuwen et al. 2010; Bellotti et al. 2012a). Moreover, intensive insecticide applications can severely impact in-field abundance, activity and fitness of locally occurring or released natural enemies (Poletti et al. 2007; Wang et al. 2008; Bellotti 2012a). The recurrent, irrational use of systemic insecticides can lead to resistance development and natural enemy suppression, and ultimately trigger increased infestations and damage by tetranychid mites or mealybugs (Sclar et al. 1998). These pest/biocontrol dynamics further exemplify the increased value and need for sustainable and environmentally sound IPM packages. IPM itself aims to reduce pest damage below economic threshold levels by relying upon multiple compatible techniques including natural enemies, plant resistance, cultural practices, and possibly minimal, targeted pesticide use, while taking into account economic, ecological and sociological consequences (Bottrell 1979; Kogan 1998). Furthermore, IPM tactics combine pest suppression and mitigation strategies, with monitoring, risk assessment and decision-support tools (Ehler 2006; Lemic et al. 2015).

In cassava, a number of most-valuable cultural and biological control options are available, almost tailor-made to smallholder production systems. Several of those practices can be readily promoted, and could prove particularly effective to address emerging pest threats (Table 2). Biological control stands out as a cost-effective and environmentally sound option for cassava pest management. The introduction of parasitic and predatory arthropods has facilitated long-term control of several key invaders, such as the cassava mealybug and the cassava green mite in Africa (Zeddies et al. 2001; Yaninek and Hanna 2003), while entomopathogens may also carry ample potential (Maniania et al. 2008;

Table 2 Overview of IPM options for managing cassava pests in Southeast Asia

Pest complex	Biological contr.	Cultural	Host resistance	Other
MEALYBUGS	paras[1,2,3], pred[1], entom fungi[3]	intercrop[1,2], clean plant material, removal[1]	low[3]	pheromone traps[2]
MITES	pred[1,2], entom fungi[3]	intercrop[1,2]	moderate[2]	botanical extracts[2,3]
WHITEFLIES	paras[2], entom fungi[2]	intercrop[1,2]	high levels[2]	-

Abbreviations: paras = parasitic Hymenoptera; pred = predatory arthropods; entom fungi = fungal entomopathogens; intercrop = intercropping; removal = physical removal and destruction of infested plant parts.
[1] implemented at some level in Southeast Asia, [2] available. Implemented in other areas and similar systems, [3] under evaluation.

Franco et al. 2009; Mascarin et al. 2013). Although largely under-researched, crop diversification and intercropping tactics could also contribute substantially to pest control and broader cropping systems resilience (Trenbath 1993; Gold et al. 1989; Delaquis et al., under review) (Table 3). While intercropping tactics could be particularly well suited for smallholder producers, increased labour demand may hinder their establishment and maintenance, in light of dropping labour availability in many of Southeast Asia's rural areas (Rigg 2013, 2015). In the section below, we present different non-chemical pest management options for each of the pest complexes, emphasizing those already implemented on Southeast Asian cassava and highlighting technologies that wait to be fine-tuned or transferred to local farmers.

3.1 Managing mealybugs

Natural enemies play a prominent role in the regulation of mealybug populations in various cropping systems (Franco et al. 2009; Daane et al. 2012). A number of parasitic or predatory insects and fungal entomopathogens are known to attack cassava mealybug pests (Bellotti 2012a,b). Encyrtid wasps (Hymenoptera: Encyrtidae) are typical mealybug parasitoids (Noyes 1994), and have been effectively used in a number of biological control programmes such as the continent-wide management of *P. manihoti* in Africa, *P. marginatus* in the Pacific and *P. hirsutus* in the Caribbean (Herren and Neuenschwander 1991; Kairo et al. 2000; Muniappan 2006). While biological control has been successfully wielded against pestiferous mealybugs, their effectiveness greatly depends upon local climate, crop management and soil conditions (Neuenschwander et al. 1990; Schulthess et al. 1997).

Initial yet deliberate steps have been taken to employ biological control against some of the mealybug invaders in Asia. Following the first outbreaks of the cassava mealybug in 2008, Thai institutions teamed up with FAO and global CGIAR centres to introduce the neotropical nymphal parasitoid *Anagyrus lopezi* (De Santis 1964) (Hymenoptera: Encyrtidae). This minute wasp causes high insect mortality through host feeding, mutilation, ovipositor probing and reproduction (Neuenschwander and Madojemu 1986). This parasitoid effectively controlled *P. manihoti* at a continent-wide scale in the 1980s in Africa. In late 2009, these same wasps were imported from Benin to Thailand, and subsequently mass-reared and released (Winotai et al. 2010). The parasitoid successfully established and lowered mealybug population levels in most Thai cassava fields

(Bellotti 2012a; Soysouvanh and Siri 2013). Next, inoculative releases were conducted in Laos, Cambodia and Indonesia, and small rearing operations were established in southern Vietnam (FAOIPM 2013, 2014). In some *P. manihoti*-infested areas in Thailand, local authorities and private sector actors continue to mass-rear and release *A. lopezi*. These (private sector supported) augmentative biological control schemes are unique and very different from past practices in Africa, where no follow-up releases were conducted (Neuenschwander 2001). Possibly, in the case of Asia, these schemes are still required, given the vastly different make-up of local cassava cropping systems, as defined by mono-cropping arrangements, increasing (broad-spectrum) pesticide use and staggered planting of fields, among others.

In addition to *A. lopezi*, three non-native wasps *Acerophagus papayae* Noyes and Schauff, *Pseudleptomastix mexicana* Noyes and Schauff, and *Anagyrus loecki* Noyes and Menezes (Hymenoptera: Encyrtidae) are successfully controlling *P. marginatus* outbreaks in India, and have possibly already made their arrival in a number of other Asian countries (Myrick et al. 2014). Also, a diverse community of naturally occurring predatory arthropods inflict mortality on mealybug taxa, with ladybeetles (Coleoptera: Coccinellidae), green and brown lacewings (Neuroptera: Chrysopidae, Hemerobiidae) often being reported (Löhr and Varela 1990; Mani and Krishnamoorthy 2008). Within Southeast Asia, the coccinellid *Sasajiscymnus quinquepunctatus* (Weise) has been identified as a candidate for biological control of *Pa. marginatus*, in Thailand (Saengyot 2012 and Burikam 2012a), and the green lacewings *Plesiochrysa ramburi* (Schneider) and *Mallada basalis* (Walker) (Neuroptera: Chrysopidae) are occasionally reared and mass-released in Thai cassava crops (Suasa-ard 2010). Lastly, the endemic apefly *Spalgis epius* (Westwood) (Lepidoptera: Lycaenidae) has also been considered for use in augmentative biological control schemes in Thailand (Saengyot and Burikam 2012b).

The use of fungal entomopathogens also carries some promise, and *Neozygites fumosa* (Entomophthorales) received some attention for use against *P. manihoti* in Africa in the 1980s (Le Ru et al. 1985; Le Ru 1986). However, the latter practice was largely discontinued following the massive success with *A. lopezi* releases. Other entomopathogenic fungi such as *Beauveria bassiana* have been used to control *Pa. marginatus*, *P. solenopsis* and *M. hirsutus* on cotton in Asia (Banu and Gopalakrishanan 2012; Surulivelu et al. 2012; Kumar et al. 2014; Ujjan et al. 2015), but their use in cassava has so far not been investigated.

In a number of arable crops, sex pheromones are employed for mealybug control (Millar et al. 2002; Franco et al. 2009), and are used either as decision-support tools to monitor field-level population levels or for mass trapping, mating disruption or volatile-mediated recruitment of natural enemies (Branco et al. 2006; Franco 2011). However, up until now, these products remain under-researched and un-used for mealybug pest control in cassava crops.

While the above tactics represent immediate, curative options to tackle pest outbreaks, an appropriate planning and implementation of a set of 'good practices' for cassava crop management can contribute substantially to preventing pest problems. For example, the use of good-quality, pest-free planting material is crucial. Non-chemical options should be used for on-farm disinfection of cassava stakes. Also, cultural control and proper field sanitation remain under-used, but could readily prevent localized mealybug outbreaks on limited extensions. In Africa, the use of maize as an intercrop drastically reduces *P. manihoti* population pressure (Agbobli 1987; Schulthess et al. 2004) (Table 3), but data from Asia and/or for other intercrop and mealybug species are not available. Work in the 1990s showed that the efficacy of biological control is directly linked to soil fertility

Table 3 Effects of intercropping strategies on arthropod prevalence in cassava. All studies were conducted in Africa

Target pest	Intercrop species	Effect (+/ 2 variation)	N studies
Cassava mealybug (*Phenacoccus manihoti)*	Maize	− 57–83% population pressure, no effect on % infested plant	2
Green mite *(Mononychellus tanajoa)*	Beans, maize, sweet potato	− 7% incidence, no effect to -21% nymph, − 40% adults on young leaves, no effects on predatory mite	2
Whiteflies (*Bemisia tabaci)*	Cotton and jatropha; maize and cowpea; bean, maize and sweet potato	− 79% eggs, − 21–58% nymphs, − 40–65% adults	3

see Delaquis et al., under review.

(Schulthess et al. 1997). However Asian cassava researchers and farmers all too often do not capitalize on the fact that a healthy and well-fertilized soil is a pre-requisite for effective pest suppression and crop health (Schulthess et al. 1997). At present, some of these crop and pest management practices are being promoted through novel extension tools, such as farmer-to-farmer educational videos (Le and Wyckhuys, unpublished).

Some of the above curative and preventative tactics can further diversify and improve the arsenal of management options that are available to Asia's cassava farmers. Certain methods, approaches and knowledge – such as sampling protocols, economic thresholds or other decision-support tools must still be developed, and additional knowledge needs to be gained on biological control options for other mealybugs. Nonetheless, we are confident that, once developed, these can further advance environmentally sound and low-cost pest management for cassava crops in Southeast Asia and globally.

3.2 Managing mites

An array of options are available for the management of cassava mite infestations, ranging from biological control and cultural practices to botanical acaricides (Byrne et al. 1983; Corbett et al. 1991; Bellotti et al. 1999; Isman 2000).

Natural enemies can effectively regulate mite populations in the world's cassava crops. The key success has been biological control using predatory mites (Acari: Phytoseiidae) (Hamamura 1986; McMurtry and Croft 1997). Cassava green mite outbreaks in Africa were effectively managed using the neotropical phytoseiid *Typhlodromalus aripo* (De Leon) (Yaninek and Hanna 2003; Mutisya et al. 2015). In Southeast Asia, the phytoseiid predator *Amblyseius* (*Neoseiulus*) *longispinosus* (Evans) has been deployed for control of *T. kanzawai*, *T. truncatus*, *T. urticae* and *E. africanus* on cassava and other crops in Thailand and the Philippines. However, this species is highly susceptible to common insecticides, which greatly impede survival, development and ultimate effectiveness (Hamamura 1986; Vasquez 1994; Nusartlert et al. 2010; Attia et al. 2013). Myriad insect and arachnid predators are known to prey upon cassava mites worldwide: the coccinellid *Stethorus gilvifrons* Mulsant and *S. siphonulus* Kapar, *Pullus* sp., *Serangium* sp., rove beetle (Coleoptera: Staphylinidae) *Oligota* sp., and spiders *Poecilochroa* sp. (Family

Gnaphosidae) and *Olios* sp. (Family Sparassidae) are described from Indonesia and India (Byrne 1983). Some of those predators could be used in augmentative biological control schemes, as exemplified by the successful use of predatory lacewings to manage *T. urticae* and *T. kanzawai* populations on Taiwanese strawberry crops (Chang and Huang 1995).

Entomopathogens are good to outstanding candidates for integrated management of spider mites (Irigaray et al. 2003). In cassava, *Neozygites* (Entomophthorales) strains can inflict high mortality to the notorious cassava green mite *M. tanajoa* (Delalibera et al. 2004). Researchers in Indonesia and the Philippines are evaluating different entomopathogenic fungi to control *T. kanzawai* (Sanjaya et al. 2013a,b). Based on these promising developments, we recommend to test those fungal strains for potential use against mites in local cassava crops.

Cultural control options are still limited, and further research is needed to evaluate different crop management tools that can help mitigate mite outbreaks. One such option is through cropping systems diversification, as intercropping with bean, maize and sweet potato has shown to drastically lower *M. tanajoa* densities in Africa (Toko 1996; Night et al. 2011) (Table 3). Although cassava varieties with high mite resistance are available, most are not the high-performance varieties preferred by farmers. Until such varieties are developed, the combined use of biological control and intermediate varietal resistance effectively controls *M. tanajoa* populations (Bellotti 2008, 2012a). In general, varieties with pronounced leaf pubescence are less susceptible to green mite infestations (Byrne 1980). While research has been conducted in Africa and Latin America on cultural control or varietal resistance against mites, much remains to be done in terms of varietal screening for resistance to Tetranychidae in Asia.

Finally, recent work reveals highly-effective repellent or acaricidal properties of different plant extracts against red mites *T. urticae*, *T. cinnabarinus* and *O. biharensis*, thus offering new options for an integrated approach to cassava mite management (Fangping et al. 2006; Zhang et al. 2008; Wei et al. 2011; Attia et al. 2013; Roh et al. 2013). Also, further work may be merited to explore opportunities to actively conserve locally occurring natural enemies, eventually through provision of non-prey (alternative) food resources such as maize pollen, for predatory phytoseiid mites (e.g. Onzo et al. 2005). In conclusion, recent research advances point toward clear opportunities for non-chemical management of mites in cassava crops, and also stress the need for continued applied research to develop, evaluate and adapt IPM technologies for some of Southeast Asia's most prominent cassava mite species.

3.3 Managing whiteflies

In the absence of recurrent population outbreaks, cassava whiteflies usually do not require control. However, this situation is starkly different in areas that are affected by whitefly-borne viral diseases (Thresh and Otim-Nape 1994). Given the recent introduction of the Sri Lanka strain of CMV into Southeast Asia (Wang et al., 2016), we consider it particularly valuable to enumerate a number of non-chemical options for management of cassava whitefly (and its associated complex of viral diseases).

Whitefly suppression can be achieved using locally occurring or introduced parasitic Hymenoptera. In West Africa, outbreaks of spiralling whitefly were effectively suppressed following fortuitous introduction and establishment of neotropical parasitoids *Encarsia haitiensis* Dozier and *E. guadeloupae* Viggiani (Hymenoptera: Aphelinidae) (D'Almeida et al. 1998). The latter species also effectively controls *A. dispersus* in India and Taiwan

(Mani et al. 2004). While predators were largely thought to be of minor importance in controlling cassava whitefly, Lundgren et al. (2014) reported a surprisingly diverse community of predators, including a netwing beetle (Lycidae), spiders, coccinellid beetles and lacewings, attacking whitefly in Colombian cassava fields. On another front, research on fungal entomopathogens is quickly advancing: different *Paecilomyces* species are associated with *B. tabaci* in India, and *Fusarium* strains have been screened for use against spiralling whitefly on vegetable crops in Asia (Faria and Wraight 2001; Aiswariya et al. 2010). None of the above fungal strains have been evaluated against *B. tabaci* whitefly in Asia, and very limited work has been done describing the natural enemy communities associated with local whitefly populations.

CGIAR institutions such as CIAT have spearheaded whitefly resistance screening and breeding, and have made available a number of cultivars highly resistant to *Al. socialis* and *B. tabaci* to cassava growers in South America and Africa (Bellotti and Arias 2001; Ceballos et al. 2004; Omongo et al. 2012). Though whitefly-resistant varieties have also been introduced into germplasm collections in countries such as Thailand, China and Vietnam, they are not deployed in the field. In addition to host plant resistance, cropping systems diversification can contribute to pest control by either inferring associational resistance to whitefly or directly favouring resident natural enemy communities (Lundgren et al. 2014; Bellotti et al. 2012a). African trials show how different intercropping schemes can drastically lower in-field abundance of whitefly (Table 3) (Fondong et al. 2002; Ewusie 2008; Night et al. 2011). Diversification schemes and the incorporation of moderately resistant varieties constitute clear mechanisms for whitefly population suppression, and are on-the-shelf technologies in case whitefly control becomes a necessity in Southeast Asia (as it may, given the recent introduction of SLCMV in Cambodia; Wang et al. 2016). Biological control may carry considerable potential as well, but critical insights into in-field abundance, activity patterns and efficacy of natural enemies urgently need to be gained.

4 Future trends and conclusion

Our work provides one of the first comprehensive overviews of the magnitude of biotic invasions and arthropod pest problems in Southeast Asian cassava crops. This 4 million ha crop, which directly underpins the livelihoods of 8 million Asian smallholder farmers and provides food security to entire nations, has recently been hit by a number of fast-spreading pest invaders and by increasing outbreaks of long-time insect colonists. While these mounting invasive pest problems can be ascribed to globalization (and the associated, cross-boundary movement of biota) and linked to climate change (Hobbs 2000; Peh 2010; Meyerson and Mooney 2007), they also elucidate inherent, critical weaknesses and deficiencies of Southeast Asia's cassava cropping systems. Grown primarily under monoculture, with inadequate use of crop rotation schemes, and with deficient soil management and fertilization, local cassava crops have proven particularly vulnerable to pest attack (see also Brown 1963; Nweke 1994; Hambäck and Englund 2005). More so, with local farmers in several Asian countries massively embracing the prophylactic use of systemic insecticides, natural defences within these cropping systems may even be further undermined. This chapter outlined a number of options for non-chemical management of some of these pest invaders, ranging from biological control to host plant resistance, through the development of IPM decision-support tools. We also point at immediate opportunities such as cropping systems diversification to enhance or restore the resilience of local cassava cropping systems (see Aye and Howeler 2012; Brooker et al. 2014) and to optimally enable natural plant defence mechanisms.

This chapter thus provides clear and immediate opportunities for low-cost, non-chemical pest management in local cassava crops, and for approaches that are fully in line with long-established principles of IPM. Only by safeguarding these valuable practices, and by recognizing, valuing and building upon past work from other continents, can we fully ensure the long-term profitability and sustainability of cassava production systems throughout Southeast Asia.

5 Where to look for further information

Additional up-to-date information on emerging cassava pest and diseases, and their management options can be found in Graziosi et al. (2016) and Delaquis et al. (in review).

6 Acknowledgements

We would like to thank Tin Maung Aye, Sok Sophearith and Lao Thao for logistical support to 2014-15 monitoring efforts. Dr. Prapit Wongtiem, Phanuwat Moonjuntha, Sirilak Lankaew and DoAE officers facilitated monitoring activities in Thailand. We are grateful to Ms. Chamaiporn Buamas (Thai DoA), Khanxay Somchanda and PPC colleagues for assistance in identifying mealybugs, and to Mr. Chhun Hy Heng (GDA) for help in orchestrating country-wide sampling efforts in Cambodia. Dr. Ngo Tien Dung (PPD) enabled country-wide surveillance activities in Vietnam. Monitoring work in Vietnam, Laos and Cambodia was made possible through respective provincial and district-level PPSD, PAFO/DAFO, and PDA offices. We thank Linh Giang and Louis Parker for developing maps on pest and cassava distribution, and Erik Delaquis for providing knowledge on intercropping included in Table 3. This initiative was conducted as part of an EC-funded, IFAD-managed, CIAT-executed programme (CIAT-EGC-60-1000004285).

7 References

Agrawal, A. A., Host-range evolution: adaptation and trade-offs in fitness of mites on alternative hosts. *Ecology* 81:500–8 (2000).

Aiswariya, K. K., Manjunatha, M. and Naik, M. I., Evaluation of fungi *Fusarium semitectum* Berk. and Ravenel and *Verticillium lecanii* (Zimm.) Viegas against spiraling whitefly *Aleurodicus dispersus* Russell on guava. *Karnataka Journal of Agricultural Sciences* 20:283–7 (2010).

Altieri, M. A. and Nicholls, C. I., Soil fertility management and insect pests: harmonizing soil and plant health in agroecosystems. *Soil and Tillage Research* 72:203–11 (2003).

Alvarez, E., Pardo, J. M., Mejía, J. F., Bertaccini, A., Thanh, N. D. and Hoat, T. X., Detection and identification of '*Candidatus* Phytoplasma asteris'-related phytoplasmas associated with a witches' broom disease of cassava in Vietnam. *Phytopathogenic Mollicutes* 3:77–81 (2013).

Ananda, K. and Rasdiman, S., White grubs identification on cassava at Kalasan region, Yogyakarta (Central Java, Indonesia) Conference Proceedings, AgrisFAO online record, http://agris.fao.org/aos/records/XB8100141 (1979).

Attia, S., Grissa, K. L., Lognay, G., Bitume, E., Hance, T. and Mailleux, A. C., A review of the major biological approaches to control the worldwide pest *Tetranychus urticae* (Acari: Tetranychidae) with special reference to natural pesticides. *Journal of Pest Science* 86:361–86 (2013).

Atu, U. G., Cultural practices for the control of termite (Isoptera) damage to yams and cassava in south) Conference Proceedings. *International Journal of Pest Management* 39:462–6 (1993).

Ayanru, D. K. G., Effects of mealybug (*Phenacoccus manihoti*) infestation on cassava yield components and plant tissue quality. *Journal of Agriculture in the Tropics and Subtropics* 88:5–10 (1987).

Aye, T. M. and Howeler, R., Cassava agronomy: intercropping systems, in *The Cassava Handbook*, edited by Howeler, R. H, CIAT, Colombia, CA, pp. 613–25 (2012).

Baker, E. W., Plant-feeding mites of Thailand (Tetranychidae, Tenuipalpidae, and Tuckerellidae). *Plant Protection Service Technical Bulletin*. Bangkok, Thailand (1975a).

Baker, E. W., Spider mites (Tetranychidae: Acarina) from southeast Asia and Japan. *Cooperative Economic Insect Report* 25:911–21 (1975b).

Banu, J. G. and Gopalakrishanan, N., Development of Formulations of a Native Entomopathogenic Fungus, *Lecanicillium lecanii* and Testing Virulence Against Mealybug, *Paracoccus marginatus* Infesting Cotton. *Indian Journal of Plant Protection* 40:182–6 (2012).

Bayubay, A. L. G. and Corpuz-Raros, L. A., Taxonomic survey of mites and sap-sucking insects associated with cassava especially in commercial plantations in Isabela Province, Philippines. *Philippines Entomology* 20:102–25 (2006).

Bellotti, A. C., Las plagas principales del cultivo de la yuca: Un panorama global. In *Simposio avances en el manejo de plagas. Memorias del XXVII Congreso de SOCOLEN* (2000, July), Medellín, Colombia, pp. 189–217 (2000).

Bellotti, A. C., Arthropod pests, in *Cassava: Biology, Production and Utilization*, edited by Hillocks, R. J., Wydra, K. and Belloti, A., CABI Publishing, Oxon, UK, pp. 209–35 (2002).

Bellotti, A. C., Cassava pests and their management, in *Encyclopedia of Entomology*, 2nd Edition, edited by Capnera, J. L., Springer, The Netherlands (2008).

Bellotti, A. C. and van Schoonhoven, A., Mite and insect pests of cassava. *Annual Review of Entomology* 23:39–67 (1978).

Bellotti, A. C., Braun, A. R., Arias, B., Castillo, J. A. and Guerrero, J. M., Origin and management of neotropical cassava arthropod pests. *African Crop Science Journal* 4:407–17 (1994).

Bellotti, A. C. and Arias, B., Host plant resistance to whiteflies with emphasis on cassava as a case study. *Crop Protection* 20:813–23 (2001).

Bellotti, A. C., Campo, B. V. H. and Hyman, G., Cassava production and pest management: present and potential threats in a changing environment. *Trop. Plant Biology* 5:39–72 (2012a).

Bellotti, A. C., Arias, B. V., Vargas, O. H., Rejes, J. A. Q. and Guerrero, G. M., Insects and mites that attack cassava, and their control, in *Cassava in the Third Millennium. Modern Production, Processing, Use, and Marketing Systems*, edited by Ospina, B. and Veballos, H., CIAT, Colombia, CA, pp. 213–50 (2012b).

Bellotti, A. C., Herrera, C. J., del Pilar Hernandez, M., Arias, B., Guerrero, G. M. and Melo, E. L., Cassava pests in Latin America, Africa And Asia, in *The Cassava Handbook*, edited by Howeler, R. H., CIAT,Colombia, CA, pp. 199–257 (2012c).

Beltrà, A., Tena, A. and Soto, A., Fortuitous biological control of the invasive mealybug *Phenacoccus peruvianus* in Southern Europe. *BioControl* 58:309–17 (2013).

Ben-Dov, Y., A systematic catalogue of the mealybugs of the world (Insecta: Homoptera: Coccoidea: Pseudococcidae and Putoidae) with data on geographical distribution, host plants, biology and economic importance. Intercept Limited (1994).

Berry, S. D., Fondong, V. N., Rey, C., Rogan, D., Fauquet, C. M. and Brown, J. K., Molecular evidence for five distinct *Bemisia tabaci* (Homoptera: Aleyrodidae) geographic haplotypes associated with cassava plants in sub-Saharan Africa. *Annals of the Entomological Society of America* 97:852–9 (2004).

Birch, A. N. E., Begg, G. S. and Squire, G. R., How agro-ecological research helps to address food security issues under new IPM and pesticide reduction policies for global crop production systems. *Phyton* 62:3251–61 (2011).

Bottrell, D. R., *Integrated Pest Management*. US Government printing office, Washington DC USA, 120p (1979).
Bouaguimbeck, H., African organic agriculture training manual. A resource manual for trainers. FiBL, Research Institute of Organic Agriculture, Switzerland (2011).
Branco, M., Jactel, H., Franco, J. C. and Mendel, Z., Modelling response of insect trap captures to pheromone dose. *Ecological Modelling* 197:247–57 (2006).
Brooker, R. W., Bennett, A. E., Cong, W. F., Daniell, T. J., George, T. S., Hallett, P. D., Hawes, C., Iannetta, P. P. M., Jones, H. G., Karley, A. J., Li, L., McKenzie, B. M., Pakeman, R. J., Paterson, E., Schob, C., Shen, J., Squire, G., Watson, C. A., Zhang, C., Zhang, F., Zhang, J. and White, P. J., Improving intercropping: A synthesis of research in agronomy, plant physiology and ecology. *New Phytologist* 206:107–17 (2014).
Brown, L. R., Agricultural diversification and economic development in Thailand: a case study. *Economic Research Service Foreign Agricultural Economic Report* 8:34 (1963).
Brown, J. K., Frohlich, D. R. and Rosell, R. C., The sweetpotato or silverleaf whiteflies: biotypes of *Bemisia tabaci* or a species complex? *Annual Review of Entomology* 40:511–34 (1995).
Byrne, D. H., Studies of resistance to the mites *Mononychellus tanajoa* (Bondar) and *Mononychellus caribbeanae* (McGregor) in cassava, *Manihot esculenta* (Crantz). *Dissertation Abstracts International B* 41:750 (1980).
Byrne, D. H., Bellotti, A. C. and Guerrero, J. M., The cassava mites. *International Journal of Pest Management* 29:378–94 (1983).
Calatayud, P. A., Tertuliano, M. and Le Rü, B., Seasonal changes in secondary compounds in the phloem sap of cassava in relation to plant genotype and infestation by *Phenacoccus manihoti* (Homoptera: Pseudococcidae). *Bulletin of Entomological Research* 84:53–459 (1994).
Ceballos, H., Iglesias, C. A., Pérez, J. C. and Dixon, A. G., Cassava breeding: opportunities and challenges. *Plant Molecular Biology* 56:503–16 (2004).
Ceballos, H., Cassava in Colombia and the world. New prospects for a millennial crop, in *Cassava in the Third Millennium: Modern Production, Processing, Use, and Marketing Systems,* edited by Ospina, B. and Veballos, H., CIAT, Colombia, CA, p. 575 (2012).
Ceballos, H., Ramirez, J., Bellotti, A. C., Jarvis, A. and Alvarez, E., Adaptation of cassava to Changing Climates, in *Crop Adaptation to Climate Change*, edited by Yadav, S. S., Redden, R. J., Hatfield, J. L., Lotze-Campen, H. and Hall, A. E., Blackwell Publishing, Oxford, UK, pp. 411–25 (2011).
Chang, C. P. and Huang, S. C., Evaluation of the effectiveness of releasing green lacewing, *Mallada basalis* (Walker) for the control of tetranychid mites on strawberry. *Plant Protection B* 37:41–58 (1995).
Corbett, A., Leigh, T. F. and Wilson, L. T., Interplanting alfalfa as a source of *Metaseiulus occidentalis* (Acari: Phytoseiidae) for managing spider mites in cotton. *Biological Control* 1:188–96 (1991).
Crowder, D. W. and Harwood, J. D., Promoting biological control in a rapidly changing world. *Biological Control* 75:1–7 (2014).
Daane, K. M., Almeida, R. P., Bell, V. A., Walker, J. T., Botton, M., Fallahzadeh, M., Mani, M., Miano, J. L., Sforza, R., Walton, V. M. and Zaviezo, T., Biology and management of mealybugs in vineyards, in *Arthropod Management in Vineyards*, edited by Bostanian, N. J., Vincent, C. and Isaacs, R., Springer, Berlin, The Netherlands, pp. 271–307 (2012).
D'Almeida, Y. A., Lys, J. A., Neuenschwander, P. and Ajuonu, O., Impact of two accidentally introduced *Encarsia* species (Hymenoptera: Aphelinidae) and other biotic and abiotic factors on the spiralling whitefly *Aleurodicus dispersus* (Russell) (Homoptera: Aleyrodidae), in Benin, West Africa. *Biocontrol Science and Technology* 8:163–73 (1998).
da Silva-Torres, C. S. A., de Oliveira, M. D. and Torres, J. B., Host selection and establishment of striped mealybug, *Ferrisia virgata*, on cotton cultivars. *Phytoparasitica* 41:31–40 (2013).
De Barro, P. J., Genetic structure of the whitefly *Bemisia tabaci* in the Asia–Pacific region revealed using microsatellite markers. *Molecular Ecology* 14:3695–718 (2005).
De Barro, P. J., Liu, S. S., Boykin, L. M. and Dinsdale, A. B., *Bemisia tabaci*: a statement of species status. *Annual Review of Entomology* 56:1–19 (2011).
De Koninck, R. and Rousseau, J. F. G., *Gambling with the Land: The Contemporary Evolution of Southeast Asian Agriculture*. Nus Press, Singapore (2014).

Delalibera, I., Hajek, A. E. and Humber, R. A., *Neozygites tanajoae* sp. nov., a pathogen of the cassava green mite. *Mycologia* 96:1002–9 (2004).

Devendra, C., and Thomas, D., Smallholder farming systems in Asia. *Agricultural Systems* 71:17–25 (2002).

Edwards, C. A., The importance of integration in sustainable agricultural systems. *Agriculture, Ecosystems and Environment* 27:25–35 (1989).

Ehler, L. E., Integrated pest management (IPM): definition, historical development and implementation, and the other IPM. *Pest Management Science* 62:787–9 (2006).

Ewusie, E., *Potentials and Prospect of Strip Cropping in the Management of Cassava Whitefly*. Texas Tech University, Lubbock (2008).

Fangping, Z., Hongfang, L. and Yongqiang, L., Repellent effects of the alcohol extracts of 25 tropical plants to Oligonychus biharensis Hirst. *Plant Protection* (2006).

FAO-IPM Asia Regional Programme, Vietnam Proceed with Mass Rearing and Field Releases of the Parasitoid, *Anagyrus Lopezi*, for the Control of Pink Cassava Mealybug Tay Ninh, Vietnam FAO-IPM online news http://www.vegetableipmasia.org (2013).

FAO-IPM Asia Regional Programme, FAO and CIAT Assist Indonesia with Biological Control for Pink Cassava Mealybug Bogor, Indonesia, Vietnam FAO-IPM online news http://www.vegetableipmasia.org/news/view/108 (2014).

FAOSTAT, Food and Agriculture Organization of the United Nations, Statistic division, http://faostat3.fao.org/home (2015).

Faria, M., and Wraight, S. P., Biological control of *Bemisia tabaci* with fungi. *Crop Protection* 20: 767–78 (2001).

Flechtmann, C. H. W., The cassava mite complex. II. New records and description of two new species in the genus *Tetranychus* from Asia. *International Journal of Acarology* 7:81–6 (1981).

Flechtmann, C. H. W., A new species of *Neotetranychus Trägårdh* (Acari, Prostigmata, Tetranychidae) from Thailand with a key to world species. *Persian Journal of Acarology* 2:35–40 (2013).

Fondong, V. N., Thresh, J. M., and Zok, S., Spatial and temporal spread of cassava mosaic virus disease in cassava grown alone and when intercropped with maize and/or cowpea. *Journal of Phytopathology* 150:365–74 (2002).

Franco, J. C., Zada, A. and Mendel, Z., Novel approaches for the management of mealybug pests. In *Biorational Control of Arthropod Pests*, Springer, Berlin, The Netherlands, pp. 233–78 (2009).

Franco, J. C., Da Silva, E. B., Fortuna, T., Cortegano, E., Branco, M., Suma, P., Torre, I. L., Russo, A., Elyahu, M., Protasov, A., Levi-Zada, A. and Mendel, Z., Vine mealybug sex pheromone increases citrus mealybug parasitism by *Anagyrus* sp. near *pseudococci* (Girault). *Biological Control* 58:230–8 (2011).

Gerling, D., *Whiteflies: Their Bionomics, Pest Status and Management*. Intercept Limited (1990).

Graziosi, I., Minato, N., Alvarez, E., Ngo, D. T., Trinh, X. H., Ayo, T. M., Pardo, J. M., Wongtiem, P. and Wyckhuys, K. A. G. Emerging pests and diseases of South-east Asian cassava: a comprehensive evaluation of geographic priorities, management options and research needs. *Pest Management Science* 72: 1071–89 (2016).

Gold, C. S., Altieri, M. A. and Bellotti, A. C., Cassava intercropping and pest incidence: a review illustrated with a case study from Colombia. *International Journal of Pest Management* 35: 339–44 (1989).

Gullan, P. J. and Kosztarab, M., Adaptations in scale insects. *Annual Review of Entomology* 42:23–50 (1997).

Hahn, S. K., An overview of African traditional cassava processing and utilization. *Outlook on Agriculture* 18:110–18 (1989).

Hamamura, T., Studies on the biological control of Kanzawa spider mite, *Tetranychus kanzawai* Kishida by the chemical resistant predacious mite, *Amblyseius longispinosus* (Evans) in tea fields (Acarina: Tetranychidae, Phytoseiidae). *Chagyo Shikenjo Kenkyu Hokoku* (1986).

Hambäck, P. A. and Englund, G., Patch area, population density and the scaling of migration rates: the resource concentration hypothesis revisited. *Ecology Letters* 8:1057–65 (2005).

Hardy, N. B., Gullan, P. J. and Hodgson, C. J., A subfamily-level classification of mealybugs (Hemiptera: Pseudococcidae) based on integrated molecular and morphological data. *Systematic Entomology* 33:51–71 (2008).

Henry, G. and Gottret, V., Global cassava trends. Reassessing the crop's future. CIAT Working document No. 157. CIAT, Colombia, CA, p. 45 (1996).

Herren, H. R. and Neuenschwander, P., Biological control of cassava pests in Africa. *Annual Review of Entomology* 36:257–83 (1991).

Herrera, C. J., Driesche, R. V. and Bellotti, A. C., Temperatureontroldent growth rates for the cassava mealybug, Phenacoccus herreni, and two of its encyrtid parasitoids, Epidinocarsis diversicornis and Acerophagus coccois in Colombia. *Entomologia Experimentalis et Applicata* 50:21–7 (1989).

Herron, G. A., Edge, V. E., Wilson, L. J. and Rophail, J., Organophosphate resistance in spider mites (Acari: Tetranychidae) from cotton in Australia. *Experimental and Applied Acarology* 22:17–30 (1998).

Hershey, C. H., Álvarez, E., Aye, T. M., Becerra, L. A., Bellotti, A. C., Ceballos, H., Fahrney, K., Howeler, R., Lefroy, R., Ospina, B. and Parsa, S., chapter 9: Eco-efficient interventions to support cassava's multiple roles in improving the lives of smallholders, in *Eco-Efficiency: From Vision to Reality*, pp. 135–60 (2012).

Hinomoto, N., Tran, D. P., Pham, A. T., Ngoc Le, T. B., Tajima, R., Ohashi, K., Osakabe, M. and Takafuji, A., Identification of spider mites (Acari: Tetranychidae) by DNA sequences: a case study in Northern Vietnam. *International Journal of Acarology* 33:53–60 (2007).

Hobbs, R J., *Invasive Species in a Changing World*, edited by Mooney, H. A. and Hobbs, R. J., Island Press, Washington, US (2000).

Howeler, R. H., Long-term effect of cassava cultivation on soil productivity. *Field Crops Research* 26:1–18 (1991).

Howeler, R. H., Cassava in Asia: trends in cassava production, processing and marketing, in *Workshop on Partnership in Modern Science to Develop a Strong Cassava Commercial Sector in Africa and Appropriate Varieties* 2020:2–6 (2006).

Irigaray, F. J. S. D. C., Marco-Mancebón, V. and Pérez-Moreno, I., The entomopathogenic fungus *Beauveria bassiana* and its compatibility with triflumuron: effects on the twospotted spider mite *Tetranychus urticae*. *Biological Control* 26:168–73 (2003).

Isman, M. B., Plant essential oils for pest and disease management. *Crop Protection* 19:603–8 (2000).

Jones, D. R., Plant viruses transmitted by whiteflies. *European Journal of Plant Pathology* 109:195–219 (2003).

Kairo, M. T., Pollard, G. V., Peterkin, D. D. and Lopez, V. F., Biological control of the hibiscus mealybug, *Maconellicoccus hirsutus* Green (Hemiptera: Pseudococcidae) in the Caribbean. *Integrated Pest Management Reviews* 5:241–54 (2000).

Kajita, H., Samudra, I. M. and Naito, A., Discovery of the spiraling whitefly Aleurodicus dispersus Russell (Homoptera: Aleyrodidae) from Indonesia, with notes on its host plants and natural enemies. *Applied Entomology and Zoology* 26:397–400 (1991).

Koh, S. H., Ahn, J., Im, J. S., Jung, C., Lee, S. H. and Lee, J. H., Monitoring of acaricide resistance of *Tetranychus urticae* (Acari: Tetranychidae) from Korean apple orchards. *Journal of Asia-Pacific Entomology* 12:15–21 (2009).

Kogan, M., Integrated pest management: historical perspectives and contemporary developments. *Annual Review of Entomology* 43:243–70 (1989).

Kumar, K. I., Kennedy, J. S. and Suresh, S., Laboratory evaluation of *Beauveria bassiana* (Balsamo) Vuillemin against crawler stage of *Paracoccus marginatus* (Williams and Granara de Willink). *Trends in Biosciences* 7: 246–9 (2014).

Lemic, D., Mikac, K. M., Kozina, A., Benitez, H. A., McLean, C. M. and Bažok, R., Monitoring techniques of the western corn rootworm are the precursor to effective IPM strategies. *Pest Management Science* early view, DOI: 10.1002/ps.4072 (2015).

Le Rü, B., Silvie, P. and Papierok, B., L'entomophthorale *Neozygites fumosa* pathogène de la Cochenille du manioc, *Phenacoccus manihoti* [Hom.: Pseudococcidae], en République populaire du Congo. *Entomophaga* 30:23–9 (1985).

Le Rü, B., Étude de l'évolution d'une mycose a *Neozygites fumosa* [Zygomycetes, Entomophthorales] dans une population de la cochenille du manioc, *Phenacoccus manihoti* [Hom.: Pseudococcidae]. *Entomophaga* 31:79–89 (1986).

Löhr, B., Varela, A. M. and Santos, B., Exploration for natural enemies of the cassava mealybug, *Phenacoccus manihoti* (Homoptera: Pseudococcidae), in South America for the biological control of this introduced pest in Africa. *Bulletin of Entomological Research* 80:417–25 (1990).

Lu, H., Lu, F. P., Xu, X. L. and Chen, Q., Potential geographic distribution areas of *Mononychellus mcgregori* in Guangxi province. *Applied Mechanics and Materials* 522:1051–4 (2014).

Lundgren, J. G., López-Lavalle, L. A. B., Parsa, S. and Wyckhuys, K. A., Molecular determination of the predator community of a cassava whitefly in Colombia: pest-specific primer development and field validation. *Journal of Pest Science* 87:125–31 (2014).

Mani, M. and Krishnamoorthy, A., Classical Biological Control of the Spiralling whitefly, Aleurodicus dispersus Russell-An Appraisal. *International Journal of Tropical Insect Science* 22:263–73 (2002).

Mani, M. and Krishnamoorthy, A., Biological suppression of the mealybugs *Planococcus citri* (Risso), *Ferrisia virgata* (Cockerell) and *Nipaecoccus viridis* (Newstead) on pummelo with *Cryptolaemus montrouzieri* Mulsant in India. *Journal of Biological Control* 22: 169–72 (2008).

Mani, M., Joshi, S., Kalyanasundaram, M., Shivaraju, C., Krishnamoorthy, A., Asokan, R. and Rebijith, K. B., A new invasive Jack Beardsley mealybug, *Pseudococcus jackbeardsleyi* (Hemiptera: Pseudococcidae) on papaya in India. *Florida Entomologist* 96:242–5 (2013).

Maniania, N. K., Bugeme, D. M., Wekesa, V. W., Delalibera, Jr. I. and Knapp, M., Role of entomopathogenic fungi in the control of *Tetranychus evansi* and *Tetranychus urticae* (Acari: Tetranychidae), pests of horticultural crops. *Experimental and Applied Acarology* 46:259–74 (2008).

Manyong, V. M., *Impact: The Contribution of IITA-improved Cassava to Food Security in Sub-saharan Africa*. IITA, Ibadan, Nigeria (2000).

Mascarin, G. M., Kobori, N. N., Quintela, E. D., Delalibera, I., The virulence of entomopathogenic fungi against *Bemisia tabaci* biotype B (Hemiptera: Aleyrodidae) and their conidial production using solid substrate fermentation. *Biological Control* 66:209–18 (2013).

Maziya-Dixon, B., Dixon, A. G. and Adebowale, A. R. A., Targeting different end uses of cassava: genotypic variations for cyanogenic potentials and pasting properties. *International Journal of Food Science and Technology* 42:969–76 (2007).

McMurtry, J. A., Croft, B. A., Life-styles of phytoseiid mites and their roles in biological control. *Annual Review of Entomology* 42:291–321 (1997).

Meyerson, L. A. and Mooney, H. A., Invasive alien species in an era of globalization. *Frontiers in Ecology and the Environment* 5:199–208 (2007).

Millar, J. G., Daane, K. M., Mcelfresh, J. S., Moreira, J. A., Malakar-Kuenen, R., Guillén, M. and Bentley, W. J., Development and optimization of methods for using sex pheromone for monitoring the mealybug *Planococcus ficus* (Homoptera: Pseudococcidae) in California vineyards. *Journal of Economic Entomology* 95:706–14 (2002).

Miller, D. R. and Miller, G. L., Redescription of *Paracoccus marginatus* Williams and Granara de Willink (Hemiptera: Coccoidea: Pseudococcidae), including descriptions of the immature stages and adult male. *Proceedings of the Entomological Society of Washington* 104:1–159 (2002).

Minafra, A. and Hadidi, A., Sensitive detection of grapevine virus A, B, or leafroll-associated III from viruliferous mealybugs and infected tissue by cDNA amplification. *Journal of Virological Methods* 47:175–87 (1994).

Miura, K., Subhasaram, T., Tawinthung, N., Noochan, N., and Shiraishi, K., Effects of termite activity on soils in northeast Thailand. *Japanese Journal of Tropical Agriculture* 34:40–7 (1990).

Muniappan, R., Meyerdirk, D. E., Sengebau, F. M., Berringer, D. D. and Reddy, G. V. P., Classical biological control of the papaya mealybug, *Paracoccus marginatus* (Hemiptera: Pseudococcidae) in the Republic of Palau. *Florida Entomologist* 89:212–17 (2006).

Muniappan, R., Shepard, B. M., Watson, G. W., Carner, G. R., Sartiami, D., Rauf, A. and Hammig, M. D., First report of the papaya mealybug, *Paracoccus marginatus* (Hemiptera: Pseudococcidae), in Indonesia and India. *Journal of Agricultural and Urban Entomology* 25:37–40 (2008).

Muniappan, R., Shepard, B. M., Watson, G. W., Carner, G. R., Rauf, A., Sartiami, D., Hidayat, P., Afun, J. V. K., Goergen, G. and Rahman, A. Z., New records of invasive insects (Hemiptera:

Sternorrhyncha) in Southeast Asia and West Africa. *Journal of Agricultural and Urban Entomology* 26:167–74 (2009).

Muniappan, R., Recent invasive hemipterans and their biological control in Asia, in *5th Meeting of the Asian Cotton Research & Development Network* (Lahore, PK, 23/25 Febuary 2011) (2011).

Mutisya, D. L., El-Banhawy, E. M., Khamala, C. P. M. and Kariuki, C. W., Management of cassava green mite *Mononychellus progresivus* (Acari: Tetranychidae) in different agro-ecological zones of Kenya. *Systematic and Applied Acarology* 30:39–50 (2015).

Myrick, S., Norton, G. W., Selvaraj, K. N., Natarajan, K. and Muniappan, R., Economic impact of classical biological control of papaya mealybug in India. *Crop Protection* 56:82–6 (2014).

Nagrare, V. S., Kranthi, S., Biradar, V. K., Zade, N. N., Sangode, V., Kakde, G., Shukla, R. M., Shivare, D., Khadi, B. M. and Kranthi, K. R., Widespread infestation of the exotic mealybug species, *Phenacoccus solenopsis* (Tinsley) (Hemiptera: Pseudococcidae), on cotton in India. *Bulletin of Entomological Research* 99:537–41 (2009).

Naing, H. H., Chandrapatya, A., Navajas, M. and Auger, P., Know more about spider mites (Acari: Tetranychidae) in Myanmar. Conference at Yezin Agricultural University, Yezin Myanmar, December 2014 (2014).

Neuenschwander, P., Spiralling whitefly, *Aleurodicus dispersus*, a recent invader and new cassava pest (2009).

Neuenschwander, P. and Madojemu, E., Mortality of the cassava mealybug, *Phenacoccus manihoti* Mat.-Ferr.(Hom., Pseudococcidae), associated with an attack by *Epidinocarsis lopezi* (Hym., Encyrtidae). *Mitteilungen der Schweizerischen Entomologischen Gesellschaft* 59:57–62 (1986).

Neuenschwander, P., Hammond, W. N. O., Ajuonu, O., Gado, A., Echendu, N., Bokonon-Ganta, A. H., Allomasso, R. and Okon, I., Biological control of the cassava mealybug, *Phenacoccus manihoti* (Hom., Pseudococcidae) by *Epidinocarsis lopezi* (Hym., Encyrtidae) in West Africa, as influenced by climate and soil. *Agriculture, Ecosystems and Environment* 32:39–55 (1990).

Night, G., Asiimwe, P., Gashaka, G., Nkezabahizi, D., Legg, J. P., Okao-Okuja, G., Obonyo, R., Nyirahorana, C., Mukakanyana, C., Mukase, F., Munyabarenzi, I., Mutumwinka, M. Occurrence and distribution of cassava pests and diseases in Rwanda. *Agriculture, Ecosystems and Environment* 140:492–7 (2011).

Noyes, J. S. and Hayat, M., Oriental mealybug parasitoids of the Anagyrini (Hymenoptera: Encyrtidae). Cab International (1994).

Nusartlert, N., Vichitbandha, P., Baker, G. T. and Chandrapatya, A., Pesticide-induced mortality and prey-dependent life history of the predatory mite *Neoseiulus longispinosus* (Acari: Phytoseiidae), in *Trends in Acarology.* Proceedings of the 12th International Congress, Springer, The Netherlands, pp. 495–8 (2010).

Nwanze, K. F., Relationships between cassava root yields and crop infestations by the mealybug, *Phenacoccus manihoti. International Journal of Pest Management* 28:27–32 (1982).

Nwanze, K. F., Leuschner, K. and Ezumah, H. C., The cassava mealybug, *Phenacoccus* sp. in the Republic of Zaire. *Pans* 25:125–30 (1979).

Nweke, F. I., Farm level practices relevant to cassava plant protection, *African Crop Science Journal* 2:563–82 (1994).

Oei-Dharma, H. P., *Use of Pesticides and Control of Economic Pests and Diseases in Indonesia*, edited by Brill, E. J. Leiden, The Netherlands, p. 199 (1969).

Oliveira, M. R. V., Henneberry, T. J. and Anderson, P., History, current status, and collaborative research projects for *Bemisia tabaci. Crop Protection* 20:709–23 (2001).

Olsen, K. M. and Schaal, B. A., Evidence on the origin of cassava: phylogeography of *Manihot esculenta. Proceedings of the National Academy of Sciences* 96:5586–91 (1999).

Omongo, C. A., Kawuki, R., Bellotti, A. C., Alicai, T., Baguma, Y., Maruthi, M. N., Bua, A. and Colvin, J., African cassava whitefly, *Bemisia tabaci*, resistance in African and South American cassava genotypes. *J. Integrative Agriculture* 11:327–36 (2012).

Onzo, A., Hanna, R., Negloh, K., Toko, M. and Sabelis, M. W., Biological control of cassava green mite with exotic and indigenous phytoseiid predators—effects of intraguild predation and supplementary food. *Biological Control* 33: 143–52 (2005).

Parsa, S., Kondo, T. and Winotai, A., The cassava mealybug (*Phenacoccus manihoti*) in Asia: first records, potential distribution, and an identification key. *PLoS ONE* 7:e47675 (2012).

Parsa, S., Hazzi, N. A., Chen, Q., Lu, F., Campo, B. V. H., Yaninek, J. S. and Vásquez-Ordóñez, A. A., Potential geographic distribution of two invasive cassava green mites. *Experimental and Applied Acarology* 65:195–204 (2015).

Peh, K. S. H., Invasive species in Southeast Asia: the knowledge so far. *Biodiversity and Conservation* 19:1083–99 (2010).

Peña, J. E. and Waddill, V., Pests of cassava in South Florida. *Florida Entomologist* 65:143–9 (1982).

Piya, S., Kiminami, A. and Hironori, Y., Sources of agricultural productivity in south and southeast Asia. *Trends in Agricultural Economics* 4:18–29 (2011).

Poletti, M., Maia, A. H. N. and Omoto, C., Toxicity of neonicotinoid insecticides to *Neoseiulus californicus* and *Phytoseiulus macropilis* (Acari: Phytoseiidae) and their impact on functional response to *Tetranychus urticae* (Acari: Tetranychidae). *Biological Control* 40:30–6 (2007).

Pretty, J. N., Morison, J. I. and Hine, R. E., Reducing food poverty by increasing agricultural sustainability in developing countries. *Agriculture, Ecosystems and Environment* 95: 217–34 (2003).

Pretty, J. N., Noble, A. D., Bossio, D., Dixon, J., Hine, R. E., Penning de Vries, F. W. and Morison, J. I., Resource-conserving agriculture increases yields in developing countries. *Environmental Science and Technology* 40:1114–19 (2006).

Prudencio, Y. C. and Al-Hassan, R., The food security stabilization roles of cassava in Africa. *Food Policy* 19: 57–64 (1994).

Radhakrishnan, A. S., Girija, S. and Anil, A. T., Organic vs conventional management in cassava: growth dynamics, yield and soil properties. *Journal of Root Crops* 39:93–9 (2015).

Ranga Rao, G. V., Giang, N. T. L., Lieu, P. and Tram, N. T. H., Occurrence of White Grubs in Groundnut Crop in Uplands of South Vietnam: A New Report. *International Arachis Newsletter* 26:29–31 (2006).

Rigg, J., From rural to urban: a geography of boundary crossing in southeast Asia. *Transactions* 1:5–26 (2013).

Rigg, J., Poverty and livelihoods after full-time farming: a South-East Asian view. *Asia Pacific Viewpoint* 46:173 (2005).

Robertson, I. A. D., The whitefly, *Bemisia tabaci* (Gennadius) as a vector of African cassava mosaic virus at the Kenya coast and ways in which the yield losses in cassava, *Manihot esculenta* Crantz caused by the virus can be reduced. *International Journal of Tropical Insect Science* 8:797–801 (1987).

Roh, H. S., Lee, B. H. and Park, C. G., Acaricidal and repellent effects of myrtacean essential oils and their major constituents against *Tetranychus urticae* (Tetranychidae). *Journal of Asia-Pacific Entomology* 16:245–9 (2013).

Saengyot, S. and Burikam, I., Development and growth ratio of predaceous coccinellid, *Sasajiscymnus quinquepunctatus* (Weise) on papaya mealybug, *Paracoccus marginatus* Williams & Granara de Willink. *Kasetsart Journal (Natural Science)* 46: 418–26 (2012a).

Saengyot, S. and Burikam, I., Bionomics of the apefly, *Spalgis epius* (Lepidoptera: Lycaenidae), predatory on the papaya mealybug, *Paracoccus marginatus* (Hemiptera: Pseudococcidae), in Thailand. *Songklanakarin Journal of Science and Technology* 34:1 (2012b).

Sakunwarin, S., Chandrapatya, A. and Baker, G., Biology and life table of the cassava mite, *Tetranychus truncatus* Ehara (Acari: Tetranychidae). *Systematic and Applied Acarology* 8:13–24 (2003).

Sakunwarin, S., Chandrapatya, A. and Visetson, S., Synergism and detoxification mechanism of crude sugar apple seed extract in *Tetranychus truncatus* Ehara (Prostigmata: Tetranychidae). *Kasetsart Journal (Natural Science)* 38:340–8 (2004).

Sanders, J. G., The identity and synonymy of some of our soft scale-insects. *Journal of Economic Entomology* 2: 428–48 (1909).

Sanjaya, Y., Ocampo, V. R. and Caoili, B. L., Selection of entomopathogenic fungi against the red spider mite *Tetranychus kanzawai* (Kishida) (Tetranychidae: Acarina). *Arthropods* 2:208–15 (2013a).

Sanjaya, Y., Ocampo, V. R. and Caoili, B. L., Transmission effect of entomopathogenic fungi on population of *Tetanychus kanzawai* (Kishida) (Tetranychidae: Acarina). *Arthropods* 2:36–41 (2013b).

Sarkar, M. A., Suasa-ard, W. and Uraichuen, S., Suitability of different mealybug species (Hemiptera: Pseudococcidae) as hosts for the newly identified parasitoid *Allotropa suasaardi* Sarkar & Polaszek (Hymenoptera: Platygasteridae). *Kasetsart Journal (Natural Science)* 48: 17–27 (2014).

Sartiami, D., Watson, G. W., Roff, M. N., Hanifah, M. Y. and Idris, A. B., First record of cassava mealybug, *Phenacoccus manihoti* (Hemiptera: Pseudococcidae), in Malaysia. *Zootaxa* 3957:235–8 (2015).

Seiedy, M., Saboori, A., Allahyari, H., Talaei-Hassanloui, R. and Tork, M., Laboratory investigation on the virulence of two isolates of the entomopathogenic fungus *Beauveria bassiana* against the twospotted spider mite *Tetranychus urticae* (Acari: Tetranychidae). *International Journal of Acarology* 36:527–32 (2010).

Sether, D. M., Ullman, D. E. and Hu, J. S., Transmission of pineapple mealybug wilt-associated virus by two species of mealybug (*Dysmicoccus* spp.). *Phytopathology* 88: 1224–30 (1998).

Schulthess, F., Neuenschwander, P. and Gounou, S., Multi-trophic interactions in cassava, *Manihot esculenta*, cropping systems in the subhumid tropics of West Africa. *Agriculture, Ecosystems and Environment* 66:211–22 (1997).

Sclar, D. C., Gerace, D. and Cranshaw, W. S., Observations of population increases and injury by spider mites (Acari: Tetranychidae) on ornamental plants treated with imidacloprid. *Journal of Economic Entomology* 91:250–5 (1998).

Shylesha, A. N., Host range of invasive Jack Beardsley mealybug, *Pseudococcus jackbeardsleyi* Gimpel and Miller in Karnataka. *Pest Management in Horticultural Ecosystems* 19:106–7 (2013).

Soysouvanh, P. and Siri, N., Population abundance of pink mealybug, *Phenacoccus manihoti* on four cassava varieties. *Khon Kaen Agriculture Journal* 41:149–53 (2013).

Sseruwagi, P., Maruthi, M. N., Colvin, J., Rey, M. E. C., Brown, J. K. and Legg, J. P., Colonization of non – cassava plant species by cassava whiteflies (*Bemisia tabaci*) in Uganda. *Entomologia Experimentalis et Applicata* 119:145–53 (2006).

Suasa-ard, W., Natural enemies of important insect pests of field crops and utilization as biological control agents in Thailand. *Proceedings of International Seminar on Enhancement of Functional Biodiversity Relevant to Sustainable Food Production in ASPAC.* 9–11 November 2010 Tsukuba, Japan (2010).

Subramaniam, T. R., David, B. V., Thangavel, P. and Abraham, E. V., Insect pest problems of tuber crops in Tamil Nadu. *Journal of Root Crops* 3:43–50 (1977).

Sugio, A., Kingdom, H. N., MacLean, A. M., Grieve, V. M. and Hogenhout, S. A., Phytoplasma protein effector SAP11 enhances insect vector reproduction by manipulating plant development and defense hormone biosynthesis. *Proceedings of the National Academy of Sciences* 108:1254–63 (2011).

Surulivelu, T., Banu, J. G., Rajan, T. S., Dharajothi, B. and Amutha, M., Evaluation of fungal pathogens for the management of mealybugs in Bt cotton. *Journal of Biological Control* 26:92–6 (2012).

Takahashi, R., Some injurious insects of agricultural plants and forest trees in Thailand and Indo-China. II. Coccidae. Report. Formosa Govt Res Inst Dept Agric Rept 81. Taihoku, pp. 1–56 (1942).

Thresh, J. M. and Otim-Nape, G. W., Strategies for controlling African cassava mosaic geminivirus, in *Advances in Disease Vector Research*, edited by Harris, K. F., Springer, New York, US, pp. 215–36 (1994).

Toko, M., Yaninek, J. S. and O'Neil, R. J., Response of *Mononychellus tanajoa* (Acari: Tetranychidae) to cropping systems, cultivars, and pest interventions. *Environmental Entomology* 25:237–49 (1996).

Trenbath, B. R., Intercropping for the management of pests and diseases. *Field Crop Research* 34:381–405 (1993).

Ujjan, A. A., Khanzada, M. A. and Shahzad, S., Efficiency of *Metarhizium* spp.(Sorokīn) strains and insecticides against cotton mealybug Phenacoccus solenopsis (Tinsley). *Pakistan Journal of Zoology* 47: 351–60 (2015).

Valentin, C., Agus, F., Alamban, R., Boosaner, A., Bricquet, J. P., Chaplot, V., de Guzman, T., de Rouw, A., Janeau, J. L., Orange, D., Phachomphonh, K., Phai, D. D., Podwojewski, P., Ribolzi, O., Silvera, N., Subagyono, K., Thiebaux, J. P., Toan, T. D. and Vadari, T., Runoff and sediment losses from 27 upland catchments in Southeast Asia: Impact of rapid land use changes and conservation practices. *Agriculture, Ecosystems and Environment* 128:225–38 (2008).

Van der Sluijs, J. P., Amaral-Rogers, V., Belzunces, L. P., van Lexmond, M. B., Bonmatin, J. M., Chagnon, M., Downs, C. S., Furlan, L., Gibbons, D. W., Giorio, C., Girolami, V., Goulson, D., Kreutzweiser, D. P., Krupke, C., Liess, M., Long, E., McField, M., Mineau, P., Mitchell, E. A. D., Morissey, C. S., Noome, D. A., Pisa, L., Settele, J., Simon-Delso, N., Stark, J. D., Tapparo, A., Van Dyck, H., van Praagh, J., Whitehorn, P. R. and Wiemers, M., Conclusions of the Worldwide Integrated Assessment on the risks of neonicotinoids and fipronil to biodiversity and ecosystem functioning. *Environmental Science and Pollution Research* 22:148–54 (2015).

Van Donge, J. K., Henley, D. and Lewis, P., Tracking development in South of neonicotinoids and fipronil to biodiversity and ecosyst. *Development Policy Review* 30:s5–24 (2012).

Van Leeuwen. T., Vontas, J., Tsagkarakou, A., Dermauw, W. and Tirry, L., Acaricide resistance mechanisms in the two-spotted spider mite *Tetranychus urticae* and other important Acari: a review. *Insect Biochemistry and Molecular Biology* 40:563–72 (2010).

Vasquez, E. A., Biological control of cassava red spider mite *Tetranychus kanzawai* Kishida. Philippine Rootcrop Research and Training Center, Baybay, Leite (1994).

Waddington, S. R., Li, X., Dixon, J., Hyman, G. and De Vicente, M. C., Getting the focus right: production constraints for six major food crops in Asian and African farming systems. *Food Security* 2:27–48 (2010).

Wang, H. Y., Yang, Y., Su, J. Y., Shen, J. L., Gao, C. F. and Zhu, Y. C., Assessment of the impact of insecticides on *Anagrus nilaparvatae* (Pang et Wang) (Hymenoptera: Mymanidae), an egg parasitoid of the rice planthopper, *Nilaparvata lugens* (Hemiptera: Delphacidae). *Crop Protection* 27:514–22 (2008).

Wang, H. L., Cui, X. Y., Wang, X. W., Liu, S. S., Zhang, Z. H. and Zhou, X. P., First report of Sri Lankan cassava mosaic virus infecting cassava in Cambodia. *Plant Disease* 100:1029 (2016).

Wang, Y., Watson, G. W. and Zhang, R., The potential distribution of an invasive mealybug *Phenacoccus solenopsis* and its threat to cotton in Asia. *Agricultural and Forest Entomology* 12:403–16 (2010).

Wei, J., Ding, W., Zhao, Y. G. and Vanichpakorn, P., Acaricidal activity of Aloe vera L. leaf extracts against *Tetranychus cinnabarinus* (Boisduval)(Acarina: Tetranychidae). *Journal of Asia-Pacific Entomology* 14:353–6 (2011).

Wetten, A., Campbell, C. and Allainguillaume, J., High-resolution melt and morphological analyses of mealybugs (Hemiptera: Pseudococcidae) from cacao: tools for the control of Cacao swollen shoot virus spread, *Pest Management Science* (2015).

Williams, D. J., Mealybugs of Southern Asia. The Natural History Museum, Kualu Lumpur, Southdene, Malaysia, 896p. (2004).

Wilson, C. and Tisdell, C., Why farmers continue to use pesticides despite environmental, health and sustainability costs. *Ecological Economics* 39:449–62 (2001).

Winotai, A., Goergen, G., Tamò, M. and Neuenschwander, P., Cassava mealybug has reached Asia. *Biocontrol News and Information* 31:10N–11 (2010).

WITS, World Integrated Trade Solution software, The World Bank http://wits.worldbank.org (2005).

Wongkobrat, A., Weerawut, T. and Prachuabmoh, O., Effect of white scale insect [*Aonidomytilus albus*] infestation on cassava yield and quality. AgrisFAO online record, http://agris.fao.org/aos/records/TH9121467 (1987).

Yaninek, J. S., Moraes, G. J. and Markham, R. H., *Handbook on the Cassava Green Mite (Mononychellus tanajoa) in Africa: A Guide to its Biology and Procedures for Implementing Classical Biological Control* IITA, Ibadan, Nigeria (1989).

Yaninek, S. and Hanna, R., Cassava green mite in Africa – a unique example of successful classical biological control of a mite pest on a continental scale, in *Biological Control in IPM systems in*

Africa, edited by Neuenschwander, P., Borgemeister, C. and Langewald, J. CABI Publishing, Oxon, UK, pp. 61–75 (2003).
Zeddies, J., Schaab, R. P., Neuenschwander, P. and Herren, H. R., Economics of biological control of cassava mealybug in Africa. *Agricultural Economics* 24:209–19 (2001).
Zhang, Y. Q., Wei, D., Zhao, Z. M., Jing, W. U. and Fan, Y. H., Studies on acaricidal bioactivities of *Artemisia annua* L. extracts against *Tetranychus cinnabarinus* Bois. (Acari: Tetranychidae). *Agricultural Sciences in China* 7:577–84 (2008).

Chapter 12

Weed control in cassava cropping systems

S. Hauser and F. Ekeleme, International Institute of Tropical Agriculture (IITA), Nigeria

1 Introduction

Weed control in cassava fields appears to have received highly variable levels of attention on different continents where cassava is grown, despite clear messages from farming communities on its importance. From Nigeria, Agumagu et al. (2008) reported major problems affecting cassava farms as identified by respondents in the following order of importance: weeds (grasses (88%), broadleaf weeds (66.6%)), rodents (64.1%), other animals (63.3%), cassava root rot (53.9%) and insect pests (48.7%). Although grasses and broadleaf weeds were perceived as the major problems, the crop is one of the few without concrete chemical and mechanical weed control options.

The preferred or most frequently used weed control approaches differ greatly among continents and countries. In southern Brazil and in other countries where cassava is predominantly a commercial crop and sold to processing factories, weed control is close to 100% herbicide-based. In Africa, most cassava is hand weeded, often by women and children. Although it is a source of employment and income for women, it tends to be drudgery and in some countries the use of child labour compromises the livelihoods and education of children. For Asia it was estimated that adequate weed control would increase yields by 7% or 3.2 million tons (Henry and Gottret, 1996; Hershey et al., 2001). For other continents such data is not available; yet considering the national average yields in Africa, there is most likely a large margin of yield and thus potential production increases.

http://dx.doi.org/10.19103/AS.2016.0014.27

However, which portion of such increases would be generated through improved weed control has yet to be estimated.

Research on weed control very often focuses on a specific weed species and its effects on a crop or on control measures to combat the species. In reality, however, farmers need to deal with a community of weeds all contributing to crop losses. Nevertheless, in some cases, the focusing on one species may be justified by its dominance; yet such situations should not lead to the impression that controlling a single species will solve weed problems in a sustainable manner.

This chapter will discuss the effects of weeds on cassava root yields, farmers' perceptions of the importance of weed control and the control methods used in different cropping systems. The focus is primarily on Africa due to the fact that the continent accounts for the largest portion of global cassava production, weed control is not as well developed and weeds pose a larger threat to livelihoods than they do in South America or Asia.

2 Effects of weeds on cassava cultivation: an overview

2.1 Labour requirements to weed cassava

Weed control consumes the second highest level of labour input among crop management operations in Asia, from a low of 13 person-days/year in Malaysia and the Philippines, to a high of 97 in Tamil Nadu, India. A CIAT survey in Asia indicated inadequate weed control in about 37% of the area planted, which would translate to an overall potential yield increase of about 7% if weed control was adequate (Henry and Gottret, 1996). Currently, most weed control is manual, but herbicide use is increasing in all countries, and is most widespread in Thailand. From India, Nedunchezhiyan et al. (2013) reported that weed control in root and tuber crops absorbs about 30% of the labour time, equivalent to 150–200 person-days ha^{-1}. In Africa cassava is largely hand weeded, consuming between 13 and 80% of the total labour required to grow cassava. It should be noted, however, that the operations being counted as part of cassava growing may be different for the different sources. Data on the actual hours or days of labour spent on weeding are rare and some reports are rather questionable.

In southern Cameroon Hauser (unpublished) measured the time spent on all operations in a cassava/maize system in large plots and found that crop establishment – including fallow slashing, burning and planting – required 309 hours ha^{-1}. The first weeding, conducted by hand and with short-handled hoes, took 197 hours ha^{-1}, or 38.8% of all labour conducted up to the time of first weeding. Labour proportions provided by Nweke et al. (2001) were much lower – ranging from 13 to 18% in five African countries (Table 1). However, the data comprised land clearing and harvesting as part of the operations, causing the proportion used for weeding to appear low.

On-station research on weeding small plots (>70 m^2) of cassava with different implements showed rather small differences among the implements used, including motorized weeders. For two weedings motorized weeders required 110–130 hours ha^{-1} and manual implements ranged between 120 and 130 hours ha^{-1}. Assuming 3–4 weedings per cassava crop these numbers might be 50–100% higher for the annual cycle. These data are much lower than those from Cameroon, yet intercrops are more difficult to weed as the distances between plants are shorter and more care needs to be taken, while in a cassava monocrop the distances are large (1 m x 0.8 m in this case) and the use of a short-handled hoe is

Table 1 Cassava production by task and days per ha in Congo, Côte d'Ivoire, Ghana, Nigeria, Tanzania and Uganda

Task	Congo	Côte d'Ivoire	Ghana	Nigeria	Tanzania	Uganda
Land clearing	66	53	44	49	54	45
Seedbed preparation	21	29	31	41	27	31
Planting	39	22	28	32	27	28
Weeding	27	28	34	38	28	32
Harvesting	48	44	53	62	46	52
Total days	201	173	191	222	182	187
Weeding % of total	13.4	16.2	17.8	17.1	15.4	17.1

Source: Nweke et al. 2001.

faster. In large plots (>1250 m^2) and at advanced weed infestation (> 1.5 Mg ha^{-1} weed dry matter) weeding with motorized roto-tiller type machines was up to 4 times faster than weeding with the hand hoe. However, time requirements alone are not a sufficient measure of weed control. In addition the human factor affecting the level of effort, care and neatness must be considered, yet it is difficult to measure and express quantitatively.

2.2 Cassava root yield losses caused by weeds

Estimates of yield losses across all major crops through interference of weeds appear largely rather old and vague, with low estimates at 5% in developed and 25% in developing countries (Vissoh et al., 2004). If no weed control measures are taken, losses across all major crops were estimated at 30% (Oerke et al., 1994). Root yield losses in cassava are highly variable due to the wide range and diversity of cassava growth types, the ecological conditions and the weed communities present. Table 2 gives an overview of the yield losses determined in trials across major cassava-growing countries. Valadao Silva et al. (2013) reported that under greenhouse conditions even the dry matter distribution within cassava plants was altered towards more shoot dry matter in the presence of certain weeds.

Weed density and composition are variable and any research comparing root yields under specific weeding regimes with un-weeded 'controls' is prone to such variability. Recent trials in Nigeria conducted on land that was under bush fallow for about five years and cropped for the second time with cassava showed that un-weeded cassava was producing root yields as high as those in plots weeded four times. This is, however, dependent on the variety, which in this case was a vigorous CMD-tolerant, erect genotype and the weed flora largely composed of small-growing species at low density.

2.3 Critical phases of weed interference

Un-weeded cassava has to be considered the worst-case scenario. In actual farming systems it is not an option to leave the cassava crop un-weeded from planting until harvest. For farmers the two essential types of information are the optimal timing of weedings and the minimum number of weedings required to keep the crop minimally or unaffected by weed interference. The timing appears to be of higher importance than the number of weedings, as most crops have so-called critical phases during which weed interference is most detrimental.

Table 2 Root yield losses in cassava across several countries

Country/region	Yield loss (%)	Notes	Reference
	60–100		Kropff and Moody, 1992
Colombia	0–60		Leihner et al., 1980
Nigeria	80		Chinaka et al., no date
Southeast Nigeria	95		Moody and Ezumah, 1974
Southeast Nigeria	25–59		Unamma et al, 1986
Southwest Nigeria	94–97	*Mimosa invisa*	Alabi et al., 2001
Uganda	75		Fermont et al, 2009
Kenya	62		Fermont et al., 2009
Cameroon	90		Ambe et al., 1992
Colombia	95		Doll and Piedrahita, 1973
Brazil, Bahia	100		Silva Soares et al., 2016
Venezuela	92		Barrios, 1973
Thailand	46		Harper, 1973
Nigeria	45		Onochie, 1975

In the rainforest zone of Nigeria, Onochie (1975) reported that the presence of weeds caused yield losses at any time, with the most damaging phase at around three MAP, when canopy closure was attained and starch accumulation in storage roots began. In a *Mimosa invisa*-infested site in Southwest Nigeria, Alabi et al. (2004) established that cassava recovered from interference if weeded by 35 DAP and did not suffer yield losses if weeding was not continued after 77 DAP. From Cameroon Ambe et al. (1992) reported highest cassava root yields when weeding was done at 28, 84 and 140 DAP; weeding at 28 and 84 DAP doubled yields compared with un-weeded plots. In Brazil, Albuquerque et al. (2008) determined the critical phase to be from 25 to 75 DAP. Delayed weeding to 50 DAP caused close to 50% root yield loss. In Brazil Costa et al. (2013) determined the critical phase starting between 66 and 88 DAP and lasting up to 71–91 days for root and starch yield, respectively, when assuming an acceptable yield loss of only 1%. When a 5% yield and starch loss was accepted, these phases started later (87 and 88 DAP) and lasted longer (up to 80 and 81 days). They established strong negative correlations between the length of time cassava was exposed to weed competition and the yield (r^2 = 0.85 for roots and 0.75 for starch). In the greenhouse Valadao Silva et al. (2013) determined the critical phase around 75 days after planting.

For farmers, knowledge of the critical phase indicates the time during which weeds need to be eliminated and thus the time when weeding needs to be conducted. The critical phase is determined by a large number of factors, for example the cassava variety, soil conditions, fertilizer use and other measures of crop husbandry, and the weed flora, and among others. The optimal timing of weeding operations depends on when the crop is most susceptible to interference of weeds on crop growth and development, which means the initiation of storage root bulking. Due to the variable number and importance of different factors affecting the critical phase, it has no fixed time window.

Therefore, the correct timing of weeding operations needs to be determined in a site- and condition-dependent manner to reliably support farmers. A further shortcoming of the approach using the critical phase is that most of the research that produced the data is rather old and therefore would not necessarily apply to the weed flora currently present, to the preferred cassava varieties and to possibly poorer soil fertility conditions that prevail today. None of the conditions can be assumed to have remained constant. Specifically, in many African cassava-growing countries, cassava varieties have been changed due to the CMD pandemic, exotic weeds have invaded fallows, and crop yields and soil fertility have declined. Land use intensification through higher cropping frequencies has likely caused larger weed seed banks, all circumstances leading to worsened growth conditions for cassava. Unlike the changes in weed flora, seed banks and soil fertility, the varietal changes triggered by the cassava mosaic pandemic may have a positive impact on the competitiveness of cassava. CMD-tolerant or -resistant varieties are usually vigorous and form dense canopies relatively early, thus are more likely to compete strongly against weeds. Recent research in Nigeria showed higher weed biomass in an older, less CMD-tolerant variety than in a highly CMD-tolerant variety (Fig. 1).

Changes in the weed flora, specifically appearances of invasive species, do not affect the critical phases, yet may require changes in the timing of weeding operations. In Nigeria the Mexican sunflower (*Tithonia diversifolia*) has established in fallows and is invading crop land. The species grows faster than cassava and maize and most common weeds. Controlling *T. diversifolia* thus requires weeding earlier than is currently recommended. The usual weeding regime starts at four weeks after planting, yet with *T. diversifolia* infestation the crops were overgrown, etiolated and suffered 'sunburn' after weeding as they were not exposed to full sunlight before the weeding (Fig. 2). Because farmers will not be able to keep the crop weed-free for any longer period and because the weed flora determines the timing and frequency of essential weeding operations, the appearance of new weed species requires the verification of the optimal weeding calendar or regime.

The following sections discuss different options to control weeds and then relate these options to approaches that would best enable farmers to remove weeds with the smallest labour requirement, at the lowest cost and with the lowest negative environmental consequences.

Figure 1 Cassava variety TMS30572 (left) and TME 419 (right) at about 2 months after planting. Both varieties were planted in the same week at Moniya, Oyo State, Nigeria. Photo: S. Hauser.

Figure 2 Cassava/maize intercrop weeded from *Tithonia diversifolia* at 4 weeks after planting at Moniya, Oyo State, Nigeria. Note the strongly etiolated maize and small cassava plants. Weeding with tools was not possible in this situation. Photo: S. Hauser.

3 Cultural control measures: land management before planting

Cassava agronomy has not received adequate attention over the last 30 years, specifically in Africa, where root yields are the lowest in the world. Considering that poor cassava crop husbandry may promote weed growth, there should be sufficient scope to reduce weed competition through improved agronomic measures, which will as well increase cassava yields. How strongly agronomic measures can affect the weeds, and if weed biomass and competition are actually reduced, will be discussed.

Farmers have several options to improve the competitiveness of a cassava crop against weeds:

1 land clearing and fallow biomass management;
2 use of cover crop fallows;
3 tillage regime before planting;
4 cassava plant density and to some extent planting pattern;
5 use of an intercrop to increase shading and competition;
6 use of fertilizer to promote cassava canopy growth;
7 choice of cassava (and intercrop) variety.

While each option may have a positive effect on cassava yield, the factors may interact and thus amplify or reduce the effects on weeds and cassava yield.

3.1 Land clearing and biomass management

In areas where fallow is still practised and where farmers have good choice of land, a set of indicators is used to identify suitable land for cropping (Norgrove and Hauser, 2016). Although in some areas farmers use crop-specific indicators, there is no information on cassava-specific indicators. Generally cassava should be planted on well-drained (not prone to waterlogging) deep soils. The soils should be stone-free or at least not have pronounced stone layers. Other criteria are listed by Hauser et al. (2014). The land clearing method may affect the level of weed infestation. However, this largely depends on the age of the fallow that is being cleared: the older the age, the lower the risk of weed infestation (de Rouw, 1995). Generally, clearing methods that do not mix or move the topsoil are better suited to prevent weed problems from seed-borne weeds, as the seeds will remain at the surface. Bulldozer clearing is thus not recommended. If trees need to be removed on land earmarked for mechanized soil preparation, other methods can be used such as uprooting with an excavator, that is, pulling trees with a chain. Although this will only work up to a certain tree diameter, the method is relatively fast and does not disturb the soil as much as bulldozer or bucket loader clearing. If the remaining biomass is burned, superficial weed seeds and vegetative propagules will be destroyed. If the entire biomass, including tree boles, is burned, the fuel load is high within small areas and the soil temperatures may get high enough to eliminate vegetative organs (Hauser et al., 2006), yet this will not be the case for larger areas within the clearing. After severe burns, such as under boles, weeds may remain absent over a long portion of the growing period (Menzies and Gillman, 2003).

Weed species usually found in the cropping phase are normally eliminated over long fallow phases (De Rouw, 1995; Akobundu et al., 1999). With shorter fallow phases the survival chances of weeds, and the risk of crop infestation, increase. In today's dominant short fallow systems, the fallow often comprises weeds of the cropping phase or of fallow species that remain troublesome in the cropping phase. In SSA *Chromolaena odorata* is such a species that forms thick fallows and is a weed in the crops.

With farmers' limited choices of fallow land, the land clearing method may not be as decisive as the fallow biomass management. In fallows that have not attained secondary forest stage, depending on the fallow age and type, the slashed material contains weed seeds at varying levels, and the soil holds root stocks and other vegetative organs from which undesired species can emerge. If the slashed material is dry and burned, it is likely that all seeds are killed, yet heat penetration into the soil is usually not deep enough to destroy underground organs across a larger area. Burning the biomass of short fallows (<15 Mg ha^{-1} DM biomass) does not damage the soil and is effective against weed seeds.

3.2 Cover crop fallow

Although cover crops never attained large-scale adoption in Africa, their effect on specific problem weeds such as *Imperata cylindrica* has been shown in cassava systems. *Pueraria phaseoloides* and *Mucuna pruriens* var *utilis* (Fig. 3) are probably the most common cover crops in the cassava-growing area of the tropics.

Figure 3 A well-developed *Mucuna pruriens* var *utilis* cover crop seeded in a grass fallow. The former grass tufts are still visible yet overgrown by the mucuna. In the front some grass blades are still visible, Kengele farm near Kiasungwa, Bas Congo, DR Congo. Photo: S. Hauser.

From Ghana, Aflakpui and Bolfrey-Arku (2007) reported the complete absence of *I. cylindrica* in three out of four sites after one and two years of *Mucuna* fallow before planting cassava. Cassava yields were higher where *Mucuna* was grown, compared with natural fallow and a treatment where *I. cylindrica* was controlled with glyphosate. In Nigeria, Ekeleme et al. (2003) established that cover crops reduced not only the weed biomass but also the weed seed banks and thus the potential of future weed infestation. Similar results were reported by Akobundu et al. (1999) from *P. phaseoloides* fallows; yet strong reductions of the weed seed bank were attained only when fallowed for three years. The main advantages of the cover crop are a combination of weed control and soil fertility improvement. However, cover crops are usually strong competitors and if they have a climbing habit, may be incompatible with cassava. In situations where herbicide use is not an option, only dead mulch systems in which the cover crop is grown and eliminated before planting cassava can be recommended. Controlling or eliminating a cover crop may require early slashing or incorporation into the soil to prevent seed production. *Mucuna*, for instance, produces large seeds that would likely germinate under a cassava crop even from deep layers after ploughing, thus becoming a weed problem. The weed climbs and can cover the crop within a short time. Weeding mucuna or any other climbing weed in cassava is labour intensive and if not done in time will cause damage to the crop, specifically if cassava is low and profusely branching.

Live mulch systems, for instance with *P. phaseoloides*, despite their advantages specifically under high rainfall conditions, require continuous control which may surpass smallholders' labour capacity. These systems were not developed for smallholders but for large-scale plantations such as rubber, to be managed by mechanical slashing. In smallholder systems, species such as *Pueraria* are generally not appreciated. An earlier review of planted fallow systems showed that cassava did not respond positively to cover crops, being the only crop with generally negative yield responses (Hauser et al., 2006). However today, with herbicides more affordable, live mulch can be managed by spraying sub-lethal doses of systemic herbicides that stop the growth for several months and permit cassava to form a canopy, further preventing the cover crop from climbing into cassava (Duindam and Hauser, 2008). The approach is labour-saving (Duindam and Hauser, 2011) and permits higher crop yields and incomes (Duindam et al., 2010). Other cover crops, especially non-climbers, may be a better option in cassava systems but they have not yet received sufficient attention to make recommendations.

3.3 The tillage regime before planting

Tillage regimes for cassava are widely variable and optimal systems to a large extent depend on soil fertility status, soil texture, availability of tractors, and weed control requirements and options. Generally, tillage – specifically the process of mixing soil when the soil is turned – will bury weed seeds and distribute them over the tilled soil profile. Weed seeds may fall dormant and remain in the soil for several years without germinating even when moisture and temperature conditions are conducive. The seeds can germinate at later stages when they are brought close to the surface by tillage. This is an evolutionary mechanism for weed seeds to remain viable in the soil. On no-till soils, contrary to tilled ones, the vast majority of weed seeds will remain on the soil surface or in the litter and mulch layer. There they are more exposed to wetting and drying and less likely to fall dormant. In the case where the biomass is burned, most of these surface layer weed seeds would be killed.

Many smallholders cannot afford or do not have access to tractors and therefore need to till manually. On soils considered sufficiently fertile and friable, farmers usually do not till unless they need to remove problem weeds such as *Imperata*. On soils of low fertility, farmers tend to till more intensively and to ridge or form mounds to concentrate topsoil around the planting stake. Weed control would be a desired side effect, since on poorer soils weed infestation may be more detrimental than on fertile ones. Fertile soils are usually those that were under fallow, and effective weed control can be done by burning; thus tillage by hand is not required. On soils prone to erosion, tillage should be minimal. If problem weeds are present and need to be controlled by manual means, tools that rake rhizomes or other vegetative parts out of the soil should be used. Biomass should be kept on the soil surface, as long as there is no risk of reestablishment, that is, only if they dry and die.

On land that can be tilled with tractors, it is common to plough and harrow before planting. Such land preparation is usually followed by herbicide-based weed control at later stages. On already degraded soils that have a large weed seed bank or perennial weed species, ploughing and harrowing are often followed by ridging (can be done by tractor) or by mounding manually. After ridging, and when cassava is planted on the crest of the ridge (Fig. 4), post-emergence herbicide application is facilitated. In this situation the furrow can be sprayed even with chemicals that would damage cassava, because the risk of contamination by herbicides taken up through the leaves is rather low.

Figure 4 Properly tilled and ridged cassava field, staying weed-free into the canopy development phase. In this example, cassava planting stakes are inserted vertically, allowing weeding tools to operate very close to the stake. Photo: S. Hauser.

On clay-rich soils the planned time of harvest needs to be considered: cassava is difficult to harvest from flat, clayey soils under dry conditions, and ridging is appropriate. Studies in Nigeria showed that ridging significantly reduced the weed biomass in cassava systems (Hauser et al., 2015). However, the effect on cassava root yield cannot be directly related to a reduction in weed growth since other edaphic factors are involved. On grass weed-infested, coarse-textured soils that are relatively easy to till manually, farmers often form low seedbeds or flat ridges by scraping topsoil with hand hoes and piling it on top of untilled soil. When forming these ridges farmers turn the soil from the furrow to the top of the soil forming the base of the ridge. They do this by turning the soil upside down and thus cover the weeds and expose the weed-free soil on top of the ridge. These manually formed flat ridges often show some weed growth along the line where the former soil surface is exposed, but less in the furrows from where the soil was removed and on the ridges where weed seeds were buried. It is not clear if the primary purpose for this tillage, from the farmer's perspective, is weed control, accumulating the more fertile topsoil or simply arranging the field to facilitate planting in lines or rows. It is, however, likely that this type of tillage has a stronger controlling effect on the weed growth than tillage conducted by tractor-drawn disc or mouldboard ridgers, because the manual ridging places the soil more precisely on top of the weeds, thus covering them more thoroughly. Tractor-drawn ridgers let the soil fall in place in an irregular manner, do not completely turn it upside down and therefore do not fully cover the weeds. Overall the intensity and benefits of the tillage regime in combatting weeds need to be balanced against the labour requirements (and thus the costs); the potential negative effects on the soil structure and consequent erosion, SOM and nutrient losses; and the ease of harvesting on clay-rich soils. Such complex interactions among the factors affecting weed growth and control may be difficult to assess and thus farmers' decision-making is often based on a few factors connected with the largest constraint.

Generally it can be recommended to plough and rogue on land infested with perennial species sprouting from vegetative underground organs, followed by harrowing and ridging. Ridging will not be necessary on light-textured soils as they do not pose as high a risk of waterlogging or problems of difficult harvest under dry conditions. On erosion-prone soils with perennial weed problems, the direction of ridging has to be across the slope to minimize run-off and gully erosion.

4 Cultural control measures: planting and crop management

4.1 Cassava plant density and planting patterns

Cassava densities vary widely in different systems, especially between cassava sole crops and intercrops (the latter is treated in the next section). From Brazil, Santos Silva et al. (2013) reported plant densities of 5000 ha^{-1} in typical local farms. They showed that root yields could be more than doubled by increasing densities to 13 000 ha^{-1}, and yet did not obtain any data on weed problems or management. Farmers in Africa have a tendency towards low cassava densities of 5000–10 000 plants ha^{-1}. Most farmers do not replace planting stakes that fail to sprout and to establish. Ayoola and Makinde (2008) reported highest cassava root yields at relatively low densities of 6666–8888 ha^{-1} from Southwest Nigeria, but with a variety no longer frequently used. They did not look into weed growth and biomass. Such low densities, combined with cassava's slow formation of a canopy, create favourable conditions for weeds in the early stages. Even with timely weeding, shading remains incomplete for a long phase (typically 3–4 months) and allows weeds to grow. Recent trials in Nigeria showed some reduction of weed biomass with higher cassava plant densities (Fig. 5), but the effects of tillage and cassava variety were stronger than the effect of increased density. The different branching types of cassava showed strong differential effects on the weed biomass, but results are not consistent. For example the low and profusely branching TMS 30572 did not outcompete weeds as well as the non- or only very late-branching variety TME 419.

Considering that cassava growers often have multiple production objectives that are regulated to some degree through plant density, this measure appears less suitable as a general weed control approach, compared to alternatives. Older sources report that yields obtained at different plant densities depended on the weeding method and timing. In weed-free cassava (herbicide application), highest yields were attained at lower plant densities, while in once or twice hand-weeded plots, yields kept increasing with increasing plant density (Doll and Piedrahita, 1973).

4.2 Intercropping

Many farmers perceive that in intercrops the density of one or both crops (or all, in case of more than two crops) needs to be reduced to provide 'space'. At higher cassava densities, the potentially positive effect of intercrops on weeds may be obliterated. Further, an intercrop plays the same role as weeds: it is a competitor. Thus, even if in an intercrop system the weeds are better controlled, the effects on cassava yields may not be positive. Another consideration in intercrops is that the use of herbicides is more complex, since both crops need to tolerate the chemicals.

In south-eastern Nigeria intercropping cassava with maize, or with maize plus lima bean, plus African yam bean and plus mucuna, did not reduce weed mass at four or eight WAP, and only marginally at 12 WAP (Ibeawuchi et al., 2007). Cassava yields were generally lower when intercropped. Unamma et al. (1986) intercropped cassava with cowpea, egusi melon, groundnut and sweet potato and measured resulting weed control and yield. Cassava root yields with cowpea and egusi melon were as high as in cassava monocrop, but were reduced when intercropped with groundnut and sweet potato. Weed control was best in egusi melon followed by groundnut and cowpea, with least control in sweet

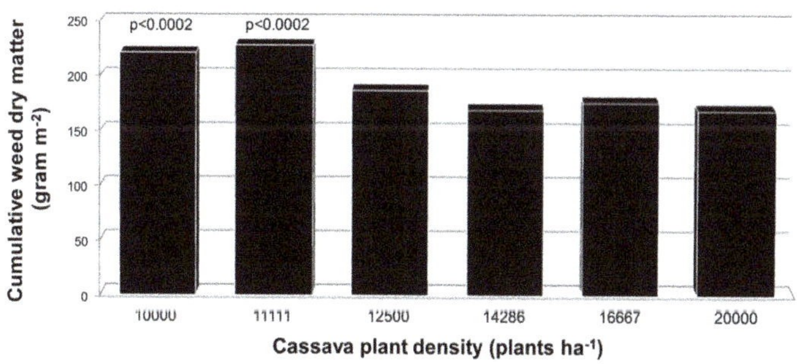

Figure 5 Cumulative weed dry matter sampled over four weedings under two cassava genotypes in two locations. *P*-values above columns indicate the level of significance compared with columns not labelled. Source: Hauser et al. (2015).

potato. Thus weed control and cassava yields were not well correlated, since intercrops that did not control weeds very well still permitted high cassava yields.

Recent research in Nigeria did not confirm that weed biomass is reduced in a cassava/maize system in which the cassava density ranged from 10 000 to 20 000 plants ha^{-1} and maize was seeded at 40 000 plants ha^{-1}. This surprising result may be partially related to the relatively difficult situation in dense crops when weeding manually. As such the weeding may not be conducted as thoroughly as in crops with ample space to manoeuvre the tools.

In India intercropping with cowpea (*Vigna unguiculata*) reduced weed biomass to a greater extent than intercropping with maize (Amanullah et al., 2006) and cassava root yields were only marginally reduced. From the same region, farmers' perception of intercropping leading to less labour for weeding has been reported (Harms, 2015).

4.3 Fertilizer application and other nutrient sources

Cassava develops slowly and requires three months or more to form a canopy that shades the soil completely. This provides a long window for weeds to germinate and grow. Fertilizer is usually applied between planting and four WAP, thus at a time that cassava has not closed the canopy. Therefore, fertilizer application needs to be scheduled in synchrony with weed control to avoid weeds accessing the nutrients and attaining more competitiveness. For the African cassava-growing countries, no recent and reliable fertilizer recommendations are available, which means that the basic information on fertilizer use to increase the competitiveness and yield of cassava is missing. Such lack of information is not surprising due to the site specificity of fertilizer recommendations which cannot be tackled with conventional trial approaches. A recently launched project is now conducting GIS-supported, structured, multi-locational research on fertilizer recommendation and the development of decision support systems for Nigeria and Tanzania.

Currently fertilizer should not be looked at as a measure to support weed control but should only be used if weeds are unlikely to interfere significantly with cassava.

Other nutrient sources, largely manure and other organic materials, might not have the same effects on weeds and the crop as does fertilizer due to slower release of nutrients. In addition there is a risk that organic materials contain viable weed seeds and lead to more weed infestation. Amanullah et al. (2006) reported lower weed dry matter production in sole and intercropped cassava when amended with poultry manure or composted poultry manure, related to faster canopy growth and closure.

4.4 Cassava varietal choice

A wide range of cassava growth types are being grown in Africa and new genotypes are being released every year. The perception in many farming communities is that early (low) and profusely branching varieties are more competitive and close the canopy earlier. Cassava varietal choice has to be looked at in connection with the farmers' production objectives and the planting density. Non-branching or late-branching varieties can be planted at higher densities and thus may attain canopy closure as early as more highly branching ones at lower densities. Manual or mechanical weed control in low and profusely branching cassava varieties may be more difficult once branching has started, since the branches will impede movement of tools and people. Erect, non- or very late-branching types allow workers to move between plants even at advanced growth stages without major risk of damaging the plants. However, the principal method of weed control may also affect the impact of branching habit (and thereby, of varietal choice). For example, herbicide-based weed control may be concluded before branching is initiated, depending on the time period the herbicide remains effective. Manual weed control, depending on rainfall and weed flora, may need to be conducted over a longer period and thus requires entering the crop at more advanced stages, thus posing problems for movement and causing damage.

Recent research in Nigeria showed that a vigorous erect cassava variety, with several main stems per planting stake, controlled weeds better than a less vigorous low branching variety (Fig. 6). Thus, apart from yield and root properties, early vigour appears to be the primary criteria, and branching habit may be a secondary consideration when choosing the variety. Similar results were presented by Bolfrey-Arku (2011) from Ghana, where a non-branching variety out-yielded a branching one by 7–12 Mg/ha, without a strong impact of either plant density or the start date of weeding. There is accumulating broad evidence that earlier recommendations to choose early and low branching cassava varieties to control weeds (Melifonwu et al., 2000) can no longer be upheld.

With the spread of the CMD pandemic in Africa, varietal choice has become more important. Resistant varieties not only prevent major yield losses, but also reduce exposure of the soil to solar radiation, allowing weeds to grow with little or no competition to produce seeds or establish vegetative organs. The failure of CMD-susceptible cassava varieties to develop a canopy (Fig. 7) may lead to aggravated weed seed production and the establishment of perennial weeds.

Figure 6 Two varieties of different vigour, branching habit and CMD tolerance: foreground right – susceptible, low branching less vigorous variety TMS 30572, and background left – highly CMD-tolerant, non-branching and vigorous variety TME 419; Moniya, Oyo State, Nigeria. Photo: S. Hauser.

Figure 7 Three cassava varieties of different CMD tolerance: foreground – highly susceptible local variety Kingabwa; background, left – highly tolerant variety Butamu; and background, right – variety Zizila; Kiyaka, Province Bandundu, DR Congo. Photo: S. Hauser.

5 Mechanical weeding options

5.1 Manual hand tools

The most common hand tools to control weeds are probably the short-handled hoe, as used in most of Africa; the long-handled hoe, more common in Asia and South America, but rather uncommon in Africa; and the cutlass. The short-handled hoe serves as a weeding tool in practically all crops and at the same time it can be used for shallow tillage such as ridging or mounding soil around a crop. This versatility may be the reason for its frequent use in Africa, where complex intercrops of three or more species are common (Büttner and Hauser, 2003). The disadvantage of the short-handled hoe is the requirement to bend down and to work in this position. The long-handled hoe allows working in an upright position, thus being less strenuous, but it requires the skill of manoeuvring the hoe blade precisely to the weeds at a large distance. It is less effective in shallow tillage, and thus serves mainly for the cutting of weeds at or marginally below the soil surface. The cutlass can be used in wide-row crops such as cassava monocrops. It may be faster than a short-handled hoe, but for good weed control the farmer needs to bend down, as with the hoe, to cut at the soil surface. Generally cutlass weeding leaves weed stems and stumps and thus there is the risk of rapid regrowth under suitable conditions.

Alternative hand tools are rotary tools on a long handle, either composed only of rotating spikes or in combination with flat blades. These tools originated from rice systems and had to be modified to work on upland soils. Recent research in Nigeria, comparing several manual tools, showed that the short-handled hoe required the least time to weed and produced highest cassava yields. The rotary weeder with flat blades and the so-called spike weeder were least effective in destroying weeds, required more time and produced lower cassava yields.

It has to be assumed that the use efficiency of any implement is strongly skill-dependent and farmers used to the short-handled hoe would work with it most efficiently. Therefore, the comparison has to be looked at with caution: the operators may have lacked the skills to handle the other implements optimally. However, the rotating tools were not suitable to remove advanced grasses and could not work close enough to the cassava stems to cut the vines of climbers. Such shortcomings are tool- and not skill-related, rendering the tools less effective.

5.2 Motorized tools

Two motorized tools appear useful in weed control in cassava: The brush cutter – a machine with a fast rotating, horizontally mounted blade or nylon string; and the roto-tiller with modified tines – a machine with vertically rotating tines mounted on a drum.

The brush cutter (Fig. 8) functions in principle in the same way as a cutlass and cuts weeds at some height above the soil surface. These machines are light and therefore allow weeding with minimal physical effort for long periods. Shortcomings include initial cost and ongoing operational and maintenance costs, such as the requirement to replace blades or nylon strings as they wear, depending on weed density and toughness (e.g. stem diameter and level of lignification) and soil texture (faster wear on sandy abrasive soils). An advantage is their quite versatile use in bush and grassland clearing. In recent trials in Nigeria the weeding quality of the brush cutter was highly operator-dependent and thus a function of skills. Many operators did not lead the blades low enough to cut at the soil surface and weeds regrew rather quickly. Skilled operators were able to cut right

Figure 8 Brush cutter mounted with a solid steel blade.

Figure 9 Narrow (left) and wide (right) Mantis® roto-tiller with original long tines to till.

at the surface, which destroyed more weeds permanently and ripped out smaller ones, leading to overall less regrowth.

The roto-tiller (Fig. 9) has to be modified since the normal tines would likely destroy storage roots at later weeding stages. The tines need to be limited in depth to about 2–4 cm to prevent rapid digging-in of the machine. In principle the action of the tines is a surface tillage, achieved by breaking the entire soil surface and mixing it, thereby uprooting and cutting up the weeds. The machine can remove older weeds and grasses by repeated passes. However, operators need to take care not to let the tines penetrate too deep into the soil, since the risk of storage root damage increases with every passage of the machine. Operating and steering the machine requires some physical strength and skills since any failure to control the machine's movement may break and destroy the cassava plants. An advantage of the roto-tiller is that it can be used for other operations such as tillage and ridging. The machine appears suitable for medium-sized farms and for weeding service providers.

5.3 Tractor-drawn machines

Due to the slow development of cassava in the early growth stages, mechanical weeding is an option as long as the tractor wheel clearance permits passing over or between the plants without causing breakage. Thus, if mechanical (or herbicide-based) weed control is intended with tractor-drawn implements, so-called 'high wheelbase' tractors should be used. Generally, weed control with tractors requires planting in straight rows at constant inter-row distances.

Cassava, unlike many other row crops, grows storage roots that may be shallow below the soil surface. These storage roots are initiated at around 6–12 WAP and any mechanical weeding needs to be very superficial to prevent their damage or destruction. As most, if not all, mechanical weeders were developed for other row crops or for other purposes such as breaking soil crusts, it is essential to adjust the working mode and depths to cassava crops. There are three basic types of tractor-drawn weeding implements: (1) blades suspended from either rigid or flexible tines being pulled horizontally through the soil; (2) rotating spike or finger discs or wheels, often rolling at an angle to the direction in which they are pulled; and (3) flexible, spring-mounted, low-diameter tines pulled through the soil surface either simply following the tractor, or powered by the PTO shaft to oscillate perpendicularly to the direction in which the machine is pulled.

The simplest weeding tools are so-called 'duck foot blades' – triangular horizontally drawn blades that run just below the soil surface, cutting off weed roots (Fig. 10). These tools only work sufficiently in the inter-row space and do not control weeds close to the cassava stem and within the cassava row unless the cassava is planted on a square pattern and at very exact distances so that the machine can be drawn both along and across the rows. The quality of weed control depends on the skills of the tractor operator and the variation of the within- and between-row distances between plants. Duck foot blades are offered in a wide range of shapes and sizes, but they are generally triangular and have a range of lifting angles, that is, to lift and crumble the detached soil to a greater or lesser degree. For weed control, the lifting angle may not matter too much, but on wet clayey

Figure 10 Duck foot weeder mounted on flexible S tines, with a parallelogram hinge to attain a large distance between the tool and the draw bar. Photo: Vitorio Fadel.

Figure 11 Left photo: rear view of rotating spike wheels (red and black) mounted behind a set of finger wheels (yellow). Right photo: front view of the same machine with two preceding duck foot blades on S tines to break the soil before being rolled over by the spike wheels.

soils, stronger crumbling through higher lift of the soil may better prevent weeds from re-rooting after cut-off. Their use is limited to soils that were not ridged.

Rotating spike or finger wheels (Fig. 11) mounted behind tractors are drawn in the inter-row space with the rotating tools' axis running at an angle to the direction of the tractor's movement. The wheels roll over the soil and the fingers penetrate the soil surface, ripping soil and weeds from the surface. Usually two or more such wheels are mounted behind one another, each with the axis at an angle balancing the angle of the previous wheel. These tools are very efficient and best used when weeds are recently emerged, are still small and have not developed a strong root system. A side effect is the multiple benefits of breaking of the soil crusts, including better penetration of rainfall. Another advantage of these tools is that they can be deployed on ridges, as the tools can be mounted at an angle that would fit the slope of the ridge.

A major advantage of these machines is that they can be used in advanced cassava crops because the bar on which the wheels are mounted can be lowered or suspended from the tractor drawbar and thus attain a larger clearance than machines on which the implements are mounted directly to the drawbar.

Flexible spring-mounted tines resembling simple harrows are pulled through the shallow soil surface, penetrating by a few centimetres, breaking the soil and ripping weeds from the soil. The tines usually vibrate from the forward movement and thus work sideways as well. The tines can be mounted in such a way that they gyrate, thus increasing the range in which they break the soil surface and rip weeds. As for the rotating tools, these implements are most effective when the weeds are still small and can be easily damaged or pulled from the soil. They can be used along and across rows if cassava is planted on a regular square pattern. These machines are limited in use by the length of the tines in relation to the height of the cassava. Low diameter tines cannot be used in already advanced cassava crops as the frames on which they are mounted would break the plants.

Today many mechanical weeding implements can be steered with sensor/camera support, thereby they can get closer to the cassava row and remove a larger proportion of the weeds without the risk of causing damage to the crop.

5.4 Other physical weed control options

In organic farming, weed control is in some cases conducted by flaming the soil surface of the inter-row space. In cassava this might be possible but does not appear to be practised.

Figure 12 Sensor/camera-controlled self-steering robot for horticultural applications. Weeding tools would be mounted under the crossbar.

In India cassava is frequently grown with plastic or textile sheets between the cassava rows, preventing most weeds from emerging or from growing to the stage of seed production. Due to the high cost of plastic mulch this option may be limited to places where cassava is a high price commodity or where yields are sufficiently high that the plastic mulch is profitable. The advantage of the system is that in combination with fertigation both weeds and evaporation are controlled.

The latest developments in mechanical weed control are robots that are self-steering through GPS and sensor/camera support, mounted with tools appropriate for the respective crops (Fig. 12). These machines were developed for high-value horticultural crops and their current costs may be prohibitive in cassava production. However, with time these technologies may find their way into 'ordinary' crops.

6 Chemical weed control

Chemical weed control in cassava involves the use of pre- and post-emergence herbicides to kill weeds, and has been shown to be very effective if applied correctly (Chikoye et al., 2006; Ekeleme, 2013). Although manual weed control is the predominant method of controlling weeds in cassava in smallholder farmers in major cassava-producing countries, chemical weed control in cassava using herbicides is becoming important. Table 3 presents a list of herbicides commonly used in cassava. In Nigeria, a recent survey of smallholder cassava farmers in one of the major cassava-producing states shows that 68.9% of farmers use herbicides to control weeds in cassava (Odoemenem and Otawa, 2011). About 9.5% of these farmers used different formulations of pre-emergence herbicides while 69.0% used post-emergence herbicides (glyphosate: 52.5%; paraquat: 16.4%) in different formulations.

6.1 Pre-emergence herbicides

Chemical weed control in cassava starts with the application of pre-emergence herbicide within three days after planting on well-prepared soil. Pre-emergence herbicides at the

Table 3 List of some herbicides evaluated and used for weed control in cassava crops in Africa, Asia and South America

Active ingredient	Country	Mode of use
Acetochlor	Brazil, Nigeria	PE
Alachlor	Brazil, Nigeria, India, Ghana	PE
Alachlor + Cyanazine	Brazil	PE
Alachlor + Atrazine	Nigeria	PE
Ametryne	Brazil, Nigeria	PE
Clomazone	Brazil, Nigeria	PE
Clomazone + Ametryne	Brazil	PE
Cyanazine	Brazil	PE
Diuron	Brazil, Nigeria, India, Ghana, Colombia	PE
Diuron + Paraquat	Nigeria, Colombia	PS
Fluazifop-*p*-butyl	Nigeria	PS
Fluchloralin	India	PE
Flumioxazin	Brazil	PE
Fluometuron	Nigeria, India, Colombia, Venezuela	PE
Glyphosate	Nigeria, India, Brazil	PS
Isoxaflutole	Brazil	PE
Isoxaflutole + Metribuzin	Brazil	PE
Metobromuron	Nigeria	PE
Metolachlor	Brazil, Nigeria, India, China	PE
Metolachlor + Atrazine	Nigeria, Ghana	PE
Metribuzin	Nigeria, Brazil	PE
Oxadiazon	Brazil, Nigeria, India	PE
Oxyfluorfen	Brazil, Nigeria, India	PE
Paraquat	Nigeria, Ghana, India, China	PS
Pendimethalin	Brazil, Nigeria, India	PE
Prometryne + Atrazine	Nigeria, Venezuela	PE
Linuron	Brazil	
2,4-D dimethylamine	Brazil	
Thiazopyr	Brazil	
Trifluralin	Brazil	
Oxadiargyl	Brazil	
Cyanazine	Brazil	

PE – pre-emergence, PS – post-emergence.
Note: This listing does not endorse the use of any chemical nor does it indicate that any chemical is registered for use on cassava in any given country.

appropriate dose and with adequate calibration work better in well-tilled (ploughed and harrowed) soil. Where the field to be cultivated is dominated by perennial and rhizomatous weeds such as *Cyperus* spp., *I. cylindrica* or *Panicum maximum* it may be necessary to kill these weeds with a post-emergence herbicide before tillage. Several factors such as soil characteristics, climate and weed species composition affect the efficacy of any given herbicide when appropriate rates are used. Although several herbicides have been evaluated for weed control in cassava (Table 3) in different cassava-producing countries, no herbicide has been developed specifically for weed control in cassava except fluazifop-*p*-butyl, which is registered and labelled for post-emergence grass weed control. Fluometuron, fluchloralin, metolachlor, atrazine, diuron and alachlor at various doses and tank mixtures have been frequently recommended for pre-emergence weed control in sole cassava in Africa, Asia and South America (Akobundu, 1980; Melifonwu, 1994; Prameela et. al., 2012). Under intercrop situations, metolachlor + atrazine and metolachlor + prometryne have been shown to be effective in controlling weeds in cassava/maize and cassava/grain legume intercrop, respectively.

6.2 Post-emergence herbicides

Because cassava is a long-duration crop staying for 10–18 months in the field before the roots are harvested, pre-emergence herbicide application alone at recommended doses often does not provide long-lasting weed control before cassava closes canopy. Complementary weed control is necessary before the cassava closes the canopy and against 'specialist' weeds that tolerate the shade, such as climbers, causing damage later in the cropping phase. This additional control can be achieved with post-emergence herbicides from four months after application of pre-emergence control. In Africa and Asia (specifically in India) complementary weed control has been achieved with application of paraquat and glyphosate when the cassava is 3–5 months old. In India (Nedunchezhiyan et al., 2013) and Nigeria (Odoemenem and Otawa, 2011) glyphosate and paraquat are commonly used to control weeds in cassava monocrop from three months after planting. Although the use of paraquat has been prohibited or de-emphasized in many counties, cassava farmers in Nigeria and India continue to use this product. The use of glyphosate cannot be recommended during the early growth stages of cassava. As long as green and functional leaves remain on the lower section of the stem, the use of glyphosate is a high-risk venture because even small amounts of the herbicide hitting cassava leaves can cause systemic damage (Fig. 13 and 14).

Figure 13 Three-month-old cassava severely infested with grasses and treated with glyphosate. Note that the weed control was very good except for the grasses within the cassava row. Some cassava plants have yellow leaves which would, at this age, not be a normal condition in cassava, and indicate possible herbicide damage. Mbalmayo, Southern Cameroon. Photo: S. Hauser.

Figure 14 Cassava damaged by glyphosate application. Note the narrow and irregular leaves, especially the younger ones. Mbalmayo, Southern Cameroon. Photo: S. Hauser.

Figure 15 Post-emergence herbicide application with a five-row hooded sprayer. Photo: Vitorio Fadel.

Glyphosate use is thus recommended once the canopy is fully closed and the lower leaves have been shed. The lignified stem of the cassava appears not to easily absorb the chemical and thus damage is less likely. Care must be taken when glyphosate is used to control weeds in cassava. Application should be directed to the weeds with a spray hood to avoid phytotoxicity on the cassava. Spray hoods can be used with knapsack sprayers as well as with tractor-carried boom sprayers (Fig. 15). At or even soon after canopy closure, herbicides can be applied using jets suspended from a boom mounted on a high wheelbase tractor (Fig. 16). The jets reach through and spray under the canopy.

Figure 16 Post-emergence herbicide application after canopy closure with directed jets, suspended from a boom mounted on a high wheelbase tractor. Photo: Vitorio Fadel.

7 Herbicide-resistant cassava

Worldwide there are very few herbicides formally registered for use in cassava. This considerably limits their use, yet many farmers are using herbicides that were found to be effective without obviously damaging the crop. Most of these herbicides are applied pre-emergence and followed by manual or mechanical weeding once the cassava has formed a canopy. Taylor et al. (2004) considered herbicide-resistant cassava as an option with the current production practices. Herbicide-resistant or -tolerant cassava would allow the use of post-emergence herbicides and reduce the labour in weeding. The *ex ante* impact of introducing herbicide-resistant cassava in Colombia was estimated to bring about 15–25% cost savings for cassava producers, with a benefit to Colombia's economy of US$300 million over a 14-year period (Pachico and Rivas, 2003). Additional benefits would be positive environmental effects through a reduction in tillage and manual or mechanical weeding, leading to less erosion and soil loss.

Sarria et al. (2000) successfully transferred a gene into cassava giving tolerance to the herbicide 'Basta' (Hoechst), thus demonstrating that the process is possible. However, the herbicide dose tolerated did not match those required to control weeds.

Today there are no cassava varieties available with such tolerance. The current rather unwelcoming public opinion on genetically modified crops and more frequent use and spread of agrochemicals are likely to affect the development of herbicide-tolerant or -resistant cassava.

8 Conclusion

With cassava gaining importance as a strategic crop, specifically in Africa, weed control will become a major issue in the near future. Agricultural land use frequency and intensity appear to increase in most developing countries, which usually leads to new problems cropping up – weeds are literally such a problem that will become more important and more difficult to tackle. All weed control options have advantages and disadvantages, which require farmers and researchers to look at solutions in the overall context of the cropping, farming and livelihood systems. Cassava growers cannot be offered a simple and safe weed control method that will be effective in all circumstances. To prevent negative consequences of farming and to ensure sustainable land use, good agronomic practice, careful soil preparation, well-informed varietal choice and timely intervention against weeds with appropriate methods need to be combined to attain high yields and high productivity. Farmers will need to acquire more skills, accumulate more knowledge and observe the crops more intensively to manage soil, crops and weeds with minimum costs and minimum damages to the larger environment and consumers. Modern weed control methods such as herbicides may appear an easy solution and yet have caused new problems such as herbicide-resistant weeds. We need to keep all weed control options open to be able to respond to new challenges in crop and specifically cassava production.

9 Where to look for further information

The journals *Weed Science*, *Weed Research* and *Weed Technology* are excellent sources of weed-related research results. *Weed Technology* is the journal with a strong focus on application of weed control methods albeit largely herbicide based. The open access *Journal of Root Crops* accessible at http://isrc.in/ojs/index.php/jrc is a good source of general information on root crops with some weed control articles. For more recent information and updates on current projects in cassava weed control, agronomy and seed systems, visit the web site http://www.cassavamatters.org/.

10 References

Aflakpui, G. K. S. and Bolfrey-Arku, G. E.-K. (2007) Integrated management of *Imperata cylindrica* (Speargrass) in yam and cassava: Weed pressure in crop, crop growth and yield. *Journal of Plant Science* 2(1): 14–24.

Agumagu, A., Adesope, O. M. and Matthews-Njoku, E. C. (2008) Perception of infestation problems on cassava farms and preference for weed management practices in humid agro-ecological zone of Nigeria. *Agriculturae Conspectus Scientificus* 73(No. 2): 115–20.

Akobundu, I. O. (1980) Weed control in cassava cultivation in the sub humid tropics. Tropical Pest Management, 26: 420–6.

Akobundu, I. O., Ekeleme, F. and Chikoye, D. (1999) The influence of the fallow management system and frequency of cropping on weed growth and crop yield. *Weed Research* 39: 241–56.

Alabi, B. S., Ayeni, A. O., Agboola, A. A. and Majek, B. A. (2001) Giant sensitivplant interference in cassava. *Weed Science* 49: 171–9.

Alabi, B. S., Ayeni, A. O., Agboola, A. A. and Majek, B. A. (2004) Manual control of thorny Mimosa (*Mimosa invisa*) in cassava (*Manihot esculenta*). *Weed Technology* 18: 77–82.

Albuquerque, J. A. A., Sediyama, T., Silva, A. A., Carniero, J. E. S., Cecon, P. R. and Alves, J. M. A. (2008) Interferencia de plantas daninhas sobre a produtividade da mandioca (Manihot esculenta) (in Portugees). – Weed Interference in cassava (Manihot esculenta) yield. *Planta Daninha* 26(2): 279–89.

Amanullah, M. M., Alagesan, A., Vaiyapuri, K., Sathyamoorthi, K. and Pazhanivelan, S. (2006) Effect of intercropping and organic manures on weed control and performance of cassava (*Manihot esculent* Crantz.). *Journal of Agronomy* 5: 589–94.

Ambe, J T., Agboola, A. A. and Hahn S. K. (1992) Studies of weeding frequency in cassava in Cameroon. *Tropical Pest Management* 38(3): 302–4.

Ayoola, O. T. and Makinde, E. A. (2008) Influence of cassava population density on the growth and yield performance of cassava – maize intercrop with a relayed cowpea. *Tropical and Subtropical Agroecosystems* 8: 235–41.

Barrios, J. R. (1973) Weed control in cassava. *In* Proceedings of 3rd International Symposium on Tropical Root Crops, Ibadan, pp. 112–14.

Bolfrey-Arku, G. (2011) In-House Review – Weed Science. 19–22 March 2011. Accessed on 29 May 2017 at http://slideplayer.com/slide/10334586/ (get more details on this web site and access date).

Büttner, U. and Hauser, S. (2003) Farmers' nutrient management practices in indigenous cropping systems in southern Cameroon. *Agriculture Ecosystems and Environment* 100: 103–10.

Chikoye, D., Ellis-Jones, J. Kormawa, P., Udensi, U. E., Ibana, S. E. and Avav, T. 2006. Options for cogongrass (*Imperata cylindrica*) control in white Guinea yam (*Dioscorea rotundata*) and cassava (*Manihot esculenta*). *Weed Technology* 20: 784–92.

Chinaka, C. C., Omeke J. O., Onu R. O. and Okoro, B. O. (no date) Cassava Production, Processing and utilization. Extension Bulletin No. 224. WAAPP – Nigeria, Agricultural Transformation Agenda, Federal Ministry of Agriculture and Rural Development, p. 31.

Costa, N. V., Ritter, L., Peres, E. J. L., Silva, P. V. and Vasconcelos, E. S. (2013) Weed interference periods in the 'Fecula Branca' cassava. *Planta Daninha* 31(3): 533–42.

De Rouw, A. (1995) The fallow period as a weed-break in shifting cultivation (tropical wet forests). Agriculture Ecosystems and Environment 54: 31–43.

Doll, J. D. and Piedrahita W. C. (1973) Effects of time of weeding and plant population on the growth and yield of cassava. Proceedings of the 3rd ISTRC symposium, 1973, IITA, Ibadan, Nigeria, pp. 399–405.

Duindam, J. and Hauser, S. (2008) Optimal herbicide application rate for pueraria fallow recalmation: Effects on vine growth, soil cover and weed suppression. http://www.tropentag.de/2008/abstracts/posters/659.pdf.

Duindam, J. W. and Hauser, S. (2011) Prioritizing research efforts to increase on-farm income generation: The case of cassava-based farmers in peri-urban Southern Cameroon. In A. Bationo, B. Waswa, J. M. Okeyo and et al. (Eds), *Innovations as Key to the Green Revolution in Africa*, Springer Netherlands, pp. 913–19.

Duindam, J., Fermont, A. and Hauser, S. (2010) Pueraria phaseoloides fallows improve labour efficiency and productivity of cassava systems in the forest zone of Cameroon. Tropentag 2010, Zurich, Switzerland. http://www.tropentag.de/2010/abstracts/links/Duindam_oMZddn1t.pdf.

Ekeleme, F. (2013) A Report on Weeds of Cassava and Management Choices for Smallholders in Africa.

Ekeleme, F., Akobundu, O. I., Isichei, A. O. and Chikoye, D. (2003) Cover crops reduce weed seedbanks in maize-cassava systems in southwestern Nigeria. *Weed Science* 51: 774–80.

Fermont, A. M., van Asten, P. J. A., Tittonell, P., van Wijk, M. T. and Giller, K. E. (2009) Closing the cassava yield gap: An analysis from smallholder farms in East Africa. *Field Crops Research*, 112: 24–36.

Harms, T. (2015) An Assessment of Intercropping and Fertilization in Cassava – (Manihot esculenta) Based Systems in the Kolli Hills, South India. MSc thesis, Department of Renewable Resources, University of Alberta, Canada.

Harper, R.S. 1973. Cassava growing in Thailand. *World Crops*, 25: 94–7.

Hauser, S., Ekeleme, F. and Dixon, A. G. (2015) Weed biomass production and cassava yields in varying cassava cropping systems. Tropentag 2015, 16–18 September, Berlin. http://www.proceedings2015.tropentag.de pp 163.

Hauser, S., Nolte, C. and Carsky, R. J. (2006) What role can planted fallows play in humid and sub-humid West Africa? *Nutrient Cycling in Agroecosystems* 76: 297–318.

Hauser, S., Wairegi, L., Asadu, C. L. A., Asawalam, D. O., Jokthan, G. and Ugbe, U. (2014) *Cassava System Cropping Guide*. Africa Soil Health Consortium, Nairobi, p. 58.

Henry, G. and Gottret, V. (1996) *Global Cassava Trends. Reassessing the Crop's Future*. CIAT Working Document No. 157. CIAT, Cali, Colombia, p.45.

Hershey, C., Henry, G., Best, R., Kawano, K., Howeler, R. and Iglesias, C. (2001) Cassava in Asia. Expanding the competitive edge in diversified markets. In *A Review of Cassava in Asia With Country Case Studies on Thailand and Viet Nam*. Proceedings of the validation forum on the global cassava development strategy, Rome, 26–28 April 2000. Available at http://www.fao.org/docrep/009/y1177e/Y1177E02.htm. Accessed on 23 December 2016.

Ibeawuchi, I. I., Dialoke, S. A., Ogbede, K. O., Ihejirika, G. O., Nwokeji, E. M., Chigbundu, I. N., Adikuru, N. C. and Oyibo P. O. (2007) Influence of yam/cassava based intercropping systems with legumes in weed suppression and disease/pest incidence reduction. *Journal of American Science* 3(1): 49–59.

Kropff, M. J. and Moody, K. (1992) Weed impact on rice and other tropical crops. Proceedings of the First International Weed Control Congress, Melbourne, pp. 123–6.

Leihner, D. E. (1980) Cultural control of weeds in cassava. In E. J. Weber, J. C. Toro and M. Graham (Eds), *Cassava Cultural Practices*. Proceedings of a workshop, Salvador, Bahia, Brazil, 18–21 March. IDRC publication No. 151e, IDRC, Ottawa, Canada, pp. 107–11.

Melifonwu, A. A. 1994. Weeds and their control in cassava. *African Crop Science Journal* 2(4): 519–36.

Melifonwu, A., Braima James, Aïhou, K., Weise, S., Awah, E. and Gbaguidi, B. (2000) Weed Control in Cassava Farms. IPM Field Guide for Extension Agents. IITA, Ibadan Nigeria, p. 23.

Menzies, N. W. and Gillman, G. P. (2003) Plant growth limitation and nutrient loss following piled burning in slash and burn agriculture. Nutrient Cycling in Agroecosystems 65: 23–33.

Moody, K. and Ezumah, H. C. 1974. Weed control in major tropical crops - a review. *PANS*, 24: 292–9.

National Root Crops Research Institute (NRCRI) (1983) *Annual report (1982)*. NRCRI, Umudike, Abia State, Nigeria.

Nedunchezhiyan, M., Ravindran, C. S. and Ravi, V. (2013) Weed management in root and tuber crops in India: Critical analysis. *Journal of Root Crops* 39(2): 13–20.

Norgrove, L. and Hauser, S. (2016) Biophysical criteria used by farmers for fallow selection in West and Central Africa. *Ecological Indicators* 61: 141–7.

Nweke, Felix, I., Spencer, Dunstan, S. and Lynam, J. K. (2001) *The Cassava Transformation: Africa's Best Kept Secret*. Michigan State University Press, East Lansing.

Odoemenem, U. I. and Otawa, B. L. 2011. Economic analysis of cassava production in Benue state, Niugeria. *Current Research Journal of Social Sciences*, 3(5): 406–11.

Oerke, E. C., Weber, A., Dehne, H. W. and Schönbeck, F. 1994. Conclusions and perspectives. In E. C. Oerke, H. W. Dehne, F. Schönbeck and A. Weber (Eds), *Crop Production and Crop Protection; Estimated Losses in Major Food and Cash Crops*. Elsevier, Amsterdam, pp. 742–70.

Onochie, B. E. (1975) Critical periods for weed control in Cassava in Nigeria. PANS 24, 292–9.

Pachico, D. and Rivas, L. 2003. A preliminary comparison of the potential welfare and employment effects of herbicide tolerant, high yielding, of mechanized cassava in different markets in Colombia. In: C. M. Fauquet and N. J. Taylor (Eds), *Cassava: An Ancient Crop for Modern Times. [Compact Disc 1]*, Proceedings of the 5th International Meeting of the Cassava Biotechnology Network (4–9 November 2001, St. Louis, MO).

Prameela, P., Menon, M. V., John, P. S. and Abraham, C. T. 2012. Evaluation of pre-emergence herbicides for cost-effective weed control in tapioca: Short communication. *Indian Journal of Weed Science*, 44(1): 58–9.

Santos Silva, T., Lima e Silva, P.S., Dantas Braga, J., Maria da Silveira, L., Pequeno de Sousa, R., 2013. Planting density and yield of cassava roots. *Revista Ciencia Agronomica*, 44: 317–24.

Sarria, R., Torres, E., Angel, F., Chavarriaga, P. and Roca, W. M. (2000) Transgenic plants of cassava (*Manihot esculenta*) with resistance to Basta obtained by *Agrobacterium*-mediated transformation. *Plant Cell Reports* 19(4): 339–44.

Silva Soares, M. R., Araujo Neto, A. C., Alcebíades Rebouças S. J., da Silva Lima, R., de Souza Moreira, E., Thiago Reis Prado, de Andrade Silva, R. and Pereira Moreira, G. L. (2016) Effect of weeds

on yield loss of cassava plants in response to NPK fertilization. *African Journal of Agricultural Research* 11: 356–70.
Taylor, N., Chavarriaga, P., Raemakers, K., Siritunga, D. and Zhang, P. (2004) Development and application of transgenic technologies in cassava. *Plant Molecular Biology* 56: 671–88.
Unamma, R. P. A., Ene, L. S. O., Odurukwe, S. O. and Enyinnia, T. (1986) Integrated weed management for cassava intercropped with maize. *Weed Research* 26: 9–17.Valadão Silva, D., Barbosa Santos, J., Pedro Cury, J., Paolinelli Carvalho, F., Barros Silva, E., Sebastião Cunha Fernandes, J., Alves Ferreira, E. and Concenço, G. (2013) Competitive capacity of cassava with weeds: Implications on accumulation of dry matter. *African Journal of Agricultural Research* 8(6): 525–31.
Vissoh, P. V., Gbehounou, G., Ahanchede, A., Kuyper, T. W. and Roling, N. G. (2004) Weeds as agricultural constraint to farmers in Benin: Results of a diagnostic study. *Netherlands Journal of Agricultural Science* 52(3/4): 305–29.

Index

www.ingramcontent.com/pod-product-compliance
Ingram Content Group UK Ltd.
Pitfield, Milton Keynes, MK11 3LW, UK
UKHW021826100926
8556IPUK00003B/182

9 781786 760043